*Coastal and Estuarine
Sediment Dynamics*

Coastal and Estuarine Sediment Dynamics

Keith R. Dyer

Institute of Oceanographic Sciences, Bidston, UK

A Wiley–Interscience Publication

JOHN WILEY & SONS

Chichester · New York · Brisbane · Toronto · Singapore

Library of Congress Cataloging-in-Publication Data:

Dyer, K. R. (Keith R.)
   Coastal and estuarine sediment dynamics.
   'A Wiley–Interscience publication.'
   Includes index.
   1. Marine sediments.   2. Estuarine sediments.
3. Coast changes. 4. Sedimentation and deposition.
I. Title.
GC380.15.D94   1986   551.3′54   85–12458

ISBN 0 471 90876 2

British Library Cataloguing in Publication Data:

Dyer, Keith R.
   Coastal and estuarine sediment dynamics.
   1. Sediments (Geology)   2. Seashore
   I. Title
   551.3′54   QE571

ISBN 0 471 90876 2

Printed and bound in Great Britain

*To Jane, Georgia, Steven, Timothy
and Frances—for putting up with
all this*

Preface

There is a vast literature on the topic of sediment transport, and, generally, the contributions to it fall into three categories: descriptive marine geological, process-oriented oceanographic, and empirical engineering. I admit this is a rather sweeping generalization but so often the engineering literature starts with dimensional reasoning and then obtains results from experiments to produce empirical relationships with little apparent attempt to relate the results to the natural environment. Conversely, the geological literature establishes some very good physical concepts but fails to quantify them sufficiently to turn them into valid theories. The oceanographic approach is to consider the driving fluid processes in the absence of a satisfactory view of the sedimentary response. Of course there are very notable exceptions to the above statements.

This book is an attempt to bring together some elements of these three approaches: to establish an overall descriptive framework from the geologists; to put in experimental results and theory from the oceanographers, where possible; and to bridge the gaps with empirical results from the engineers where the theory is inadequate. Because of the complex interactions between the fluid, the bed, and the particles in the fluid and on the bed, it is very difficult to achieve a complete flowing logic through the book. It is impossible to avoid a certain amount of repetition, or dropping a topic at a certain point, to take it up again later in a slightly different context. Waves are almost ubiquitous, yet it is simpler to treat tidal currents and waves separately; in most cases of importance, of course, the two act together. Nevertheless without understanding them separately, there is little chance of understanding their interaction. Similarly cohesive and non-cohesive sediments are considered separately, even though there are large areas of the sea bed where the sediments are cohesive until they move, and are non-cohesive until they are deposited.

My objective was to provide a text book with the minimum of complex mathematics and derivations, but one which explains the concepts, which can then be carried forward by the reader to a further level of understanding by reference to the literature. The expert reader will probably note that some of their favourite references are missing. This is partly because I have had to choose in order not to produce an inordinately long and complex book, and partly because I defy any one person to have actually read and understood all the literature. Unfortunately, much of the engineering literature is not readily applicable to the sea and has been omitted as a consequence.

This book is aimed at complementing the many other textbooks in the field rather than replacing them. Since writing the manuscript the excellent book by Sleath (*Sea Bed Mechanics*, Wiley–Interscience) has been published, which I think actually adds to this book, rather than detracting from it, and vice versa.

This book has required considerable effort from many other people besides myself. As the reader will be aware, I have drawn extensively on the published results of work carried out in the last twelve years in the Sedimentation Group of the Institute of Oceanographic Sciences at Taunton—the group that I had the good fortune to lead. That Group has now unfortunately dispersed (not of their own volition) but hopefully it will reform with the same zeal elsewhere. I would like to thank all the members of the Group for their contributions over the years but specifically to several for their comments on drafts of the manuscript: Alan Carr, Alan Davies, Tony Heathershaw, Nick Langhorne, Richard Soulsby and Tim Smith. Also I would like to thank Paul Komar and Ian Robinson for their comments, and Dorothy Croston for all the typing and collation. The mistakes and misunderstandings remaining are mine. Last, but not least, I wish to thank British Rail for space and time for the background reading, and thank the many hotels in whose bedrooms the writing was done.

KEITH DYER

Taunton

June 1985

Contents

Symbols

Symbols used throughout the text. Other symbols are defined where they are used.

a	Amplitude of surface wave motion
A	Area
A_b	Amplitude of near bed oscillatory water motion (Equation 3.49)
B	$= w_s/\beta\kappa u_*$
C	Suspended sediment concentration
$\overline{C}$	Turbulent mean concentration
c'	Turbulent deviation of concentration
c	Phase velocity of waves
c_g	Group velocity of waves
C_a	Reference concentration
C_0	Particle concentration in the bed
C_v	Solids volume concentration
C_D	Drag coefficient
d	Pipe diameter
d_0	Wave orbital diameter $= 2 A_b$
D	Grain diameter: subscript a, b, c, diameter on grain axes; subscript n, nominal diameter
e_b	Bedload efficiency factor
E	Wave energy
E	Erosion rate
f	Friction factor (Equation 3.8)
f_w	Wave friction factor (Equation 3.57)
f	Coriolis parameter
F	Froude number (Equation 3.13)
F_i	Interfacial Froude number
g	Gravitation acceleration
h	Water depth
H	Wave height
H_r	Ripple height
I_1	Immersed weight longshore transport rate of sediment
k	Wave number
k_s	Bed roughness under waves
k	Friction coefficient

K	Proportionality coefficient in Bagnold's sediment transport formula (Equation 7.8)
K_s	Eddy diffusion coefficient for sediment
K_m	Eddy diffusion coefficient for fluid
K_{wc}	Eddy viscosity under combined waves and currents
l	Mixing length
L	Monin–Obukov length
M	Mobility number (Equation 5.5)
n	Frequency
n	Number concentration of particles
N_z	Coefficient of vertical eddy viscosity
P_l	Longshore wave thrust
q	Transport rate of sediment
q_s	Suspended sediment transport rate
Re	Reynolds number $= ud/\nu$ or uh/ν
Re_*	Grain or boundary Reynolds number $= u_* D/\nu$
Re_w	Wave Reynolds number (Equation 3.54)
Re_s	Settling Reynolds number (Equation 4.3)
R_f	Flux Richardson number (Equation 3.36)
Ri	Gradient Richardson number (Equation 3.37)
T	Wave period
u	Horizontal velocity component
$\bar{u}$	Turbulent mean horizontal velocity
u'	Turbulent horizontal velocity deviation
$\hat{U}$	Depth mean horizontal velocity
u_{100}	Velocity 100 cm above the bed
U_∞	Mean velocity outside the boundary layer
u_m	Near bed maximum orbital velocity (Equation 3.50)
u_*	Friction velocity $= \sqrt{\tau/\rho}$
u_{*c}	Threshold friction velocity
U_r	Ripple migration rate
v	Lateral velocity component
$\bar{v}$	Turbulent mean lateral velocity
v'	Turbulent lateral velocity deviation
$\bar{v}_l$	Longshore current velocity
w	Vertical velocity component
$\bar{w}$	Turbulent mean vertical velocity
w'	Turbulent vertical velocity deviation
w_s	Particle settling or fall velocity
w_0	Settling velocity of single particles in hindered settling formulation
W	Stream power $= \tau u$
x	Horizontal coordinate direction
y	Lateral coordinate direction
z	Vertical coordinate direction
z_0	Bed roughness length

α	Dynamic friction angle (Figure 4.5)
β	$= K_s/K_m$ (Equation 6.3)
β	Bed slope angle, beach slope, resuspension rate (Equation 6.19)
γ	$= (\rho_s - \rho)/\rho$
δ	Boundary layer thickness
δ_L	Viscous sublayer thickness
θ	Dimensionless shear stress (Shield's Entrainment Function) (Equation 4.18)
θ_c	Threshold value of Shield's Entrainment Function
θ	Water surface slope angle (Chapter 3). Angle of wave approach (Chapter 11)
κ	Von Karman's constant (Equation 3.17)
λ	Wavelength of surface wave or bedform
λ_c	Beach cusp wavelength
λ_e	Edge wave wavelength
λ_r	Ripple wavelength
λ_D	Dune wavelength
μ	Coefficient of molecular viscosity
ν	Kinematic viscosity $= \mu/\rho$
ρ	Water density
ρ_s	Density of sediment particles
ρ_m	Bulk density of sediment
σ	Angular frequency $= 2\pi/T$
τ	Shear stress, bed shear stress
τ_0	Bed shear stress
τ_B	Bingham shear stress, or residual stress
τ_b	Bingham shear stress, or residual stress
τ_c	Threshold shear stress
τ_d	Critical shear stress for mud deposition
τ_e	Critical erosion shear stress
φ	Phi unit
φ	Angle of repose, or angle of static friction
Φ	Dimensionless transport rate of sediment

CHAPTER 1

Introduction

Sediment movement occurs as the mobile bed tries to readjust its shape or texture in order to resist better the forces causing movement. In this way sand grains move from a flat bed to cause ripples, ripples move to create sandbanks, and erosion and deposition alters the coastal outline. To be able to describe and predict these changes is essential for successful coastal defence, the prevention of coastal flooding, for navigation, for dredging and the stability of structures such as oil rigs and pipelines placed on the sea bed. Despite many years of research we are still quite a way from being able to do this wholly satisfactorily.

There have been many descriptive studies by geomorphologists and geologists and these have provided the conceptual models for the cause-and-effect relationships. Geomorphologists have mainly restricted their view to the land, but have had a great impact in the understanding of beaches and coast erosion. Geologists have mainly been interested in describing present-day sediments in order to understand the conditions under which similar deposits were laid down in the geological past. Many of these studies have concentrated on the internal structure of the sediments, their textures, fabrics and depositional sequences, in particular concentrating on the accessible intertidal deposits. Reviews of these topics are contained in Allen (1982) and Reineck and Singh (1980). Geologists have also used acoustic surveying techniques. Since the 1930s echo-sounders have provided the means for examining the detailed shape of the sea bed, and since the 1960s side scan sonar has been available for mapping the distribution of bed features. The side scan results show gradations of bedform that can be related to differences in current velocity, and variations of shape have been diagnosed as characteristic features of the tidal current oscillation. These have been summarized by Stride (1982). When coupled with current measurements and sea bed samples, sonar studies have led to regional descriptions of sediment circulation; the sources, sinks and transport paths, based on two simple concepts. One is that sediment becomes finer in the direction of transport and the other is that the asymmetry of the bedforms indicates the direction of transport.

However, these studies give no indication of how much sediment is moved, or when, but they provide an obvious framework into which more quantitative studies must fit.

The greatest advances in quantifying the relationships between water flow and the sediment transport rate have been made by engineers in the relatively simple situations found in rivers, irrigational channels and in flumes. The classical description of the sequence of bedforms observed in a flume with increasing velocity was made by Gilbert in 1914, but it was not until 1928 that Shields formalized non-dimensional parameters that described the threshold of grain movement over a range of grain sizes and densities, in terms of the bed shear stress. This development was helped by the advances in fluid mechanics made in the 1920s by Prandtl, and von-Karman and others. The advances enabled formulations of the characteristics of the boundary layer flow near the bed and of the turbulence, and laboratory experiments investigating these aspects eventually gained impetus with the development of hot film anemometers and laser doppler anemometers. Much of this work has been summarized by Hinze (1959) and Schlichtling (1968). More recent work has investigated the intermittent structure of the turbulent shear stresses and traced the movement of individual particles within the flow; thus getting closer to the physics of grain movement.

The pragmatic engineering approach has continued, however, investigating in more detail, for instance, the transition conditions between different bedforms. Additionally engineers such as Rouse, Kennedy and Einstein developed theoretical relationships which were calibrated using laboratory and field data. Several dozen different sediment transport formulae have been developed, each of which is suited to particular data sets or experimental conditions, but they often differ by an order of magnitude, or more, when applied to the same situation. Consequently many problems remain. These approaches have been summarized by Raudkivi (1967), Yalin (1977) and Graf (1971).

One of the most significant contributions to sediment transport studies has been that of Bagnold. Starting in 1940 with his book on desert dunes he has written a series of papers which developed an approach to sediment transport based on sound physical principles. This has resulted in focussing interest in the physics of moving grains, and his approach has been adopted by many oceanographic workers for application to the sea. However this approach has been less readily accepted by the engineers.

Over the last 10 to 20 years there has been increasing interest in combining the engineering, geological and physical approaches and making direct physical measurements of sediment movement in the sea. This has been made possible by improved instrumentation, by better ways of recording large quantities of data, and by computer analysis techniques. Nevertheless, progress has still been hampered by the lack of means of reliably measuring sediment transport. Some of this work has been summarized by Stanley and Swift (1976).

There are a number of differences between flumes and the sea, which means that one has to be careful when applying to the sea theory which has been derived from flume investigations. These differences are summarized in Table

Table 1.1 Comparison between sediment transport in flumes and in the sea

Variable	Flume	Sea
Flow	Uniform in space Steady in time	Non-uniform Tidally varying
Water depth	Centimetres	Metres, to hundreds of metres
Bed shear stress	Calculated from water surface slope	Calculated from velocity profiles, or from turbulent shear stresses
Bed shape	Initially flat	Bed normally rippled
Sediment	Single grain size	Multimodal and multi-component
Waves	Monochromatic, regular	Irregular spectrum

1.1. The most fundamental difference is the presence of random waves in the sea. Though many studies of wave induced sediment motion have been carried out in flumes, there are distinct limits on the water depths and wave periods that are possible because of the size of the flumes, and because the waves are usually regular and of a single period. Also any studies of combined waves and currents are restricted to co- and contraflow situations. Whereas in flume experiments it is possible to calculate the drag on the bed by measuring the surface water slope, in the sea this cannot be done, and the stress measurements are intimately entwined in problems of bedform induced form drag and the effects of moving sediment on the near-bed velocity profiles. Because the tidal flow is oscillatory, the bed surface bears the effects of the previous half-cycle modified by the sometimes rapid deceleration. The bedforms will not grow until these effects have been reversed, and consequently lag effects between shear stress and bedform development are greater in the sea than in rivers.

In many areas the sediment distribution has been acted upon for hundreds or thousands of years by the tidal and wave processes. Present day movement is therefore often minimal, and sediment is only likely to move at spring tides and even then the rates of transport are likely to be small, as the threshold of movement will only just be exceeded. In other areas sediment movement may only occur when there is storm wave activity to enhance the tidal streams. Because the forces have been acting for so long it is also possible that the original source of the sediment moving on the shelf may have been exhausted and the amounts of material in transport may be the merest shadow of that in the past. The present day currents may therefore be competent to move more than the measured load of sediment if the sediment were available.

These days there are a number of workers who have developed instrumental rigs that can be placed on the sea bed to remotely record sediment movement during storms, but the results from them are meaningful only when placed in the overall regional context. This requires conventional marine geological investigations to determine the sediment circulation patterns, as well as widespread measurements of the current field, both tidal and non-tidal. Once the physics of sediment movement is adequately known, then the flow field can be mathematically modelled and the sediment transport calculated, using the direct measurements as validation. Though this is done at the moment, there are usually inadequate measurements to prove the modelling and this, coupled with deficiencies in the sediment transport theory, make the errors unacceptably large. However, the modern instrumentation developments lead one to expect that these problems will be overcome.

When it comes to using the results of these studies in the geological context, there are additional difficulties. In the ancient sediments one does not see the sediment transport, but only a fraction of the sediment originally deposited. In situations where it is possible to estimate the original rates of deposition of sand, it is obvious that sedimentary beds a few tens of metres thick would have been deposited in a few months or years. This contrasts with the millions of years that elapse for the formation of units a few thousand metres in thickness. As a result we see perhaps only one percent of the potentially deposited sediment. Obviously large quantities of sediment have been intermittently re-eroded, or there are horizons during which nothing was deposited for long periods. This dilemma has been highlighted by van Andel (1981). Eventually we need to know what controls the final preservation of the deposits that we see being formed today. This knowledge will aid the search for oil and other sedimentary deposits of economic importance.

Table 1.2 Variables important in sediment transport

Sediment size, shape, density and mineralogy of grains
Sediment settling velocity
Sediment availability
Flow depth
Water density, viscosity
Bed shear stress
Bedform wavelength, height, steepness
Maximum tidal velocity
Residual tidal velocity
Wave period, amplitude

It is apparent from the above that there are many variables that have to be considered in sediment transport studies in the sea. These are listed in Table 1.2. The influence of these various factors will be considered in the following chapters. Since the sediment movement is driven by the near-bed flow, and the moving sediment itself affects this flow, we have to devote considerable space to an examination of the fluid forces and boundary layer flow.

CHANGING SEA LEVEL

The sediments on the shelves and coasts, and in the estuaries of the world, are derived from the land via river discharge, from coast erosion and from redistribution of sediments dumped on the shelf during glacial periods or during low sea level stands. The present low relief of the continental shelves and coastal plains is thought to be the result of continual transgression and regression of the sea, and the consequential migration of the shoreline. During periods of sea level retreat shorelines are abandoned and the waves are able to attack the newly-exposed rock of the sea bed. The rejuvenated landscape will cause the rivers to incise their courses, and they will discharge large amounts of sediment increasingly further out onto the shelf. Because of the changing length and depth of the embayments the magnitudes of both the tidal and residual currents would change. However, they could either increase or decrease depending on local conditions.

During transgressions, beaches would be gradually pushed shorewards. Longshore transport would cause bays to become infilled and the submerged valleys of the rivers would become buried under sediment. The river discharge of sediment would diminish, however. If the rate of sea level rise is fast then banks, beaches and other sedimentary features would become relict beneath the sea.

Estuaries are likely to be more numerous during transgressions than during regressions. They are ephemeral features being fairly rapidly altered and destroyed, having an average life of probably only thousands of years. The ephemeral nature of estuaries has been discussed by Schubel and Hirschberg (1978).

The sediment response to the changing sea level is unlikely to be immediate and, as a consequence, the sediment distribution and characteristics are likely to be modified for some time after the sea level becomes static. In some areas, however, the sediment will fairly quickly achieve a reasonable equilibrium with the mean currents and wave forces, even though, because of the natural variability, considerable amounts of transport can occur. In many circumstances it is difficult to tell whether there are any long term trends because the large variations mask them, and it may take only a small variation of sea level to destroy a local equilibrium.

Sea level can alter because of a number of causes. Changing ocean currents and variation in the temperature of the ocean water can cause minor fluctuations, but there are two main effects: eustatic and isostatic variation.

Eustatic variations arise because of alterations in the volume of the world's oceans due either to changes in the elevation of the ocean floor or to the variation in the actual volume of water present in the oceans. The ocean floor may vary in its elevation due to plate tectonic activity, since the width and elevation of the mid-ocean ridges changes with slight variations in the rate at which sea floor spreading occurs. A rise of the ridge would cause the water to rise onto the coastal plains. These movements are of long time scale ($\sim 10\,\text{my}$) and can account for the major geological transgressions and regressions. Alternatively the volume of water present in the oceans can be changed by the amount trapped as ice in the glaciers on the continents. If the present glaciers were to melt, a sea level rise of about 60 m would result. Over the last half a million years or so, during the Pleistocene glaciation, many glacial periods and interglacials have

6

caused large and quite rapid (~ 100,000 y) fluctuations of sea level, largely
between present day sea level and about 100 m below.

Isostatic variations are caused by the alteration in elevation of the continents
due to erosion of mountains or because of tectonic activity. This can be quite
localized, as well as covering large areas. A good example of isostatic effects is
the rebound of the Scandinavian region because of the removal of the Pleistocene ice sheet. As a consequence the land at the northern end of the Baltic Sea is
rising at a rate of 1 cm y^{-1}.

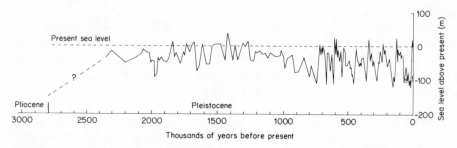

Figure 1.1 Variation of sea level during the last two million years. *From Stienstra, 1983,
Marine Geol., 52, 27–37. Reproduced by permission of Elsevier Publishing BV*

Since the chronology of sea level fluctuations is determined mainly from the
elevation of relict beach features, eustatic and isostatic effects are difficult to
separate. Nevertheless examination of worldwide patterns, and relying on tec-
tonically stable areas, has enabled a fairly consistent picture to emerge. This has
been extended backwards in time by the study of oxygen isotope ratios in deep
sea cores, which define temperature stages relating to glacials and interglacials.
Figure 1.1 shows the postulated sequence over the last two million years.

Of particular interest to modern sedimentologists is the most recent period,
the Holocene interglacial period. About 18,000 years before present there

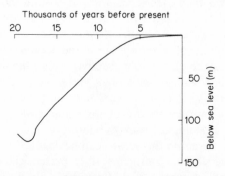

Figure 1.2 The rise of sea level during
the Flandrian Transgression

appears to have been a low sea level of about -120 m (Figure 1.2), which would place the shore line near the edge of the present continental shelf. The level rose at a rate of about 1 m per century (1 cm y^{-1}) as the glaciers retreated. This sea level rise is known as the Flandrian transgression. As can be seen from Figure 1.2 sea level reached the present day elevation about 5,000 years ago. Since then various authors suggest that minor fluctuations have occurred, which are probably related to slight climatic changes, minor advances and retreats of glaciers, and to local isostatic effects. At present there appears to be a general rise in sea level of about 1 mm y^{-1}. This will of course exacerbate coast erosion, if only temporarily, until the next advance of the glaciers. However it has been predicted that sea level will continue to rise and may even accelerate in the next few decades. This is a result of the global warming caused by the increasing levels of carbon dioxide in the atmosphere.

SEDIMENT DISCHARGE FROM RIVERS

The amount of sediment eroded from a drainage basin, the sediment yield, depends on the geology, the topography and the climate. However, the yield does not usually equal the amount of sediment discharged into the estuary because a great deal is deposited in the lower reaches of the river. It has been estimated that the total annual discharge of sediment is about 7×10^9 tons worldwide (Milliman and Meade, 1983). Examination of the annual rates shows that the main sediment discharge is in the Far East, the monsoon areas and the tropics (Figure 1.3), where the sediment yield can exceed 1000 tons $km^{-2} y^{-1}$. Though the discharge of water can be high in temperate latitudes and glaciated areas, the sediment concentrations are relatively low.

The Asian rivers carry in excess of 75 per cent of the world sediment discharge, and two thirds of that is carried by the Hwang-Ho (Yellow) River and the Ganges/Brahmaputra. The former drains an area of very easily eroded glacial loess silts, sediment concentrations can reach 60 per cent by volume at times of flood, and of the annual discharge of about 1.1×10^9 tons, between a quarter and a third can be transported in only two or three days of flood (Wan, 1982). Since the mouth of the Yellow River changed its outlet into the Yellow Sea by about 400 km in the mid-19th century, the effects on the coastal sedimentation pattern in both areas is likely to have been drastic. Chen et al. (1982) have traced the development over the last 2000 years of the mouth of the Yangtze River. This river is fourth in sediment discharge in the world, discharging nearly 5×10^8 tons y^{-1}, and, in one area, the coast is building outwards at 1 km in 40 years. Satellite measurements have shown 3.8×10^6 tons of sediment suspended within the top metre of the water column over an area off the mouth reaching out to 20–30 km from the coast (Yun and Wan, 1982).

During the year there is considerable variation around the annual mean sediment discharge. The maximum discharge can occur in winter for temperate latitudes, the spring for areas with winter snow cover, and the summer in semitropical areas. The discharge is the product of the flow rate times the

8

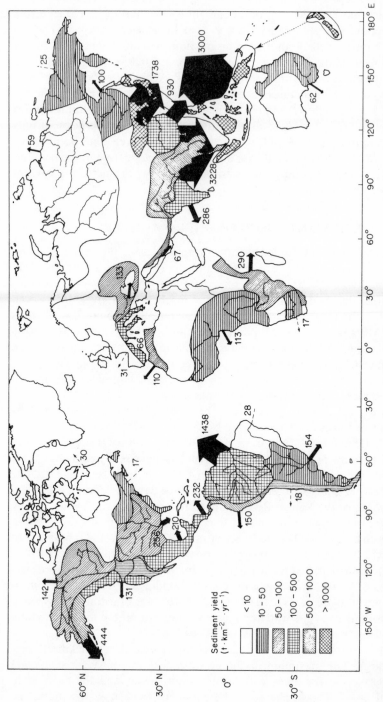

Figure 1.3 Annual discharge of suspended sediment from the drainage basins of the world. Numbers are average annual input in 10⁶ tons. *From Milliman and Meade, 1983, J. Geol., 91, 1–21. Reproduced by permission of the University of Chicago Press*

suspended sediment concentration and many studies have considered the relationship between these variables. The relationship can be represented by a rating curve of the form

$$C = a\,Q^b \qquad\qquad 1.1$$

where C is the suspended sediment concentration in mg l^{-1} and Q is the flow rate in $m^3\,s^{-1}$. The constant a ranges between 0.004 and 80,000 and the exponent b varies between 0.0 and 2.5 (Müller and Förstner, 1968).

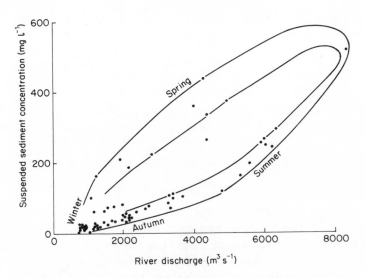

Figure 1.4 Monthly river discharge against sediment concentration for the Fraser River, B.C. *From Milliman, 1980,* Estuar. Coastal Mar. Sci., *10, 609–633. Reproduced by permission of Academic Press Inc. (London) Ltd.*

In some rivers, however, there is considerable hysteresis in the rating curve. Figure 1.4 shows an example from the Fraser River in Canada. Because much of the catchment is snow covered in winter the sediment concentration is low, and once most of the sediment weathered by frost action is eroded in the spring the mobile material becomes exhausted. Consequently for a given river discharge there are two possible suspended sediment concentrations. Hysteresis is unlikely to occur when there is no paucity of readily mobile material.

Within each limb of the hysteresis curve considerable variation of suspended sediment concentration can be produced for the short period individual floods, and for a given river discharge the suspended sediment concentration can cover a wide range. Therefore some factor other than discharge is still important.

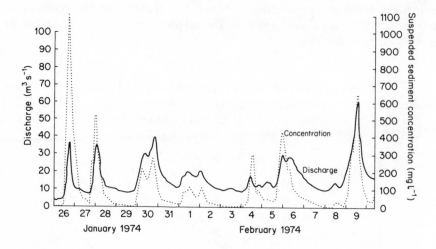

Figure 1.5 Discharge and suspended sediment concentration variation in the River Creedy, Devon, UK. *From Walling and Webb, 1981,* IAHS Publication No. 133, *177–194. Reproduced by permission of IAHS Press, Oxon*

Examination of Figure 1.5 shows two characteristics: When two flows of similar magnitude occur close together, the second has a lower sediment concentration. Also for prolonged peak flows the sediment concentration falls more rapidly than does the flow. Both effects are related to availability of mobile sediment, with a period of drying out and weathering between storms necessary to release new material. An example of these effects is given by Wood (1977).

It is obvious by extrapolation of the above comments that most sediment discharge occurs on the occasional extreme events. As the erosive power of the stream rises very rapidly with discharge, the material deposited on other less extreme flows becomes re-mobilized. Figure 1.6 shows the discharge of the Susquehanna River during 1972, the year of Tropical Storm Agnes, and the associated suspended sediment concentration. The sediment concentration during the storm was more than 40 times greater than any previously recorded, and the sediment discharge was equivalent to between 30 and 50 years of normal flow. Further discussion of the effect of this storm on the Chesapeake Bay estuary will be found on page 258.

The importance of extreme events is shown by the cumulative curve in Figure 1.7. In this example 90 per cent of the sediment discharge occurs in 5 per cent of the time, and for 80 per cent of the time the base load contributes virtually nothing to the overall discharge. This intermittency causes considerable problems for sampling as well as in estimating the effect of riverine sediments on estuarine and coastal sediment budgets.

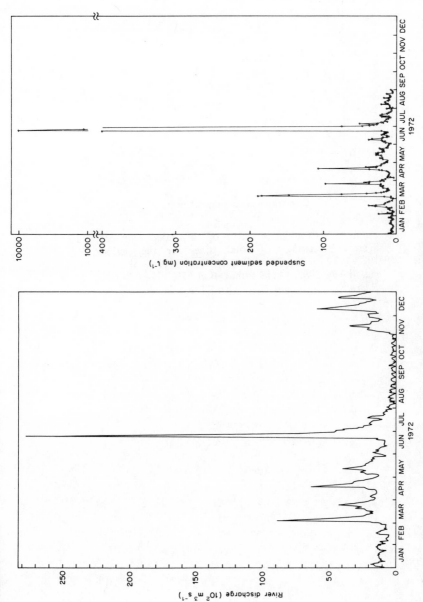

Figure 1.6 Discharge and suspended sediment concentration in the Susquehanna River during 1972, showing the effect of Tropical Storm Agnes. *From Schubel, 1974, in Gibbs, R. J. (Ed), Suspended Solids in Water. Plenum Marine Science, 4, 113–132. Reproduced by permission of Plenum Press*

12

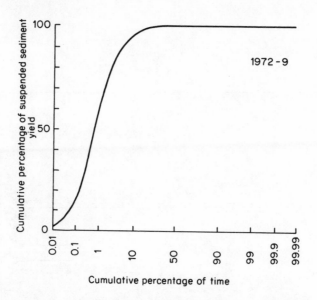

Figure 1.7 Cumulative curve of suspended sediment against percentage of total time, for the period 1972–1979 for the River Creedy, Devon. *From Walling and Webb, 1981,* IAHS Publication No. 133, *177–194. Reproduced by permission of IAHS Press, Oxon*

CHAPTER 2

The Sediment

MINERALOGY OF SEDIMENT PARTICLES

The basic unit from which all of the important sedimentary particles are made is the silica tetrahedron. In this, four oxygen atoms are arranged at the corners of a tetrahedron with a silicon atom at the centre. In quartz, the oxygen atoms are shared between adjacent tetrahedra and a strong closely-packed structure with the chemical formula SiO_2, results. Consequently quartz is hard, chemically stable and very resistant to erosion. Quartz is an abundant mineral in granites and other acidic igneous rocks and, when these are weathered, forms separate crystalline sand grains generally less than about 1 mm in size. The non-crystalline, amorphous form of silica is known under a variety of names depending mainly on colouring produced by impurities. The most common are chert and flint.

In the feldspars a proportion of the silicon atoms are replaced by aluminium, which has a very similar atomic radius. However, since aluminium has only a $3+$ charge, instead of the $4+$ of silicon, the charge is balanced by the addition of potassium. Orthoclase feldspar consequently has the formula $KAlSi_3O_8$ and, as can be seen, a quarter of the silicon atoms are replaced. Other feldspars include sodium or calcium instead of potassium to balance the ionic charges. Because of the relative ease of chemically exchanging these balancing atoms, feldspars weather more readily than quartz. Also they are physically weaker. Feldspars are found in a wide range of igneous rocks and form a minor proportion in the sediments formed from them.

The other major group of minerals is the clay minerals. In this group the SiO_4 tetrahedra form sheets by being linked by three of their corners. This gives a silicon–oxygen ratio of 4 to 10. The sheets are then linked in different ways, generally involving a layer of alumina or aluminium hydroxide, and substitution of one element by another can give a large number of varieties.

13

14

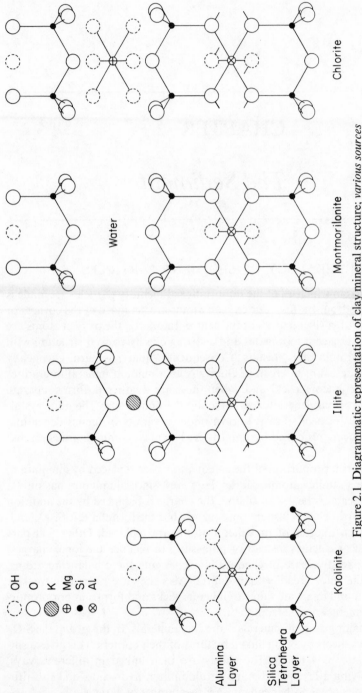

Figure 2.1 Diagrammatic representation of clay mineral structure; *various sources*

There are four main clay minerals, whose structures are shown diagrammatically in Figure 2.1. Kaolinite has a two-layer structure of a silica tetrahedra sheet and an aluminium hydroxide layer. It is formed from alkali feldspars under acid conditions and is the constituent mineral of china clay. The other minerals have a three-layer structure with the aluminium hydroxide layer sandwiched between two silica tetrahedra sheets. These sandwiches are then joined to each other in different ways. In illite, the three-layer structures are joined by potassium atoms and, additionally, aluminium substitutes for silicon in the tetrahedra, and iron and magnesium for aluminium in the hydroxide layer. With montmorillonite, the sandwiches are joined by varying amounts of water. This means that the lattice is of variable size and can expand, and that the strength of the bonds between the layers is rather small. Consequently, montmorillonite can be fairly easily split down into very small grains. Magnesium and iron can replace aluminium in the lattice giving a variety of forms. The montmorillonite clays as a whole are also known as smectites. In chlorite, the three-layer sandwiches are joined by a layer of magnesium hydroxide. Illite is formed from micas and alkali feldspar under alkali conditions. Montmorillonites are formed from basic rocks or other low potassium silicate rocks when calcium and magnesium are present.

All of the clays, to a varying degree, can undergo base exchange whereby ions in solution can exchange with others in the clay. Thus the character of the clays can change, particularly when they are introduced into sea water where there is a high concentration of ions in solution. Illite and chlorite are likely to form from kaolinite and montmorillonite in the sea, but much of the illite is of a 'mixed-layer' form with layers of varying composition.

There have been many attempts to use the distributions of the clay minerals as indicators of geological processes and of sediment provenance. For instance, Goldberg and Griffin (1964) found a latitudinal distribution of clay minerals in the Atlantic Ocean and from this it was interpreted that montmorillonite was produced from degradation of volcanic debris, illite from arid areas, kaolinite from tropical weathering and chlorite from glacial flour. In estuaries a number of studies have shown clay mineral ratios as reasonable indicators of land or marine derived material (page 255).

The clay minerals can all be fairly easily broken down to small fragments. They form an increasingly important proportion of the particles with decreasing sediment grain size, and they are almost exclusive in the fraction $< 2\,\mu m$. However, their mineralogy is difficult to determine and their sedimentation characteristics are dominated by the ionic charges on their surfaces. These charges interact with the ions in solution and result in the particles flocculating when in suspension. On the bed the charged particles stick together giving cohesive forces that are much stronger than the gravitational forces on the particles. Consequently there is a fundamental difference in sedimentary behaviour between sands and clays or muds.

GRAIN SIZE

As a result of the finer sediment grains being more mobile than the coarse ones, a large amount of information can be obtained from a study of the size of the grains and the proportions of grains in different size fractions. The range of

Table 2.1 Grain size scales for sediments

Grade scales

Wentworth (1922) after Udden		Phi $\varphi = -\log_2$ (mm)	(mm)	Microns (μm)
Boulder				
		−8	256	
Cobble		−7	128	
		−6	64	
		−5	32	
Pebble		−4	16	
		−3	8	
		2	4	
Granule				
	Very coarse	−1	2	
Sand		0	1	1000
	Coarse			
		+1	$\frac{1}{2}$	500
	Medium			
		+2	$\frac{1}{4}$	250
	Fine			
		+3	$\frac{1}{8}$	125.0
	Very fine			
		+4	$\frac{1}{16}$	62.5
Silt	Coarse			
		+5	$\frac{1}{32}$	31.3
	Medium			
		+6	$\frac{1}{64}$	15.6
	Fine			
		+7	$\frac{1}{128}$	7.8
	Very fine			
		+8	$\frac{1}{256}$	3.9
Clay	Coarse			
		+9		1.95
	Medium			
		+10	$\frac{1}{1024}$	0.98
	Fine			
		+11		0.49
	Very fine			
		+12	$\frac{1}{4096}$	0.24
	Colloid			

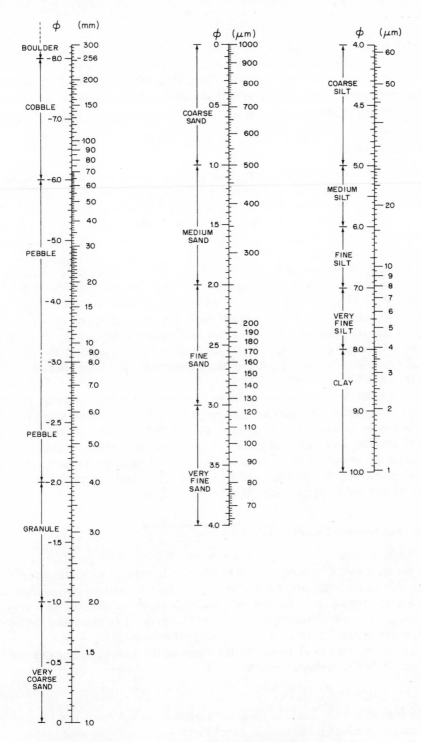

Figure 2.2 Conversion chart for phi units φ to microns μm

grain size may reflect some part of the variability of the forces causing movement. There is a large literature in the geological field on the characterization of different sedimentary environments by certain properties of the sediments. These are then applied to the geological column to determine the environment of deposition of ancient sediments.

Grain Size Scale

A grade scale of size that is most commonly used is the Udden–Wentworth scale (Wentworth, 1922). This is shown in Table 2.1. As can be seen, the base is a 1 mm size and the other grades follow by dividing or multiplying by two. To avoid the awkward situation of using a large number of decimal places in describing the smaller sand and silt grades, a smaller unit is used. This is the micron (μm) where 1 micron = 1/1000 mm.

Because of the use of powers of two in the Wentworth series, Krumbein (1934) proposed a logarithmic scale, the phi scale, on which the size in phi (φ) units, $\varphi = -\log_2$ (size in mm). As can be seen in Table 2.1, the various size grades divide at whole numbers of phi units. The negative logarithm was used in order that the grains smaller than 1 mm had positive values, which are easier to use. The larger grains are relatively infrequently analysed and the negative values are less inconvenient. A complete conversion chart from phi units to μm is shown in Figure 2.2.

To be able to use these size scales we need to be able to measure reliably the sizes of the sediment grains and the proportions in the different size intervals. This is done generally either by sieving or by sedimentation techniques, though sophisticated electronic instruments are now becoming more commonly used. The details of many of these techniques are described in books such as Carver (1971). Initially, however, it is useful to examine the results of physical measurements of individual grains and the definitions of grain size.

Size and Shape of Grains

Because most grains are fragments of crystals or chips of amorphous minerals, it would be most unusual for them to be spheres. Generally the grains are triaxial ellipsoids and have a long D_a, a short D_c and an intermediate D_b diameter. Additionally, the nominal diameter D_n is the diameter of a sphere having the same volume and weight as the grain. For ellipsoidal particles $D_n = (D_a.D_b.D_c)^{1/3}$, and it has been found that on average $D_n \approx D_b$.

Using the three axial diameters, the shape of the grains can be expressed in terms of the Corey shape factor S:

$$S = D_c / \sqrt{D_a D_b} \qquad (2.1)$$

which gives a shape factor of unity for a sphere. Though many other measures of shape have been proposed, the Corey factor seems to be the most useful.

The overall shape of the grains can also be thought of in terms of sphericity.

Sphericity is defined as the cube root of the volume of the particle divided by the volume of the circumscribing sphere. It is unlikely to change greatly by abrasion during transport and, consequently, is a fairly fundamental property of the grain. On the other hand, the roundness can alter significantly during transport. Roundness, and its inverse, angularity, refers to the outline of the grain, and roundness can be defined as the average radius of corners and edges divided by the radius of the maximum inscribed circle. These measures are very tedious to determine even on large gravel size particles and it is usual to derive values of roundness and sphericity from a visual comparison with standard charts, such as Figure 2.3. A descriptive roundness scale can also be applied to the grains.

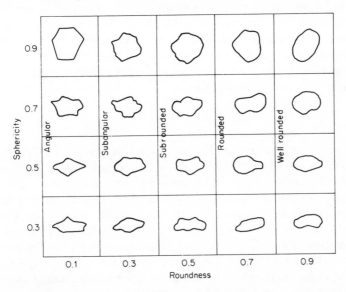

Figure 2.3 A descriptive roundness and sphericity chart for sand and gravel grains. After Krumbein and Sloss, 1963

An interesting alternative way of taking account of shape variations in sand size sediment is by considering their 'rollability' (Winkelmolen, 1969). This is done by passing the various sieve fractions through a rotating drum whose axis is inclined at an angle of 2° 30' to the horizontal. As the grains cascade within the drum, the most rollable grains are sorted out and appear first at the collector. There is a good correlation between rollability and shape, the near-spherical grains being the most rollable.

GRAIN SIZE ANALYSIS

Normally the sediment samples have to be treated in bulk, since only a few particles can be examined and sized using a microscope. A certain amount of pre-treatment is necessary, depending on the character of the sediment and the type of analysis proposed. The choice of analysis technique depends largely on the overall grain size of the sediment, with the different methods being restricted to certain ranges of size. Consequently several methods often have to be used and the results patched together in the overlap regions. There is increasing use made nowadays of various electronic techniques, such as the Coulter Counter, in which a very much wider range of sizes can be treated using the same technique. However these instruments are expensive and the traditional techniques of sieving and pipette analysis are still widely used.

Wet Sieving

If the sediment sample contains any silt or mud, this must be removed while the sample is still wet. Otherwise it will form a hard, dry, compact mass, and it will be necessary to grind the sample to break it down to its original grains. Wet sieving is normally carried out using (4φ) stainless steel sieves. The mud is washed through the sieve into a container for later analysis, and the sand and coarser sediment dried for sieve analysis.

Sieve Analysis

In this technique the sample of clean, dry sediment is passed through a series of standard sieves. These are arranged with aperture sizes at $\frac{1}{2}$ phi intervals. However the actual size needs to be checked at intervals using a microscope. As the apertures are more or less square they tend to separate particles depending on their intermediate diameter. Though sieves smaller than 63 μm (4φ) can be bought, this size is normally taken as the lower limit of effective sieving.

In operating the sieving procedure, loading on the sieves is particularly important. McManus (1965) states that an initial thickness of only 4 to 6 grains is the maximum on the sieve to avoid the apertures becoming clogged with grains only slightly larger than the holes. When this happens, obviously some of the small grains will not get through and a distorted grain size distribution curve results. The maximum load on a stack of 20 cm sieves at $\frac{1}{4}$ phi intervals, the coarsest being 0 phi, is consequently 20 gm (Carver, 1971), though this is probably rather excessive. In order to obtain samples of this size the sediment has to be split into representative subsamples.

Pipette Analysis

In order to measure the size of the material finer than 4 phi, the Andreason pipette technique is commonly used. In this method the sediment washed off in the wet sieving is made up to a known volume (generally 1 l) with water, and is placed in a tall cylinder. After stirring, the fluid is allowed to settle and during the settling 20 ml samples of the suspension are drawn off in a repeatable manner from a fixed distance beneath the surface. The samples are then evaporated and weighed, and the residue represents the total amount of sediment finer than a particular size in suspension. This size is related to the settling velocity of the grains and the time of sampling. The settling velocity is sensitive to the water temperature which must be controlled, especially if analysis is being carried out into the clay range. The velocity is calculated from Stokes Law assuming that the grain density is 2.65 gm cm^{-3}. This can be summarized in the simple formula

$$\text{Size (}\mu\text{m)} = F \sqrt{\frac{\text{Depth of sampling (cm)}}{\text{Time (min)}}} \qquad (2.2)$$

The factor F has values 12.99 at 24°C, 13.30 at 22°C and 13.60 at 20°C. For a sampling depth of 10 cm the variation of particle size with time is shown in Figure 2.4

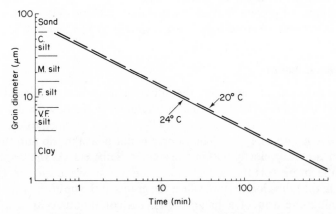

Figure 2.4 Sampling time against grain size for pipette analysis with an extraction depth of 10 cm

The initial concentration is obtained by taking a sample while still stirring, and the percentage weights are then calculated relative to this concentration, which immediately gives the cumulative percentage finer than the size appropriate to the sample time. The results can be combined with sieve analyses of the sand fraction to give a complete grain size distribution.

The lower limit of analysis is at about 0.5 μm below which Brownian motion of the water molecules begins to affect the particles settling. For grain size greater than about 30 μm, the time of cessation of stirring, or rather the time at which the turbulence of stirring dies away and the particles settle undisturbed, is critical. Also the time taken to withdraw the sample is significant. Consequently reproducibility is not good at these larger sizes.

High particle concentrations can also produce variations in the expected settling velocities. With concentrations in excess of about 10 per cent 'hindered settling' occurs. At this concentration, the particles only have a separation of about 2 grain diameters and it is seen that columns of sediment and water fall quite rapidly, but are separated by similar columns rising. This has the effect of decreasing the fall velocity, and causes a series of fairly distinct interfaces in the sedimenting suspension and these settle at a constant rate. Within each layer the concentration remains fairly constant with time. To avoid this problem concentrations need to be kept low, certainly less than $10 \, gm \, l^{-1}$. At smaller concentrations vertical density currents can occur if the suspension is not uniformly distributed, causing an increase in settling velocity. However the effect of this settling convection on the settling velocity is equivalent to only a small temperature change (Keunen, 1968).

Flocculation of the particles can also be a problem when there is a lot of clay present. Because of ionic charges on the particle surface they tend to stick together in large, low density flocs, which will fall very slowly for their size. This can be solved by adding small quantities of a dispersing agent, such as sodium hexametaphosphate (Calgon) or sodium oxalate.

The techniques of pipette analysis are descibed in Carver (1971), and British Standard and several other techniques are reviewed by McCave (1979a).

Sedimentation Columns

The same principle of measuring the fall velocities was first applied to sand size material by Emery (1938). Modern instruments are described by Schlee (1966), Channon (1971) and Rigler et al. (1981). Most sedimentation columns have an active length of 1–2 m. The sample is put in at the top and the grains attain a constant fall velocity within a few grain diameters. As the larger grains fall faster, the grains sort themselves into a continuous size sequence. One way of measuring the fall velocity is to collect the grains at the bottom of the tube on a pan forming one arm of a balance giving a continuous output of weight against time. Another technique is to measure the differential pressure between the active tube containing the falling grains and a passive stand pipe. The pressure alters with time as the grains fall below the lower orifice of the transducer connection. In the first case the weight gives the fraction coarser than a particular size, in the other the output of the sensor will be proportional to the total weight of material falling in the column, i.e. a cumulative value smaller than the size appropriate to the time since injection of the sample. A calibration of size is obtained either from fall velocity tables (see Chapter 4), or by passing sieve fractions of natural grains or glass ballotini through the column.

Similarly to pipette analysis, there are again difficulties as the result of grain interaction. To reduce this, and to assist sample introduction, a wetting agent is often used. Hulsey (1961) has shown that fall velocities can be 20 per cent higher for a concentrated dispersion of grains, because of density current effects, than for single particles. He has studied the variation in fall velocity with sample load and concludes that the greatest changes in velocity occur with loads in the 0.25–1 gm range. Since the grains only achieve fall velocities characteristic of the system in which they fall, and this is not reproducible from one system to another, Keunen (1968) suggests that no satisfactory measurements are possible on grain sizes less than 100 μm, the size below which settling convection becomes significant.

It is particularly useful to be able to convert the grain size obtained by sieving to that obtained by settling, and vice versa. Many comparisons between the two techniques have shown wide discrepancies, particularly in the higher moments of the grain size distribution. However, recent papers by Baba and Komar (1981) amd Komar and Cui (1984) have indicated how these discrepancies may be resolved.

Komar and Cui (1984) carried out measurements of the sizes of a number of grains by microscopic examination, by sieving and by settling analysis of single grains. They found that because the grains can pass diagonally through the sieve holes, the mean intermediate diameter of the grains D_b was slightly greater than the nominal sieve size, or that of a sieved series of spheres. The comparison yielded $D_b = 1.32 D_s$, where D_s is the sieve aperture size. In terms of the phi units $\varphi_b = \varphi_s - 0.40$. With this relationship we can convert the sieve size into a representative size of the actual grains by moving the grain size distribution curve towards the coarser sizes by 0.40 φ units. In order to determine the relationship between the settling velocity and the intermediate diameter D_b, they utilized a relationship developed by Baba and Komar (1981). Baba and Komar showed that the measured settling velocities, w, can be converted to the settling velocity w_s of an equivalent sphere of diameter D_b by $w = 0.977 w_s^{0.913}$. A complex empirical formula of Gibbs et al. (1971) (Equation 4.8) was then used to convert the values of w_s to the diameters D_b. The values thus obtained gave a good comparison with the adjusted sieve values.

The settling technique has the advantage of speed and bears more relation to the hydraulic behaviour of the grains, and this is of most interest in studies of sediment movement. However, there are obviously drawbacks to all of the established ways of measuring sediment grain size distributions. Sedimentologists must be aware of these in applying statistical techniques to the resulting data, and in trying to interpret them. There is obviously a need to standardize on techniques of measurement, and on methods of presenting the results. In particular the intermediate diameter appears to be the dimension most valuable in analysing grain size.

There is a particularly noticeable lack in the literature in replicate determinations of size and size distributions on several sub-samples of the same sample. Great weight is placed on small differences from sample to sample when it comes to environmental interpretation of the results, but few analyses make any assessment of whether these differences are larger than the error limits of the analysing technique.

PRESENTATION OF GRAIN SIZE ANALYSIS

Grain Size Distribution

The percentages of the total weight retained on the sieves of a uniformed sand, such as a beach sand, when plotted against grain size as a histogram shows results similar to Figure 2.5A. The result is an asymmetrical distribution with the largest percentage (the mode) being retained on a sieve on the fine side of the

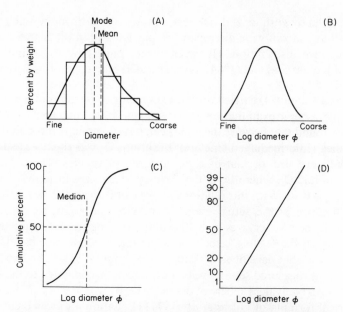

Figure 2.5 Various methods of displaying grain size distributions.
(A) Histogram of frequency distribution (B) Frequency distribution versus log diameter (C) Cumulative frequency distribution
(D) Cumulative frequency distribution as probability plot

mean diameter. The mean diameter is the size corresponding to the centre of gravity of the area under the curve. Such a curve is difficult to quantify in statistical terms, but it is useful for defining the presence of several modes in a mixed sediment.

When the same results are plotted in semi-logarithmic terms against the logarithm of the diameter, the curve closely approximates the symmetrical, bell-shaped curve typical of a normal (Gaussian) distribution (Figure 2.5B). Thus the ideal uniform sediment is log-normal, and deviations of the curve from the ideal form can be described by statistical techniques. It should now be apparent why the phi scale is so useful, since it immediately presents the results in terms of the logarithm of the diameter. Because the histogram gives misleading results if the interval between the sizes is not constant, quarter phi intervals are normally chosen.

In order to compare one sample with another it is usual to present the results as a cumulative curve presenting the cumulative percentages finer than a particular size (Figure 2.5C). This is generally a smooth curve, so that any faulty sieves show up as a kink in the curve. The mode of the distribution now occurs at the inflection point in the curve. The size given by the 50th percentile is the median grain size.

An additional assessment of how true the concept of log-normality is for any sample can be achieved by plotting the cumulative percentages on arithmetic probability paper. This effectively stretches the ordinate axis so that a normal sediment will plot as a straight line (Figure 2.5D). As we will see later this presentation is used in interpreting the sediment distribution in environmental terms.

The reason for the log-normal distribution of the sediment is unknown. It has been speculated that it may be a response to the turbulent velocity and shear stress fluctuations near the bed. There are a few periods of high stress imposed on a more usual lower intensity background in a steady flow. Consequently, the larger grains will only be infrequently moved and the smaller ones will move very much more frequently. However, the explanation is likely to be complicated by the effects of packing of the coarse and fine grains, and the fact that the larger grains may shelter the smaller ones. It is conceivable then that both coarse and finer grains would move together at about the same velocity.

The properties of the grain size frequency distribution curves can be used for characterizing the samples and allowing comparisons between them. Deviations from the normal distributions may also be environmentally sensitive. There is a vast literature on this which has been reviewed, amongst others, by Folk (1966).

As we will see, an enormous number of concepts have been built up around the assumption of log-normality, but there are alternatives. Bagnold (1937) plotted the log-histogram in which the frequency axis in Figure 2.5B, as well as the grain size axis, is plotted in logarithmic terms. When this is done the curve often approximates to a hyperbola, rather than to the parabola which would arise from a Gaussian distribution. Therefore it is argued that most sediments are not log-normal. Various characteristic parameters for the hyperbola have been developed and are reviewed in Bagnold and Barndorff-Nielsen (1980), together with several examples of their application. However this mode of presentation and description has not achieved wide usage.

Multimodal Nature of Sediments

Naturally, on a world-wide basis, there appear to be gaps in the sediment size distribution in which relatively fewer grains are found. This results in a multimodal average sediment. There is generally a paucity of material in the range 0–2 phi and also in the range 3.5–5 phi. The former appears to be the result of weathering of rocks, either into joint and cleavage fragments of larger size, or into separate crystalline grains in the sand size. The latter division may well be the result of the change in the technique of analysis, coming, as it does, in the awkward interval between sieving and pipette analyses. However it does tend to divide the fairly common sand from muds that are mineralogically distinct. Some workers have also stated that there is another zone of scarcity between the silts and clays.

Ternary Diagrams

Because of the multimodal nature of sediments, easy and useful classification can be achieved by plotting the percentages occurring in each mode on a triangular diagram (Figure 2.6). Each apex is 100 per cent of that particular

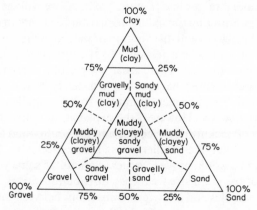

Figure 2.6 Ternary diagram for mixtures of clay, sand and gravel. *After Shephard, 1954, J. Sediment. Petrol., **24**, 151–158. Reproduced by permission of the Society of Economic Paleontologists and Mineralogists*

model size and the triangle can be divided into a series of areas, such as sandy-gravel, gravelly-sand, etc. The most commonly-used triangle is that for gravel–sand–clay, but sand–silt–clay is also useful. This technique forms a good means of comparative description, and can illustrate trends in suites of samples taken from a particular environment, or divide samples from different environments, in a very simple way.

Statistical Parameters

Mean Size

For a normal distribution, the mean, median and mode all coincide and can be represented by the size of the 50th percentile. Slight deviations from normality, however, can make these three measures different. To represent accurately the mean we need to calculate the centre of gravity of the area beneath the grain size curve, or the centre of gravity of the curve itself on the cumulative distribution. This calculation can be done mathematically in the same way as determining the centre of gravity, by considering the moment of the percentages in each of the constant size increments and, consequently, this method is known as the moment method. Seward-Thompson and Hails (1973) compare programs for doing this.

The mean size $\bar{x}_\varphi = \sum \dfrac{\text{Percentage} \times \text{grain size}}{100} = \sum \dfrac{f\,m_\varphi}{100}$ (2.3)

where f is the percentage in each grade of size m_φ.

Because of the open-ended nature of the grain size curves, an assumption has to be made concerning the mean size of the percentages passing the finest sieve, and also of that retained on the coarsest sieve. These assumptions are important since, though the weights may be small, the distance from the mean is large, and their moments significant. The relevant size is taken as the centre point of each grade. Consequently, the interval between sieve sizes must be maintained at $\frac{1}{4}$ phi throughout the range to keep errors to a minimum.

Sorting

Within the grain size distribution we need some measure of the spread of grain size. This is given by the standard deviation (also called the sorting), which is calculated as the second moment about the mean.

$$\sigma_\varphi = \sqrt{\frac{\Sigma f(m_\varphi - \bar{x}_\varphi)^2}{100}} \qquad (2.4)$$

For a normal curve 68.3 per cent of the distribution lies within $\pm \sigma_\varphi$ of the mean (i.e. between the 84th and 16th percentiles), 95.4 per cent lies within $\pm 2\sigma_\varphi$ and 99.7 per cent within $\pm 3\sigma_\varphi$ (Figure 2.7).

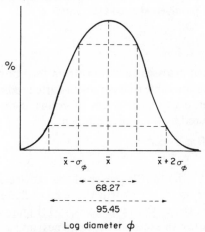

Figure 2.7 Sorting of a grain size distribution

A low value of sorting characterizes a sharp-peaked curve with few grains very much larger or smaller than the mean—in other words a uniform sample. A classification of sorting can be set up using the numerical value of the sorting.

Skewness

The skewness can be defined in terms of the third moment of the distribution:

$$\alpha_{3\varphi} = \frac{1}{100} \frac{\Sigma f(m_\varphi - \bar{x})^3}{\sigma_\varphi^3} \qquad (2.5)$$

In this formula the sum of the cube of the moments is normalized by the cube of the sorting. This makes the skewness independent of the sorting. The skewness

28

decreases the symmetry of curve about the mean. In a normal distribution the mode, mean and median coincide and the skewness is zero. In a positively-skewed distribution, the median and the mode are on the coarse side of the mean (Figure 2.8). There is then a tail in the fine sizes. The opposite occurs for negative skewness.

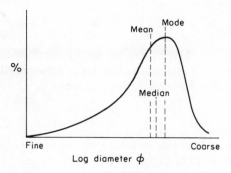

Figure 2.8 Grain size distribution for a positively skewed sediment

Kurtosis

This is a measure that assesses deviations from the normal at the extremities of the distributions. It is possible to have a well-sorted sediment with zero skewness, but which has an excess or a deficit of grains at the extreme sizes. This tendency can be measured by the fourth moment

$$\alpha_{4\varphi} = \tfrac{1}{100} \frac{\Sigma f(m_\varphi - \bar{x})^4}{\sigma_\varphi^{\;4}} \tag{2.6}$$

Again normalization by the fourth power of the skewness makes the kurtosis independent of the sorting. A platykurtic curve has a flatter peak than the normal (or a deficit in the extremes) and kurtosis < 3 (Figure 2.9). The leptokurtic curve has a sharp peak (or an excess in the extremes) and a value > 3. A normal or mesokurtic curve will have a kurtosis value of 3.

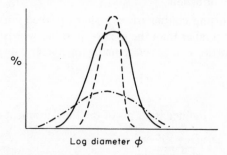

Figure 2.9 Kurtosis: ——— Normal distribution; – – – Leptokurtic curve; –·–·– Platykurtic

Graphic Measures

Before the advent of computers made calculation of the statistical parameters by the moment method easy, it was necessary to calculate them by graphical approximations. These are still widely used, partly because they can be derived from the plotted curves, and the interpretation of the results in terms of the distribution is simpler.

Trask (1930) proposed the use of the median (the 50th percentile) as a measure of the sediment size, and the ratio of the 25th and 75th percentiles as a measure of the sorting. As we have seen, neither is an accurate representation of the statistical distribution, but due to their simplicity they are still occasionally used.

Inman (1952) proposed the average of the size of the 84th and 16th percentiles, $M_\varphi = \varphi_{84} + \varphi_{16}/2$ as a measure of the mean. This is reasonable for nearly normal curves but fails to represent accurately badly skewed curves. For sorting he proposed the Phi Standard Deviation

$$\sigma_\varphi = \frac{\varphi_{84} - \varphi_{16}}{2} \tag{2.7}$$

since these two percentiles are two standard deviations apart on a normal curve. This measure, however, does not cover very much of the curve. McCammon (1962) has calculated the efficiencies of these measures by comparing them with the moment mean and sorting for normal distributions. The mean M_φ was 74 per cent efficient and the sorting σ_φ 54 per cent.

Inman (1952) suggested two measures for the skewness, one for the central part of the curve and the other for the extremes. These are

$$\alpha_\varphi = \frac{\varphi_{84} + \varphi_{16} - 2\varphi_{50}}{\varphi_{84} - \varphi_{16}}$$

and

$$\alpha_{2\varphi} = \frac{\varphi_{95} + \varphi_5 - 2\varphi_{50}}{\varphi_{84} - \varphi_{16}} \tag{2.8}$$

The second of these measures can have an absolute value greater than 1.00 since it is normalized in a different way from the first. Also, since in comparing the skewness in the extremes with the sorting in the centre, the skewness is not independent of sorting.

Inman also proposed a measure of kurtosis

$$\beta_\varphi = \frac{(\varphi_{95} - \varphi_5) - (\varphi_{84} - \varphi_{16})}{\varphi_{84} - \varphi_{16}} \tag{2.9}$$

Normal curves would have a β_φ of 0.65 using this index.

Folk and Ward (1957) proposed an index for the mean

$$M_z = (\varphi_{16} + \varphi_{50} + \varphi_{84})/3 \tag{2.10}$$

which McCammon (1962) rated as 88 per cent efficient. They also developed an Inclusive Graphic Standard Deviation by including two more points on the curve, the 5th and 95th percentiles which are 3.3 standard deviations apart, and taking the average of this and the Inman measure σ_φ they obtained

$$\sigma_I = \frac{\varphi_{84} - \varphi_{16}}{4} + \frac{\varphi_{95} - \varphi_5}{6.6} \tag{2.11}$$

McCammon (1962) rated this as 79 per cent efficient. Folk and Ward proposed the descriptive scale of sorting based on this measure shown in Table 2.2. For

Table 2.2 Descriptive scales of sorting, skewness and kurtosis and (Folk and Ward, 1957)

Sorting σ_I		Skewness S_{KI}		Kurtosis K_G	
Very well sorted		Very negative	−1.0	Very platykurtic	
	0.35				0.67
Well sorted		Negative	−0.30	Platykurtic	
	0.50				0.90
Moderately sorted		Nearly symmetrical	−0.10	Mesokurtic	
	1.00				1.11
Poorly sorted		Positive	0.10	Leptokurtic	
	2.00				1.50
Very poorly sorted		Very positive	0.30	Very leptokurtic	
	4.00				3.00
Extremely poorly sorted			1.00	Extremely leptokurtic	

skewness, they combined the two parameters of Inman and obtained an Inclusive Graphic Skewness

$$S_{KI} = \frac{\varphi_{16} + \varphi_{84} - 2\varphi_{50}}{2(\varphi_{84} - \varphi_{16})} + \frac{\varphi_5 + \varphi_{95} - 2\varphi_{50}}{2(\varphi_{95} - \varphi_5)} \tag{2.12}$$

This measure is independent of sorting and has absolute limits of ± 1.00. The value S_{KI} is approximately 0.23 of $\alpha_{3\varphi}$, the skewness obtained by the method of moments.

For kurtosis they introduced the Inclusive Graphic Kurtosis

$$K_G = \frac{\varphi_{95} - \varphi_5}{2.44 \ (\varphi_{15} - \varphi_{25})} \tag{2.13}$$

In a normal curve the spread between the 5th and 95th percentile is 2.44 times the spread between the 25th and 75th percentiles. Thus a normal curve has a Graphic Kurtosis of 1.00. Mathematically the minimum value is 0.41, and the highest possible seems to be about 8.00. Descriptive scales of skewness and kurtosis are shown in Table 2.2.

From arguments based on the packing of bimodal mixtures, in which the 75th and 25th percentiles are significant (see page 31), Dyer (1970a) developed indices for sandy gravels by adapting Folk and Ward's formulae. The sorting

$$\sigma_D = \frac{\varphi_{84} - \varphi_{16}}{4} + \frac{\varphi_{75} - \varphi_{25}}{2.67} \tag{2.14}$$

the spread between the 25th and 75th percentiles being 1.33 standard deviations. For skewness

$$S_{KD} = \frac{\varphi_{84} + \varphi_{16} - 2\varphi_{50}}{2(\varphi_{84} - \varphi_{16})} + \frac{\varphi_{75} + \varphi_{25} - 2\varphi_{50}}{2(\varphi_{75} - \varphi_{25})} \qquad (2.15)$$

The proposed kurtosis was

$$K_D = \frac{\varphi_{85} - \varphi_{16}}{1.5(\varphi_{75} - \varphi_{25})} \qquad (2.16)$$

In a normal curve the spread between the 16th and 84th percentile is 1.5 times that between the 25th and 75th. Thus the normal curve has a K_D of 1.0.

SEDIMENT PROPERTIES

Shell Content

A large proportion of many marine sediments is composed of broken and complete shell material, in some instances forming almost the complete sediment. There will be two fractions in the shell: that which was transported with the sediment and that which was not. The latter are most likely to be either alive or recently dead and should be eliminated. However there are no general criteria to separate the two fractions, except that the transported shell is most often broken or abraded.

If shelly samples are sieved, the shell material is generally retained on the coarser sieves and the mineral grains pass through to the finer ones. The result will be a positively skewed distribution curve. Sieving the mineral grains separately shows that they are much better sorted. However, if the total sediment sample is passed through a sedimentation column, the resulting curve shows a more log-normal distribution. The shell fragments, though larger than the mineral grains, fall at much the same velocity because of their shape and density. They exhibit hydraulic equivalence to the quartz sand grains.

The normal way of treating shelly sediments is to remove the shell by digestion with dilute hydrochloric acid and to consider the mineral grains as an entirely separate and coherent population. This may not be adequate in many circumstances, particularly when measurements of threshold of movement or of sediment transport rates is attempted. Some useful information, however, can be obtained from maps of shell content when used in conjuction with other information. It can be expected for instance that with transport away from the source, the percentage of shell in the sediment will decrease, and because of abrasion it will be of decreasing size.

Packing of Grains

The way in which the grains in a sediment are arranged must affect the stability of the surface grains—it is harder for the fluid flow to move a grain out of a dense packed surface layer than if it were loose. Also the state of packing must be related to the way in which the sediment was deposited. Denser packing appears where the bed is actively subject to wave and current activity and looser packing occurs where sediments settle from suspension in still water. In order to attain stability in the water flow, the sediment probably attains a minimum porosity, a maximum degree of packing.

The various ways of systematically packing spheres have been investigated by Graton and Fraser (1935). They show that there are basically four ways of stacking the grains (Figure 2.10). In the cubic arrangement the sphere centres form the corners of a cube. This is obviously the most unstable arrangement. The packing fraction, or packing density (the ratio of solid occupied space to total space), for this is 0.52. In other words the porosity is 48 per cent. By displacing the upper layer slightly, the spheres will rest in the hollows formed by the four lower spheres. This is the tetragonal arrangement which has a packing density of 0.70.

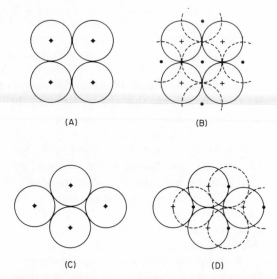

(A)   (B)

(C)   (D)

Figure 2.10 Packing of spherical grains: (A) Cubic (B) Tetragonal (C) Orthorhombic (D) Rhombo- hedral
· Centres of upper layer of spheres, full circles.
+ Centres of lower layer, dashed circles

In the orthorhombic packing the centres of the lower grains form a diamond and the centres of the upper layer are directly over them. This has a packing density of 0.60. Displacing the upper layer into the hollows gives the rhombo- hedral arrangement which has the greatest packing density of 0.74, a porosity of 26 per cent.

Most sediment grains, however, are non-spherical and may be better visu- alized as prolate spheroids (rugby footballs). Allen (1970a) has shown that the packing density is the same as for spheres except in the cubic packing, in which the packing is a function of the spheroid orientation and axial ratio.

Measurement on random packing of spheres has shown that there are quite large natural variations in packing which depend on the rate of introduction of the particles into the container. The porosity increases with deposition rate to some limiting value, as not all grains have time to roll into a minimum energy position in a hollow before being supported from the sides by other grains. The porosity can subsequently be considerably reduced by simply tapping the container, the particles moving into a configuration with a lower potential energy. This phenomenon has been investigated by Scott (1960), who has shown that there are two limiting random packings of uniform spheres: a dense one with a packing density of 0.64 and a loose one with a packing density of 0.60.

Obviously non-uniformity of grain size will have a large effect on packing density and porosity, because small grains can occupy the large void spaces. As a consequence, packing density increases and porosity decreases. The effect of size mixing on porosity obviously declines rapidly as the particle sizes approach each other.

There is a certain diameter ratio at which a smaller sphere can pass through the passages between the larger voids. This is the 'critical ratio of entrance' (Fraser, 1935). For cubic packing this ratio is 0.414 and for rhombohedral packing it is 0.154. Thus, for a grain size less than about 1/7th of the larger grains, more or less free passage through the pore spaces can occur and the smaller particles can flow into the coarser lattice as a separate deposition stage. Also there is a 'critical ratio of occupation', which gives the diameter of the largest grain that can sit in the voids without disturbing the coarse grain lattice. In this instance the particle must have been emplaced at the same time as the coarser grains. For the cubic case the ratio is 0.732, and for the rhombohedral case where there are two sizes of voids it is 0.414 and 0.225

Fraser (1935) has shown that in a binary mixture with a diameter ratio of 1 : 6.3, there is a minimum porosity when the proportion is 25 per cent small spheres to 75 per cent large. In the limiting case when the diameter ratio is 1 : ∞, the small spheres can occupy 0.64 of the void between the large spheres. This gives a maximum packing density of 0.87 at a relative composition of 26.65 per cent to 73.35 per cent. Yerazumis et al. (1962) found a maximum packing fraction of 0.684 per cent at a relative composition of 26.5 per cent to 73.5 per cent for a diameter ratio of 1 : 3.8. It appears from these results that the maximum packing fraction varies with the diameter ratio much more than with the relative composition.

Ternary mixtures have been examined by Dexter and Tanner (1971), and Statham (1974) has investigated compositions in the intermediate diameter ratio range. He obtains the results summarized in Figure 2.11, with minimum porosities at about 70 per cent large: 15 per cent medium: 15 per cent small grains. These results also indicate maximum packing densities for binary mixtures at about 70 per cent large to 30 per cent small spheres.

Theoretically the absolute sizes of the grains should have no effect on the porosity. However, for natural sediments, the porosity tends to increase with decreasing grain size. This is the result of the increased importance of friction,

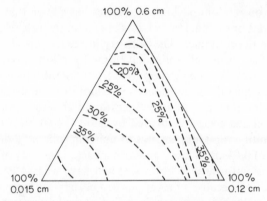

Figure 2.11 Porosity % for a mixture of three sizes of sphere. *From Statham, 1974,* Sedimentology, *21, 149–162. Reproduced by permission of the International Association of Sedimentologists*

ionic forces and angular grains as the size decreases. Platy grains in particular can lead to the formation of bridging structures giving larger than average void spaces. Typical values of porosity for some natural sediments are listed in Table 2.3. For shell material, Allen (1974) has shown packing densities in the range 0.1–0.2. Obviously badly-sorted sediments are likely to be less porous than well-sorted ones.

Table 2.3 Typical porosity values

Grain size	Porosity (%)
Coarse sand	39–41
Medium sand	41–48
Fine sand	44–49
Fine sandy silty clay	50–54

An additional measure of packing is the linear concentration λ, where λ is the ratio of the mean diameter of the grains to the mean free separation distance. This is related to the packing density for equal size spheres by the equation

$$\lambda = \frac{1}{\dfrac{C_*}{C}^{\frac{1}{3}} - 1} \tag{2.17}$$

where C_* is the maximum static packing density $C_* = 0.74$. Consequently $\lambda = \infty$ when the spheres are packed rhombohedrally. When the particles are non-spherical and non-uniform the mean diameter of the grains may not be an adequate representation of the packing state. This measure is not widely used, however.

To move the upper layer of grains over the lower layer by a relative shearing motion requires the particles to move upwards out of the hollows, but only in the tetragonal and rhombohedral arrangements. This causes an increase in porosity. The phenomenon is called dilation or dilatancy. It is most easily exemplified by the observation that when standing on a wet beach a zone of drier sand forms around one's feet. This is caused by the pressure dilating the sand and the pore spaces becoming undersaturated with water. Once the pressure is removed the reverse happens. Dilatancy also occurs when grains avalanche down a slope.

Angle of Rest

Grains can be piled on each other until they reach a critical angle at which failure takes place. An avalanche of grains then occurs down the slope until a stable angle is attained, an angle slightly less than that for failure. The definition and determination of the angle of rest is not quite as straightforward as one might suppose. There appear to be large discrepancies between the angle of repose of individual particles on a fixed bed, and of a layer on a loose bed. For single grains values exceeding 50° have been found, which compare with 32–35° for the mass angle of repose. In general, the angle increases with decrease in size and with decreased sphericity. Allen (1970b) has investigated theoretically the values of the angle of initial yield φ_i, from the point of view of the angles of tilt necessary to roll spheres out of the hollows in the regular packing states. The cubic packing gave $\varphi_i = 25°$, and rhombohedral $\varphi_i = 38°$, when surface roughness is neglected. Friction between the grains would increase these angles considerably. For haphazard arrangements with packing densities between 0.64 and 0.60, there is an upper possible limit in the range $32° > \varphi_i > 22°$ and a lower possible limit in the range $30° > \varphi_i > 15°$. For a very fine sand grade beach sand experimental values were between 31° and 43°, the values rising with packing density.

Miller and Byrne (1966) carried out experiments with single spheres and grains of size D resting on a bed of a different size. For natural grains they found that the angle of repose was represented by

$$\varphi = 61.5 \left(\frac{D}{D_B}\right)^{-0.3} \qquad (2.18)$$

where D_B is the bed particle diameter. When the upper particles are larger than the bed grains the angle of repose is small, and when the upper grains are small they hide in the spaces and are difficult to roll out. However, Equation 2.18 states that for equisize particles the angle of repose is 61.5°. In comparison with the other results, it may be more appropriate to use a value of about 32 as the coefficient in Equation 2.18.

The angle of final repose, the angle of residual shear φ_r, is also variable and a function of porosity. For a series of glass spheres, sand and gravel of different sizes Statham (1974) showed values of φ_i varies from 24°–51° and φ_r from

$22°-35°$. The largest values were for the coarsest material, φ_i showing more variation than φ_r. The values of $(\varphi_i-\varphi_r)$ varied between $2°$ and $16°$. Statham also investigated binary and ternary mixtures and found φ_r rising with decreasing porosity. It appeared that tests including natural sand particles gave higher values than those using only glass spheres, pointing to the probable importance of the surface roughness of the particles. Results for a ternary mixture of glass spheres is shown in Figure 2.12. They give a maximum value of φ_r of about $28°$, and there is a noticeable inverse correlation with porosity when compared with Figure 2.11.

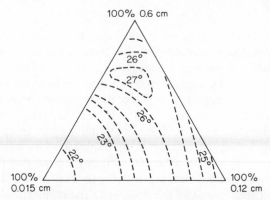

Figure 2.12 Angle of repose φ_r for a mixture of three sizes of sphere. *From Statham, 1974,* Sedimentology, *21, 149–162. Reproduced by permission of the International Association of Sedimentologists*

Observations in nature on the avalanche slope in the lee of desert dunes show angles generally of about $30°-33°$.

Grain Fabric

In a sediment, the grains are seldom deposited in a manner that results in a homogeneous sediment with uniform properties throughout. The non-uniformity gives the sediment a grain fabric. An X-radiograph of the sediment core would show laminations of varying thickness, sometimes of sub-millimetre scale in fine sediments (Figure 2.13). When these laminae have been formed from a settling suspension, they are likely to show a size grading, coarsest grains at the bottom and becoming finer towards the top. This structure is called graded bedding. Obviously sampling the sediment to obtain sensible grain size curves causes problems when laminae are present. When a grain fabric is not present

Figure 2.13 X-radiograph of a mud core showing fine-scale lamination of clay and fine silt. Length of core 18 cm

other disturbing factors such as animals, which produce bioturbation, or vertical water movements producing fluidization must be invoked.

Elongated particles tend to roll along the bed with their long axes normal to the direction of movement. However, their most stable position on the bed is with their axes aligned with the current, because in this attitude they present the least resistance to the flow. Additionally, they tend to have their nose down in the upstream direction so that they do not get flipped over by the current getting underneath. This attitude is called imbrication and the angle of upstream inclination, the imbrication angle, is generally $8°-10°$.

Because grains tend to align themselves with the current flow, there is considerable interest in measuring their orientation as a means of determining the current direction at the time of deposition. To do this requires an oriented, undisturbed sample of the sediment which then has to be impregnated with resin or plastic so that the grain orientation is frozen. The resulting 'rock' can then be sliced and thin sections taken in three orthogonal planes can be examined under a microscope. By selecting a large number of grains randomly, a rose-diagram can be constructed to show a histogram of the frequency that grains are oriented in particular angular segments. This technique was used by Curray (1956) in studies of recent coastal sands.

In order to achieve a more rapid measurement of grain orientation Winkelmolen et al. (1968) developed a method of optically scanning a black and white photograph of the prepared sample surface. Carrying this out in several directions gave a variation in reflected light intensity with the grain orientation. There are now commercial beam scanning devices which give similar results.

The three-dimensional grain orientations can best be done, however, using the anisotropy of magnetic susceptibility. The technique has been described by Rees (1965) and Hamilton and Rees (1970). It relies on the fact that any ferromagnetic grain which is put in an induced magnetic field has free poles on its surface which oppose the inducing field and tend to demagnetize it. The strength of the demagnetizing field depends on the geometry of the grain, being least when it is applied across the longest dimension and greatest when the inducing field is applied across the shortest dimension. This effect is only significant in minerals with a high susceptibility, such as magnetite, but only a minimum number of about 100 such grains within a 10 cm^3 specimen is required. Consequently the technique is best applied to reasonably fine sands. The anisotropy of susceptibility is measured by placing a small sample within a known magnetic field in a torsion magnetometer, and deriving the forces on the sample by measuring its deflection within the magnetic field. When these measurements are repeated with the specimen aligned on three orthogonal axes, the direction of maximum and minimum anisotropy, and consequently the mean grain orientation, can be calculated. Of course the original sample has to be oriented with respect to magnetic north on collection. The measurement with the torsion magnetometer takes only 15–20 minutes per specimen. Criteria have been evolved with this method to determine whether disturbance to the depositional fabric has occurred.

ENVIRONMENTAL INTERPRETATION OF GRAIN SIZE CHARACTERISTICS

Relating the grain size distributions to the fluid flow relies on a definition of the different ways in which sediment grains are transported. Bedload comprises grains that are rolling, sliding or undergoing short hops (saltation). They are moving within a few grain diameters of the bed and are travelling more slowly than the fluid. The bedload is sometimes also called the traction load. Suspended load are those grains travelling with the water, at the same velocity, and which are supported by the turbulence. These grains settle towards the bed when the flow velocity diminishes. Wash load are those grains which are perpetually in suspension, even when the fluid has been flowing at a low velocity for some time. Other characteristics of these modes will be discussed later.

The fact that fine grains are much more easily moved than coarser grains and will tend to travel faster and further has led to one of the basic maxims of marine geology, that of 'fining down the transport path': the mean sediment size decreasing with distance from the source. Near the source the fine fraction would be transported in suspension, but the coarser fraction would be transported on or near the bed. Consequently, a sample of the mobile fraction of the bed surface would have a relatively large mean size, would not be very well sorted and would be skewed towards the fine fraction (positively skewed). Under extreme conditions where all of the fine material has been winnowed from the surface materials, a coarse lag deposit a few grains thick is left on the surface protecting the material beneath. If this is removed then further erosion of fine grains will occur. Flume experiments on 'aging' of the bed shows that the mean diameter coarsens with time and the sorting improves (Rana *et al.*, 1973).

Further from the source, where the currents are lower, most of the coarser material will have been left behind. The bed now becomes skewed towards the coarse grades (negatively skewed), but should be better sorted. The coarsest material would only be moved occasionally as surface creep, the intermediate sizes would be both by saltation and suspension, and the fine material by suspension.

Even further downstream, the coarsest material is no longer present. The intermediate sizes predominate but still move in saltation. The finer material is transported in suspension. The mean grain size has thus decreased, the sorting has improved, and skewness has decreased.

Eventually the velocity is only capable of moving the intermediate sizes occasionally as surface creep, and there is obviously very much more of the fine material in the bed. The sorting will have worsened and the skewness will again be positive. In the extreme only the fine material will be present. The sediments will then be well sorted and symmetrical.

The above sequence is well described by Inman (1949) and he highlighted the role of surface creep, saltation and suspension in the transport and in the grain characteristics.

Obviously, when the coarse, intermediate and fine fractions correspond to the three modes of gravel, coarse sand and fine sand, the sorting and skewness can be described in terms of a mixture of modes. The observed grain size distributions can be produced either by several modes reacting together, or, alternatively, by selective sizes being removed from a much broader distribution, by transportation processes. Because several modes occur naturally, most people choose the former way of interpreting the size distribution. This was done for sand and gravel by Folk and Ward (1957) as an application of their statistical parameters, and they found that the relative proportions of each mode defined a systematic series of changes in the parameters. The mean size, sorting and skewness were linked in a helical trend (Figure 2.14). Kurtosis also consistently varied along the helix. Thus for pure sand or gravel, sorting was good, the distributions were symmetrical and mesokurtic. For equal mixtures of the two modes sorting was poor, and though skewness was small, the distributions were very platykurtic. They suggest that further loops of the helix would occur for finer modes.

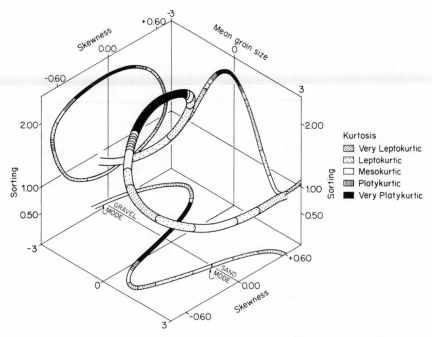

Figure 2.14 The relationship between mean grain size, sorting, skewness and kurtosis for a sand–gravel mixture in varying proportions. *From Folk and Ward, 1957, J. Sediment Petrol., 27, 3–26. Reproduced by permission of the Society for Economic Paleontologists and Mineralogists*

The relationship between mean size, sorting and skewness has been used by many people to distinguish different sedimentary environments. Mason and Folk (1958), for example, used a skewness/kurtosis plot to distinguish beach, dune and aeolian flat environments. They found the beach sands were negatively skewed and mostly leptokurtic. Also Friedman (1961) showed that beach sands tended to be negatively skewed and dunes positively skewed.

Doeglas (1946) has considered the effect on the grain size distribution curves of mixing two or more components. By plotting the curves of many sediments on probability paper he showed that they were composed of two or more almost straight lines, each of which represented a normal population. The more or less horizontal part of the curve gives the relative proportions of the components (Figure 2.15). This complex structure was considered to be the result of variations in the capacity of transporting medium. He shows that similar results can be obtained by removing the ends of a symmetrical distribution, but comparison of the material with that at the source would make it apparent if this was the case. The central half of the distribution is not considered as distinctive as the variations at the extremes, and for the fine grades the content depends not

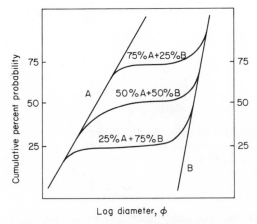

Figure 2.15 Cumulative probability plots of the grain size distributions of two normal populations, populations A and B, and mixtures of them in different proportions

only on the current conditions at the place of observation but much more on those upstream. He considers that the transported material has three components: (1) Bottom deposits of local origin; (2) Fine grades of material from far upstream; and (3) Intermediate sizes of material from less far upstream. It was recognized that the coarse particles generally rolled along the bed, the fine fraction was carried in suspension and the middle grades moved by saltation. However, these were not specifically related to the curve shapes, though three types of curve were recognized: resulting from sediments where the fines had been removed, where the coarse fraction had already been deposited, and where both had been removed in equal proportions.

The three methods of transport were also recognized by Moss (1962, 1963) and related to separate populations intermixed within the same sample. He used shape and size of the particles to distinguish the populations. By plotting the elongation function (D_a/D_b) against the long diameter D_a a curve was obtained

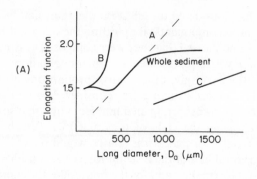

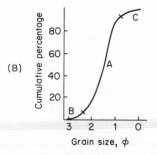

Figure 2.16 (A) Elongation function D_a/D_b plotted against the long diameter of the grains D_a, showing three component populations. (B) The three components within the cumulative grain size distribution. *From Moss, 1972,* Sedimentology, *18, 159–219. Reproduced by permission of Elsevier Science Publishers BV*

for particles longer than 100 μm (Figure 2.16A) which could be separated into three basic populations. Population A formed the bulk of the sediment, B the fine tail and C the coarse tail. Moss equated population A with the saltation load; grains roughly equal in size to those in the bed, and arranged so that the larger and slightly elongated end of the population has the same hydraulic characteristics as the smaller more equant ones. The fall velocities of the larger elongated grains and the smaller spherical ones are the same, and the elongated ones start to move on the bed at the same threshold velocity as the smaller more spherical ones. The B population is the part of the suspended load that appears near the bed and is equivalent to the interstitial population. This requires a particular size relationship to the main grain framework as required by packing criteria. Long thin grains will be able to penetrate between the grains on the bed surface as easily as the smaller spherical ones. As a consequence these grains show higher elongation functions in the larger sizes, which overlap the smaller sizes of the A population. The C population is equated with the surface creep, those grains considerably coarser than the bed grains, which are rolled along the bed. The more equant grains roll better and more consistently than the elongated ones, and consequently the C population does not have high elongation functions.

Moss proposed that the three populations could be recognized in the cumulative grain size distribution curve, as shown in Figure 2.16B.

The importance of shape in the transportability of grains was also emphasized by Winkelmolen (1969). He distinguished two types of deposit; receiving deposits show low rollability values for the coarser grains and increasingly better rollabilities for the fine grains, lag deposits have the opposite character, high relative rollability values for the coarser grains and low values for the finer grains. Poor rollability apparently promotes transport by suspension and saltation. Better rollability favoured bottom transport.

Moss's ideas were applied to grain size distributions plotted on arithmetic probability paper by Visher (1969). With this presentation natural samples are composed of several straight line segments. It is suggested that each of these is a separate log-normal sub-population which may be related to one of the three transport modes (Figure 2.17). In most of the curves the saltation population covers at least 98 per cent of the total distribution and the break between the saltation and traction population commonly occurred at about 2φ. The break between saltation and suspension often was 3φ or greater. From a large number

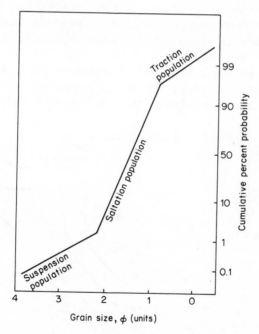

Figure 2.17 Example showing the separation of a grain size distribution into three populations relating to the mode of transport. *Following Visher (1969),* J. Sediment Petrol., *39, 1074–1106. Reproduced by permission of the Society of Economic Paleontologists and Mineralogists*

of samples Visher shows that significant differences in the grain size distributions can occur by the different transport processes acting in different relative magnitudes. From the log-probability curves it is possible to distinguish the processes which include currents, swash and backwash, wave, tidal channel, fallout from suspension, turbidity current and aeolian dune.

Middleton (1976) has considered the hydraulic interpretation of the breaks in the size distribution. He considers that a criterion for suspension is approximately $w_s/u_* = 1$ where w_s is the settling velocity and u_* is the friction velocity, and that this defines the break point between the surface creep, or traction population, and the intermittent suspension, saltation population. He also shows that the size distribution of the suspended sediment can be calculated providing the friction velocity is known, and this is confirmed by comparison with data for rivers. Consequently it might be possible to calculate the friction velocity of ancient deposits by determining the size break between the two components. However Middleton's approach considers only the mean shear and neglects the effects of turbulence in intermittently producing more extreme conditions. This has been considered by Bridge (1981a). Taking a turbulent flow with a known distribution of fluctuating velocities, he assumes that a small increment of instantaneous shear stress will transport sediment of sizes smaller than the appropriate threshold values according to a cubic transport rate. The sediment distribution curve is then given by the weights of each sediment fraction transported, when integration both over time and the shear stress range is carried out. The resulting cumulative curves compare very well with measured ones, and a distinct break is visible between the suspension and traction populations. This break corresponded to the criterion of $w_s/u_* = 0.64$, rather than the value of unity.

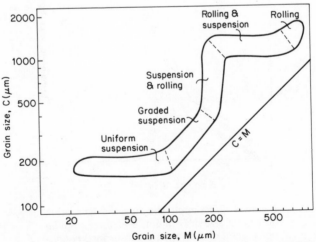

Figure 2.18 Example of a C–M diagram, and its interpretation in terms of modes of transport. *After Passega, 1964,* J. Sediment Petrol., *34, 830–847. Reproduced by permission of the Society of Economic Paleontologists and Mineralogists*

Passega (1957) used two percentiles to describe the grain size distribution curve rather than the complete curve. As a measure of the overall size the median M is used. The coarse fraction was considered more representative of the depositional agent than the fine fractions, and the coarsest 1 percentile C is taken as being representative of the maximum grain size. This is a measure of the 'competency' of the current, whether it can move the sizes that are available. These two parameters are plotted on a C–M diagram for suites of samples from different environments, as in Figure 2.18. The results show several distinct trends, a constant M section, a constant C section and a C = M section, which are interpreted as due to traction, suspension and saltation respectively. It is often difficult, however, to measure the 1 and 99 percentiles with any accuracy.

It is obvious that it is necessary to consider the areal variation in the grain size frequency diagrams to assess the overall pattern of erosion, transport and deposition, and to remove some of the effects of variations caused by provenance. So far the interpretation of size distribution curves has demanded a knowledge of the environment in which the sediment was found. The interpretation of the size analysis is therefore biased by what the analyser thinks are the processes. Extension of the results to ancient sediments in an attempt to hindcast their depositional environment is consequently hazardous. It would be better to have some quantitative but independent judgement of grouping of the samples on their grain size characteristics, and then consider the results in the light of their environment.

A technique for grouping data from samples is factor analysis. This method may be used to study the relationships between variables (R-mode factor analysis), or between objects or samples (Q-mode factor analysis). The difference is one of objective, the same data and techniques being used in both applications. In analysis of sediment data the percentage of the sediment retained on each sieve would be the variables, and could include such additional information as shell content. R-mode analysis would then consider the relationships between the variables and could define, for instance, that large grain size associated with low shell content is a common factor, and that good sorting together with low shell content is another. Q-mode analysis on the other hand, would define a series of factors, sample groupings, one of which could contain those samples having large grain size, poor sorting and high shell content, while another could have intermediate grain size, good sorting and low shell content. R-mode analysis thus would be suitable for examining common features that might relate to the processes involved in forming the sediment, whereas Q-mode analysis would be more suitable for mapping the distribution of sediments showing the same attributes, which could then be interpreted in terms of their environment.

The techniques of factor analysis are described by Harbaugh and Merriam (1968).

Q-mode analysis was applied to samples from Barataria Bay, near the mouth of the Mississippi, by Klovan (1966). Three factors were derived and these were interpreted in terms of the sample environments as surf energy dominated, current energy dominated, and gravitational energy dominated where settling from suspension was active. Stubblefield *et al.* (1975) have also used Q-mode factor analysis on a set of samples from the New Jersey Shelf. Three groupings resulted, each characterizing a particular part of the ridge topography. The environments in the three regimes were interpreted using C–M analysis.

R-mode analysis was applied to samples from the Gironde Estuary in France, by Allen, G. P. *et al.* (1972). They found three factors explained 73 per cent of the grain size variation in the sediment. By comparison with Visher's curves the sample distribution suggested that sediment fractions finer than 3φ form a suspension population, that between 3φ–1.3φ were the saltation population, and that coarser than 1.3φ–0.6φ surface creep. They also used Passega's C–M diagram, but with the coarsest 5th percentile rather than the coarsest 1st percentile. Sediments containing a significant amount > 5 per cent of the surface creep population plotted either as 'graded suspension plus rolling' or 'rolling'. The C–M pattern seemed to indicate a transport intensity gradient which was a function of this factor in the samples. When all these results were combined, four transport mechanisms were suggested; $>3\varphi$ suspension, 3φ–2φ graded suspension, 2φ–0.6φ saltation, $< 0.6\varphi$ surface creep.

All of the above techniques have been applied to regional sediment studies in many parts of the world and have formed a good basis for interpretation of patterns of sediment transport and circulation. On the basis of these models it is possible to define the sources and sinks of the sediment, where it is eroding and depositing, and the transport paths between them. However the problem remains of being able to say under what conditions the transport takes place and how fast the sediment moves. This will be examined in the following chapters.

CHAPTER 3

The Fluid Flow

PROPERTIES OF FLOW

The measurement of fluid movement can be considered in two ways. An Eulerian measurement is one where the speed and direction is recorded through time at a point fixed in relation to external coordinates. Conventional current meters measure an Eulerian velocity. A Lagrangian measurement is one where a particle trajectory is followed. Its speed and direction at any one time can be reconstructed, but they refer to different positions with respect to the external reference. Where the Eulerian velocities do not depend on time the flow is steady, and where the Lagrangian velocities are also constant the flow is additionally uniform. Thus Eulerian and Lagrangian velocities are the same only if the flow is both steady and uniform.

If a short time exposure is taken of a number of marked particles in a flow, each of them will appear as a short line, individually defining segments of a stream-line. In a non-turbulent flow the stream-lines will be fixed particle trajectories and there will be no flow across them. Let us consider a space in the flow that is surrounded by stream-lines. A_1 and A_2 are the areas of the two ends normal to the flow direction and u_1 and u_2 are the velocities through the ends (Figure 3.1).

In a small time increment the volumes flowing through the ends must be equal, provided the water can be considered incompressible. Thus

$$A_1 u_1 = A_2 u_2$$

This gives

$$Au = \text{constant} \tag{3.1}$$

Equation 3.1 is the equation of volume continuity and states that when a stream-tube narrows, the velocity of the flow increases. The flow is thus non-uniform.

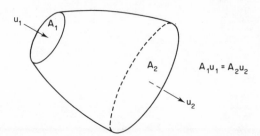

Figure 3.1 Continuity of flow through a stream
tube

Let us assume that there are pressures p_1 and p_2 acting at the ends of the stream-tube which are at depths of h_1 and h_2. The net work done moving the fluid equals the force times the distance. The work done causes a change in energy. The potential energy of the fluid increases by the height difference times the weight of fluid, and the kinetic energy of the fluid increases by half the mass times the velocity squared.

By the rules of conservation of energy the work done equals the energy gained, thus

$$(p_1 - p_2) = \rho g(h_2 - h_1) + \rho(\frac{u_2^2}{2} - \frac{u_1^2}{2})$$

or

$$\rho gh + p + \rho u^2/2 = \text{constant} \tag{3.2}$$

Equation 3.2 is known as Bernoulli's equation. It states that the depth, pressure and velocity cannot all increase simultaneously along a stream-tube, nor can any two if the third remains constant. Contraction of a stream-tube implies an increase in velocity and probably also a decrease in pressure.

It can be seen that the third term in equation 3.2 has the dimensions of a pressure, and it is called the dynamic pressure. It is readily understandable that a fluid exerts a pressure in addition to the normal hydrostatic pressure by virtue of its motion.

Bernoulli's equation is really only valid for ideal fluids. An ideal fluid is one that exerts no drag on the boundaries and which will have no internal velocity gradients when it is moving. In other words it is a fluid with infinitesimally small viscosity. Obviously ideal fluids are very unreal, but flow problems analysed using the ideal fluids concept have been shown to give reasonable enough approximations to the true flow in many situations.

Fluid Drag

A real fluid moving past a body will exert a drag force on it. The stream-lines are distorted around the moving body creating a pressure differential between the upstream and downstream sides of the body. Each element of the body will have a dynamic pressure directed normal to the surface. The sum over the body of the horizontal component of the dynamic pressures gives the form drag contribution to the total drag. In addition, there will be the frictional drag due to the shearing of the fluid past the surface of the body and this will depend on the viscosity of the fluid, the roughness of the surface and on the detailed form of the near surface flow. If the fluid is moving past the body with sufficient speed to cause turbulence then the energy used in creating the turbulence is taken from the mean flow by additional drag. The frictional drag is thus the sum of the shear stresses over the body surface. The total drag is written as

$$F_D = C_D A \frac{\rho u^2}{2} \qquad (3.3)$$

where C_D is a drag coefficient, and A is the projected area of the body normal to the flow.

In some instances it is possible to calculate the drag and the drag coefficient from theoretical considerations, but in most circumstances values for C_D have had to be found by experiment. The results have shown that the drag coefficient depends on the shape of the body and on the Reynolds number, which reflects the flow conditions in the zone close to the surface of the body. The problem of drag on spheres falling in still water will be considered later. It is perhaps the simplest of drag problems, providing the spheres are smooth, not rotating and not interacting with one another. However, much of sediment transport theory hinges on calculation or measurement of the drag on stationary or mobile grains at the bed. This problem is one of extreme complexity.

Laminar Flow

When a fluid flows slowly past a boundary the drag is related to the viscosity of the fluid. There is a velocity gradient near the wall and this produces a shear stress on planes parallel to the wall whose magnitude depends on the viscosity and the velocity gradient. The shear stress is the result of a momentum transfer normal to the wall caused by the fluid molecules in their movement passing between higher velocity layers and lower velocity layers and gaining or losing momentum as a consequence. The shear stress

$$\tau = \mu \frac{du}{dz} \qquad (3.4)$$

where μ is the coefficient of molecular viscosity which has a value about 0.01 poise (cgs units) for water. Fluids that conform to Equation 3.4, where the stress is proportional to the rate of deformation, are called Newtonian fluids.

In steady laminar flow all particles passing a point will follow the same path and that will be the same as a stream-line. There may be intermittency in the flow, but it is either stationary with respect to the boundary or it affects the whole flow giving pulsations in the mean velocity. However, flows in the sea are hardly ever laminar.

Turbulent Flow

Turbulence in the flow causes random movements of small eddies within the fluid, the instantaneous situation being complex and quickly changing. It is only by considering time average conditions that one can observe any coherence in the flow, and the stream-lines are drawn following average velocity vectors rather than the instantaneous ones. Consequently, particles passing a point in the flow do not necessarily follow the same path. The eddying movements around the average stream-line are much larger than the molecular ones and the momentum exchanges and the shear stresses are larger. The resulting turbulent shear stress has been experimentally observed to be proportional to the square of the time averaged velocity, thus

$$\tau \propto u^2 \tag{3.5}$$

In order to obtain an equation for the turbulent shear stresses analogous to that for viscosity Boussinesq linearized Equation 3.5 and introduced the approximation

$$\tau = N_z \frac{d\bar{u}}{dz} \tag{3.6}$$

where N_z is a coefficient of eddy viscosity which has a magnitude several orders larger than the molecular one, but which has the same dimensions of L^2T^{-1}. This concept has the disadvantage that the eddy viscosity is not a constant related to the fluid properties as is the molecular viscosity. It will vary with the mean velocity, as well as with other things, and it is not isotropic. Consequently the eddy viscosity is likely to vary throughout the water column. However, despite its unsound physical meaning, the Boussinesq approximation is one that is often used since, in linearizing the equations of motion, it makes them more tractable.

It is often convenient to express the boundary shear stress τ_0 in a form that has the dimensions of a velocity. This is done by defining a friction velocity u_* such that

$$\tau_0 = \rho u_*{}^2 \tag{3.7}$$

The total shear stress in the fluid is the sum of the viscous and turbulent contributions, but the viscous contribution can be considered negligible above a certain Reynolds number, except in some cases very near the wall where turbulence is restricted.

Reynolds Number

When Osborne Reynolds injected a dye streak into laminar flow in a tube he found that it kept a constant width and distance from the wall, until the flow rate was increased to a point at which the dye threads became contorted and broke up into eddies. The contortions were random with time and also with distance along the pipe. The flow had become turbulent. When this happened the resistance to flow through the pipe increased, the extra resistance being caused because of the energy of the mean flow absorbed in creating the turbulence. Experiments with different tube diameters and fluids showed that the onset of turbulence occurred at a constant value of a dimensionless combination involving the flow velocity u, pipe diameter d and fluid kinematic viscosity $v = \mu/\rho$.

This combination is now known as the Reynolds number

$$Re = \frac{ud}{v}$$

For pipe flow Re had a value of about 2000 when conditions ceased to be laminar, but full turbulence was not developed until an Re of about 4000.

The Reynolds number compares the relative importances of inertial and viscous forces in determining the resistance to flow, and a Reynolds number can be constructed for many situations by choosing the appropriate velocity and length scales. The critical value of Re defining the transition from laminar to turbulent conditions will be different for each case, however.

The Reynolds number concept is particularly useful in modelling flow phenomena. When the lengths, velocities and kinematic viscosities are chosen such that the Re is the same in prototype and model, then the characteristics of the flow are comparable and the resistance to flow is scaled accurately.

Turbulence

Measurement of the velocity vectors in a turbulent fluid averaged over a sufficiently long period will give very much more consistent results than the instantaneous ones. The random fluctuations become averaged out with time and one can distinguish a steady background water flow which moves the turbulence along and on which turbulence is superimposed. The velocities can be separated into three orthogonal components, u, v and w, which act respectively in the x streamwise horizontal direction, the y lateral horizontal direction and z vertically upwards.

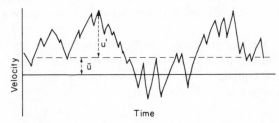

Figure 3.2 Decomposition of a turbulent velocity into mean ū and fluctuating u' parts

The horizontal velocity u at a point at any instant can be considered as being composed of a time mean flow ū and a turbulent deviation u' (Figure 3.2). Thus $u = \bar{u} + u'$. Similarly $v = \bar{v} + v'$ and $w = \bar{w} + w'$. There will be a very large number of occurrences of velocities near to the mean and much smaller numbers at higher or lower velocities. The distribution can thus be treated by statistical methods. The mean of the turbulent deviations is of course by definition zero, but the mean of the deviation squared is not zero. The root mean square values are used as an indication of the magnitude or the intensity of the turbulence, similarly to a standard deviation. It is better to consider them normalized against the mean flow or against the friction velocity, in order to be able to compare magnitudes at different levels or in different flows.

Consequently
$$\frac{(\overline{u'^2})^{\frac{1}{2}}}{\bar{u}} \text{ or } \frac{(\overline{u'^2})^{\frac{1}{2}}}{u_*}$$

The first of these expressions is known as the turbulence intensity. Near the boundary the intensities of turbulence are generally of the order 10 per cent of the local mean flow. Away from the wall the values are somewhat less. The turbulent kinetic energy is equivalent to half the sum of the variances of the turbulent fluctuations in the three coordinate directions, times ρ (see p. 199).

Let us consider what happens to particles of water involved in turbulent motion when there is a mean velocity gradient. Particles moving upwards arrive at a layer in which a higher mean velocity prevails. Since they preserve most of their original velocity they create a negative u' in the faster layer. Conversely, the particles arriving from higher velocity layers above give rise to a positive fluctuation in the horizontal velocity in the slower layers. Therefore, on average, a positive (upward) w' is associated with a negative u' and a negative w' is associated with a positive u'. Consequently the time average $\overline{u'w'}$ is not only non-zero but also it is negative. This can be expressed as an inverse correlation between the longitudinal and vertical fluctuations of velocity at a point. Though there is no net exchange of water in this mixing process, there is an exchange of momentum across a horizontal plane parallel to the mean flow direction. This momentum exchange is the shear stress τ acting on the plane. Thus $\tau = -\rho\overline{u'w'}$. There are also, similarly, stresses on two vertical planes due to interactions between the streamwise horizontal, the vertical and the lateral fluctuations i.e. $-\rho\overline{v'u'}$ and $-\rho\overline{v'w'}$.

These turbulent stresses are called the Reynolds' Stresses. By far the most important one from our point of view is that involving a vertical momentum exchange, for at the boundary it is this component of stress that is involved in moving the sediment grains.

Flow in Pipes

Though flow in pipes does not have any direct application to the sea, there are a number of general similarities that justify brief consideration.

In a smooth pipe the velocity gradient and consequently the shear stress must be zero at the centre, and the gradient and shear stresses are maximum at the wall. For viscous flow in narrow tubes the velocity profile is parabolic and the shear stress is linear. The velocity profile at the wall is also almost linear.

Flow in the pipe obeys the law

$$\tau_0 = \tfrac{1}{8} f \rho <\overline{u}>^2 \tag{3.8}$$

where $<u>$ is the cross sectional mean velocity, τ_0 is the shear stress on the wall and f is the Darcy–Weisbach friction coefficient.

For laminar flow, Equation 3.8 can also be written as f = 64/Re. (3.9)
This equation represents experimental data up to an Re $\sim$ 2000.

Above about 4000 and up to an Re of about 10^5, Blasius determined the formula

$$f = \frac{0.3164}{Re^{\frac{1}{4}}} \tag{3.10}$$

The Reynolds and Blasius experiments were both completed with smooth pipes. Nikuradse extended the measurements by glueing sand grains to the insides of the tubes. This resulted in the resistance coefficient becoming constant at high values of Reynolds number when the turbulence was fully developed. The value that the drag coefficient attained under these conditions, however, varied according to the roughness of the bed, being greater for the larger grains. For the smaller roughnesses the drag was in accordance with the Blasius formula at intermediate Reynolds numbers. The results are shown in Figure 3.3 and it is obvious that there are three flow regimes with transition zones between them; the laminar regime where the drag is given by Equation 3.9, the smooth turbulent regime given by Equation 3.10 and the rough turbulent regime when the drag depends on the scale of the roughness. As we shall see later the smooth turbulent regime has a thin layer close to the wall where the stress is transmitted by viscosity. This layer is called the viscous or laminar sublayer. In the rough turbulent regime this sublayer is disrupted by the roughness elements. As the Reynolds number rises the viscous sublayer thins until even tubes with small roughness elements tend to have rough turbulent flows.

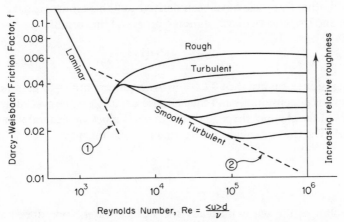

Figure 3.3 Resistance coefficient for rough and smooth pipes. Curve 1, Equation 3.9. Curve 2, Equation 3.10. d is the pipe diameter

Flow in Channels

Flows in natural channels with a free surface are rarely laminar. Because of the large water depths and high velocities the Reynolds numbers are large.

Uniform steady flow of a liquid down a sloping channel where, by definition, the water depth h is constant, is a balance between the driving gravitational forces and the resisting frictional forces. Thus

$$\tau_0 = \rho g h \sin \theta \qquad (3.11)$$

where $\sin \theta$ is the water slope.

The bed shear stress τ_0 can be related to the mean velocity by introducing a friction factor according to the quadratic friction law $\tau_0 = \rho k <u>^2$. The friction coefficient k will obviously vary with Reynolds number, with the shape of the channel and the roughness of the channel bed. $<u>$ is the cross-sectional mean flow.

Substituting in Equation 3.11 gives

$$<u> = \sqrt{1/k} \cdot g h \sin\theta \qquad (3.12)$$

A number of flow formulae have been developed by engineers engaged in irrigational channel design and most practical flow formulae are thus of the form $u \propto R^p S^q$ where $S = \sin\theta$.

R is the hydraulic radius, the ratio of the cross-sectional area to the wetted perimeter, which is used in preference to the water depth h because of the variable depth across the section of most channels.

In 1769 the engineer Chezy developed the formula $<u> = C\sqrt{RS}$ where C is the Chezy coefficient. However C is neither a constant nor a dimensionless number since it has the dimension $L^{\frac{1}{2}}T^{-1}$, (acceleration$^{\frac{1}{2}}$), as can be seen by comparison with Equation 3.12.

Several other formulae have been derived to fit experimental data from flow in rivers and irrigation canals. Manning proposed the formula

$$<u> = \frac{1}{n} R^{\frac{2}{3}} S^{\frac{1}{2}}$$

The dimensions of n are then $TL^{-\frac{1}{3}}$. Most engineers prefer to assume that the Manning n is dimensionless and then choose a suitable value according to visual estimates of the channel shape and roughness. Values are generally in the range 0.01–0.05. The Chezy coefficient is related to Manning n by

$$C = \frac{1}{n} R^{\frac{1}{6}}$$

Since flow in the sea is unsteady the surface water slope does not directly relate to the bed shear stress. Also the hydraulic radius is difficult to define. Therefore these engineering approaches are not applicable to the sea.

The velocity of flow in a channel cannot continue increasing indefinitely without a further drastic change in the character of the flow. This change occurs when inertial forces become comparable with gravitational forces. The Froude number compares the inertial and gravity forces and is expressed as

$$F = \frac{<u>}{\sqrt{gh}}$$ (3.13)

When $F < 1$ the flow is subcritical or tranquil; when $F > 1$ the flow is supercritical or shooting. At and above the critical Froude number ($F = 1$) a hydraulic jump occurs at which the flow suddenly increases in depth and conditions become tranquil again. The critical Froude number can also be thought of as defining the point at which a long surface wave can no longer propagate upstream, since the velocity of shallow water waves $c = \sqrt{gh}$.

Many of the other flow properties in channels, such as the velocity profiles, are similar to those observed in pipes.

THE BOUNDARY LAYER

Because of the friction between the flowing water and the solid boundary, the flow velocity is reduced. This reduction is greatest near the bed but gets less further away into the body of the flow. The velocity consequently deviates from the free stream value within a layer near the wall. This layer is called the boundary layer, but its thickness δ is difficult to measure, as the free stream velocity is approached asymptotically. In the laboratory the thickness is usually defined as the height where the velocity is within 1 per cent of the free stream velocity, but this is difficult to apply in the sea.

In a pure oscillatory current the boundary layer thickness

$$\delta = u_{*m}/\sigma$$ (3.14)

where u_{*m} is the maximum current, and σ is the angular frequency of the oscillation (Soulsby, 1983). For a semidiurnal M_2 tide, $\sigma = 1.4 \times 10^{-4} s^{-1}$, and the boundary layer for typical u_* values would be of the order of a 100 metres thick. In the sea, however, the rotation of the earth needs to be taken into account, and the tidal flow forms an ellipse with a major axis a, and minor axis b (see page 92). For a reasonable value of the sea bed drag coefficient (Soulsby, 1983)

$$\delta = 0.0038\left(\frac{\hat{U}_a\sigma - \hat{U}_b f}{\sigma^2 - f^2}\right)$$ (3.15)

where $\hat{U}_a$ and $\hat{U}_b$ are the depth mean flows in the direction of the ellipse axes, and f is the Coriolis parameter.

In shallow water the boundary layer can fill the whole water depth, and at maximum flow in some circumstances the velocity can be logarithmic up to the surface. In deeper water, however, the top of the boundary layer is often limited by thermal stratification. Reviews of the marine boundary layer have been given by Wimbush and Munk (1970), Bowden (1978) and Soulsby (1983).

As we have already seen there are three flow states: 1. Laminar flow;

56

2. Smooth turbulent flow; 3. Rough turbulent flow. All three have a different structure within the boundary layer. We will confine our attention to the last two, as laminar flows are infrequent in natural flows.

Within the boundary layer there is a zone close to the wall in which the shearing stress is assumed to be constant with height. This has a thickness of 0.1–0.2 δ. The flow in this inner or wall region is not directly affected by external conditions, such as variation in mean pressure gradients in the flow direction. In conditions of smooth turbulent flow (Figure 3.4A) there is a thin viscous or laminar sublayer next to the wall, which is of the order of a few millimetres thick. The motion within it is definitely not laminar, but the profile of mean longitudinal velocity is linear. Because viscosity is important, the term viscous sublayer is the most appropriate. Above the viscous layer is a transitional, or buffer, layer to the overlying fully turbulent logarithmic layer, in which the velocity profile is logarithmic with height.

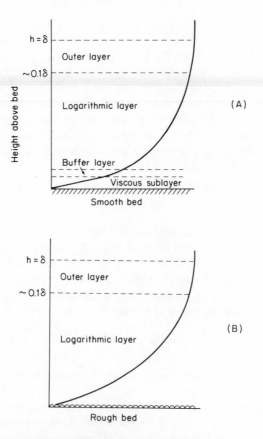

Figure 3.4 Diagrammatic representation of the velocity profiles for (A) Smooth turbulent, and (B) Rough turbulent flow. The thicknesses of the layers are not to scale

In the fully turbulent part of the layer, the scale of the larger eddies, which must be anisotropic and elongated in the flow direction, is determined by distance from the wall. The turbulent eddies influence the viscous sublayer down to the wall, causing wave-like disturbances, though the motions are of an entirely viscous nature. The spatial and time variations of velocity, and the resulting pressure gradients, plus pressure fluctuations due to turbulence outside the sublayer, cause local instabilities and even local separations. The sublayer consequently periodically grows and disintegrates.

In rough turbulent conditions the sublayer and transitional zone are absent, the fully turbulent layer extending right down below the tops of the roughness elements (Figure 3.4B).

The outer layer comprises the remaining 80–90 per cent of the boundary layer. There the flow is independent of viscosity, but is dependent on the wall shear stress and is highly affected by external conditions. These are determined by the wall shear stress far upstream, for the flow has a reasonably long memory. In the outer layer the shear stress and the turbulence energy diminishes towards the top of the boundary layer. Outside the boundary the velocity is constant, the flow is non-turbulent and the shear stress is zero.

Velocity Profiles in the Boundary Layer

Viscous sublayer

Within the viscous sublayer the shear stress is constant and the motions are dominated by viscous forces. Equation 3.4 holds, therefore the velocity gradient is constant and the velocity profile linear. Inserting the definition of the friction velocity, Equation 3.5 gives

$$u_*^2 = v\frac{d\overline{u}}{dz}$$

Integration gives

$$\frac{u}{u_*} = \frac{zu_*}{v} \tag{3.16}$$

As $u = 0$ at $z = 0$ the constant of integration is also zero. The left hand side gives the velocity profile normalized against the friction velocity and the right hand side is a distance or frictional Reynolds number which is sometimes written as $z +$. The ratio of kinematic viscosity to friction velocity gives a length scale that is used to define the thickness of the laminar sublayer $\delta_L = v/u_*$. Thus as the velocity gradient and the shear stress increase, the viscous sublayer becomes thinner.

Turbulent layer

The velocity profile in the turbulent boundary layer can be simply derived by dimensional reasoning. In a zone close to the bed where the shear stress can be considered constant, and equal to the bed shear stress, the magnitude of the velocity gradient can only be governed by the shear stress and the distance from the boundary. The friction velocity is related to the shear stress and has the dimensions of a velocity. Therefore

$$\frac{d\bar{u}}{dz} \propto \frac{u_*}{z} \quad \text{or} \quad \frac{d\bar{u}}{dz} = \frac{1}{\kappa}\frac{u_*}{z} \tag{3.17}$$

The constant κ is known as von Karman's constant, which we will see later has a value of 0.4. The length κz is known as the mixing length, and represents the mean eddy size contributing to momentum transfer at the height z.

Integration of Equation 3.17 will then give

$$\frac{\bar{u}}{u_*} = \frac{1}{\kappa}\ln z + C \tag{3.18}$$

Implicit in the derivation of Equation 3.18 is that the shear stress is virtually constant with height in a 'constant stress layer' near the boundary. The constant of integration in Equation 3.18 must be found from the conditions at the wall. Its value will depend on whether the flow is rough or smooth turbulent, and on the sizes, forms and positions of the roughness elements. The easy method of determining the constant of integration is to define a height z_0 above the wall at which the mean velocity is zero. For rough turbulent flow z_0 is known as the roughness length and can be written in terms of the grain size D, if the grains are considered as standing on the boundary. Thus we have the von Karman-Prandtl equation

or

$$\frac{u}{u_*} = \frac{1}{\kappa}\ln\frac{z}{z_0} \tag{3.19}$$

$$\frac{u}{u_*} = \frac{1}{\kappa}\ln\frac{z}{D} + B' \tag{3.20}$$

When there is a viscous sublayer present, the distance z_0 is proportional to the sublayer thickness. Thus $z_0 = \beta\, v/u_*$. Putting this in Equation 3.19 gives

$$\frac{u}{u_*} = \frac{1}{\kappa}(\ln\frac{z\,u_*}{v} - \ln\beta)$$

or

$$\frac{u}{u_*} = A\ln\frac{z\,u_*}{v} + B \tag{3.21}$$

This equation is known as the Universal logarithmic velocity distribution, since Equations 3.20 and 3.21 can be made compatible if

$$B' = B + A\ln\frac{Du_*}{v} \tag{3.22}$$

The combination Du_*/v is known as the grain or boundary Reynolds number Re_*, and is useful in expressing the structure of flow near the boundary. Thus Equation 3.21 can represent the velocity distribution in the fully turbulent layer over both rough and smooth boundaries with the appropriate value of B. Equation 3.16 and Equation 3.21 can also be combined in a formula expressing the 'law of the wall'

$$\frac{u}{u_*} = f\left(\frac{z\,u_*}{\nu}\right)$$

This applies over the constant stress layer, out to between about 0.1 and 0.2 δ.

The values for the constants A, B and B' can be determined from the pipe flow measurements of Nikuradse. The results are shown in Figure 3.5. For large grain diameters or for boundary Reynolds numbers in excess of about 70 the value of B' is 8.5. For values of the boundary Reynolds numbers less than about 5 the intercept of the curve gives a value of B of 5.5 and the slope of the line a value of A of 2.5. Thus von Karman's constant $\kappa = 0.4$. The value of B' is likely to depend on the geometry of the roughness distribution and, for anything other than a uniform distribution of equal sized grains, the right hand end of the curve on Figure 3.5 could be transposed upwards or downwards. This would also involve an alteration in the boundary Reynolds number defining the lower limit of rough turbulent conditions.

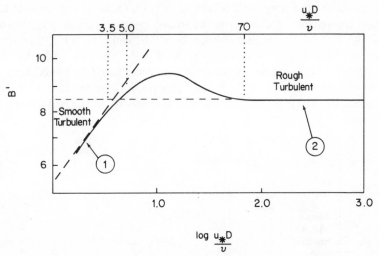

Figure 3.5 Roughness function B' versus grain Reynolds number. Curve 1, Equation 3.23. Curve 2, Equation 3.24. Solid line observed values

The velocity distribution for smooth turbulent flow will therefore be

$$\frac{u}{u_*} = 2.5 \ln \frac{z\,u_*}{\nu} + 5.5 \qquad\qquad \frac{u_*D}{\nu} < 5 \qquad\qquad (3.23)$$

and for rough turbulent flow

$$\frac{u}{u_*} = 2.5 \ln \frac{z}{D} + 8.5 \qquad\qquad \frac{u_*D}{\nu} > 70 \qquad\qquad (3.24)$$

In the transition zone between $5 > \dfrac{u_* D}{v} > 70$ there is a viscous sublayer, but it must be of equivalent thickness to the grain heights. In this range the appropriate values for the constants should be derived from the results on Figure 3.5 when necessary. Sometimes rough turbulent flow conditions are taken to start at $\dfrac{u_* D}{v} = 3.5$ (Inman, 1949).

Comparison between Equation 3.19 and 3.23 suggests that when the wall is hydraulically smooth $z_0 = \tfrac{1}{9}\dfrac{v}{u_*}$ so that $\beta = \tfrac{1}{9}$. Also at the beginning of the transitional region it is apparent the grain diameter $D = 5\dfrac{v}{u_*}$. At the intersection of the profiles given by Equations 3.16 and 3.23 the viscous sublayer thickness is $\delta_L = 11\dfrac{v}{u_*}$. Thus the transition region commences where the sublayer thickness is $\delta_L = 11 D/5$; the grains begin to disrupt the viscous sublayer when they are about a half of that layer's thickness.

Comparison between Equations 3.19 and 3.24 suggests that when the flow is rough turbulent the roughness length $z_0 = \tfrac{1}{30} D$. Other measurements suggest that the zero velocity level between the grains occurs at a distance 0.20 to 0.27 D below the top of the grains (Blinco and Partheniades, 1971; Einstein and El-Samni, 1949). These imply values of B' of 0.56–0.79, but it is rather unrealistic in these circumstances to consider that Equation 3.20 holds when extrapolated right down between the grains. This point is exemplified if we consider Equation 3.19. According to this formula the velocity at the bed when $z = 0$ is minus infinity!

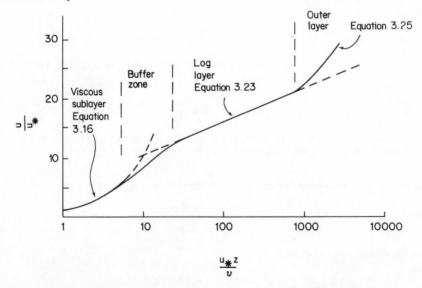

Figure 3.6 Velocity profile measured over a hydraulically smooth boundary

Various velocity profiles measured experimentally over smooth boundaries in the laboratory (Figure 3.6) show that a viscous sublayer exists out to about $z_+ = \dfrac{u_* z}{\nu} = 5$, and then a transitional, or buffer, zone out to about $z_+ = 30$. The region from $z_+ = 30$ to $z_+ = 1000$ gives good agreement with the logarithmic profile of Equation 3.23. However, at $z_+ > 1000$ when $z/\delta > 0.15$ and when we are in the outer region of the boundary layer, the observed points diverge from the logarithmic profile and can be better represented by a power law distribution such as

$$\frac{\bar{u}}{u_*} = 8.3 \left(\frac{u_* z}{\nu} \right)^{1/7} \tag{3.25}$$

This curve also fits reasonably well the observations at $z_+ < 1000$. The exponent of the power law distribution varies with the total flow Reynolds number from 1/5 to 1/10, the smaller value being for the larger Reynolds numbers. For rough boundaries, Equation 3.25 can also be written as

$$\left(\frac{\bar{u}_1}{\bar{u}_2} \right) = \left(\frac{z_1}{z_2} \right)^{1/n} \tag{3.26}$$

where $n = 5$ to 10. In this form the profile of velocity, which is a straight line on a log velocity–log height plot, can be reconstructed from a measurement at one height, if the exponent is known, but this approach gives no information on the bed shear stress. Since Equation 3.26 gives a velocity continually increasing with height, the power law profile would only be appropriate in areas where the boundary layer fills the whole water depth.

In the outer region the divergence of the velocity profile from the universal logarithmic distribution can also be considered as a velocity defect law since the velocity asymptotically approaches the free stream velocity U_∞ as the edge of the boundary layer is approached. Thus

$$\frac{U_\infty - \bar{u}}{u_*} = f \left(\frac{z}{\delta} \right) \tag{3.27}$$

The formulation

$$\frac{U_\infty - \bar{u}}{u_*} = 9.6 \left(1 - \frac{z}{\delta} \right)^2 \tag{3.28}$$

has been proposed as a reasonable representation of the velocity profiles obtained by several experimenters.

An alternative form of the universal velocity defect law has been presented as

$$\frac{U_\infty - \bar{u}}{u_*} = -5.6 \log \frac{z}{\delta} + 2.5 \qquad \text{for } \frac{z}{\delta} < 0.15 \tag{3.29a}$$

and

$$\frac{U_\infty - \bar{u}}{u_*} = -8.6 \log \frac{z}{\delta} \qquad \text{for } \frac{z}{\delta} > 0.15 \tag{3.29b}$$

The velocity defect law is most applicable in situations where the boundary layer does not occupy the whole depth, and where there is a nearly constant velocity for some depth below the water surface.

Experiments have shown that to a good approximation the bed shear stress is proportional to the square of the velocity in the boundary layer. Thus

$$\tau_0 \propto U^2$$

or

$$\tau_0 = \rho C_D U^2 \tag{3.30}$$

where C_D is a drag coefficient. Equation 3.30 is known as the quadratic stress law. Commonly in the sea, current measurements are taken at 1 m above the bed and therefore $\tau_0 = \rho C_{100} U_{100}^2$ or $u_*^2 = C_{100} U_{100}^2$. It can be seen from Equation 3.19 that C_{100} is related to the roughness length z_0, by

$$C_{100} = \left(\frac{\kappa}{\ln 100/z_0} \right) \tag{3.31}$$

By making an assumption about the drag coefficient, it is possible to estimate the bed shear stress from a near bed velocity measurement.

Velocity Profiles in the Sea

The only measurements of the viscous sublayer in the sea appear to be those of Caldwell and Chriss (1979) and subsequent publications. They profiled a heated thermistor through the viscous sublayer and into the sea bed, and found a layer with a linear velocity gradient which was 0.7 cm thick. Above it a logarithmic profile was measured. The shear stress calculated from both sections agreed to within 5 per cent. In other measurements, however, they obtained complicated profiles exhibiting a viscous sublayer, but the velocity profiles gave shear stresses only a quarter of those obtained in the overlying logarithmic layer (Chriss and Caldwell, 1982). They explained the discrepancy as due to the influence of biologically-induced surface roughness. Chriss and Caldwell (1984) obtained values for the viscous sublayer thickness δ lying between $8 \, v/u_*$ to $20 \, v/u_*$, rather than the laboratory value of $11 \, v/u_*$. This spread of values they also explained as due to roughness changes upstream of the measurement position.

The power law profile (Equation 3.26) is occasionally used in the sea. Van Veen (1938) fitted a large number of profiles from the Dover straits to this curve and obtained an exponent of 1/5.2. Dyer (1970b) found from measurements in a tidal channel that the exponent increased from about 1/10 at low flow rates to about 1/5 at high flow rates at one position. At another, however, the exponent tended to decrease slightly with increasing flow.

The power law profile has been used for estimating the bed shear stress from mid-depth current measurements. An assumed exponent allows extrapolation of the velocity to a height of 1 m above the sea bed. This value can then be used with a logarithmic profile, or the quadratic stress law, to obtain the bed shear stress, by assuming the roughness length or the drag coefficient. The errors in this procedure are large but often no better alternative is available. Davies, C.

M. (1980) has carried this process out for measurements near a sand bank in the Bristol Channel, using an exponent of 1/8.

Ludwick (1974) has compared velocity profiles measured in an area of sand waves and shoals at the entrance to Chesapeake Bay with the velocity defect law (Equation 3.28). Only a small proportion of the profiles gave a good match, mainly because of topographic effects.

The von Karman–Prandtl logarithmic profile (Equation 3.19) has been used extensively in the sea. Equation 3.19 represents a straight line when u is plotted against ln z. The slope of this line will be the value κ/u_* and the intercept will give the value of the roughness length z_0 (Figure 3.7). Consequently the bed shear stress can be calculated quite easily from the measured velocity profile in the turbulent part of the constant stress layer.

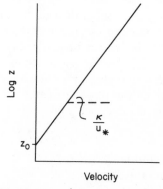

Figure 3.7 Von Karman–Prandtl
velocity profile, Equation 3.19

Measurements in the sea have verified that the von Karman–Prandtl logarithmic profile holds within the bottom 1–2 m of the flow for some of the time. This was first shown by Lesser in 1951 and subsequently by Charnock (1959), Sternberg (1966), Dyer (1970b), and Channon and Hamilton (1971) amongst others. The profile is normally fitted to data using a least squares technique, but the data has to be averaged over a reasonable length of time in order to decrease random errors to a minimum (see page 84). Some aspects of this problem have been considered by Heathershaw and Simpson (1978) and by Lesht (1980). The latter author found that, provided the data were averaged for a period in excess of 20 times the characteristic timescale of the flow, no further improvement in the logarithmic profile resulted from longer averages. In practice this requires averaging over about 10 min. The errors in the profiles also depend rather critically on the numbers of current meters used in obtaining the profile. Wilkinson (1984) provides a method for calculating the statistical errors in fitting a logarithmic profile with current meters at six heights. Using 95 per cent confidence limits, errors in τ_0 of ± 35 per cent and in z_0 of ± 77 per cent were obtained for one data set. These may be unusually small in that many authors use fewer current meters, with correspondingly worse errors.

Gross and Nowell (1983) have measured the profile at 10 levels within the bottom 3.6 m of a tidal flow about 15 m deep. Averaging over 10 minutes reduced the uncertainties in the mean values to ± 8.8 per cent for u_*, ± 50 per cent for z_0 and ± 40 per cent for τ_0. They found, however, that averaging for longer periods did not further reduce the errors. Consequently it appears that there are fundamental difficulties in obtaining values of the bed shear stress with errors of less than ± 30 to 40 per cent.

Various measurements including those of Charnock (1959), Caldwell and Chriss (1979) and Soulsby and Dyer (1981) have confirmed that a von Karman constant of 0.40 is appropriate to the sea.

Though the logarithmic profile is widely observed, there are several reasons why the velocity profile should not necessarily be logarithmic. There are also difficulties in assuming that the shear stress estimated from the slope of the profile should be that moving sediment at the bed. Slight curvature of the profile is sometimes obvious but is more often hidden in the random errors. Most of the effects cause the profile to be convex upwards. Thus, since the lowest current observation is seldom closer than 15 cm to the bed, this curvature, if extrapolated towards the bed, would give a very much steeper slope, a much smaller shear stress than that which would be calculated from the profile higher in the flow. Consequently, considerable care must be taken in fitting a logarithmic profile to observed data.

The disturbing effects are:
1. Acceleration or deceleration.
2. Variations in upstream roughness.
3. Bedforms.
4. Stratification in the water or due to suspended sediment.
5. Errors in determining the zero datum of the current meter array.
6. Waves.

These will be considered in turn, except for the effect of waves which will be dealt with in Chapter 7.

Accelerating Flow

Away from the bed the relative importance of inertia to frictional effects is greater than near the bed. Consequently in an accelerating current the flow well away from the boundary will retain a 'memory' of the preceding driving forces longer than that near the bed. The flow profile will therefore be concave upwards (Figure 3.8) and the current is smaller than the logarithmic value by an amount that increases with z. Conversely, in a decelerating current the profile is convex upwards. These profiles reflect the fact that the bed shear stress slightly leads the velocity in an oscillating flow. Soulsby and Dyer (1981) have characterized the acceleration by du_*/dt and an acceleration length scale

$$\Lambda = \frac{u_* |u_*|}{\dfrac{du_*}{dt}}$$

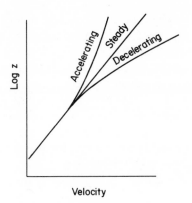

Figure 3.8 Curvature of the log-
profile produced by acceleration
and deceleration

The modulus sign ensures that the bed shear stress retains its appropriate direction. By a derivation similar to that for stratified flow, the general profile form is obtained as

$$\frac{u}{u_*} = \frac{1}{\kappa} \left[\ln \frac{z}{z_0} - \frac{z - z_0}{\gamma \Lambda} \right]$$

(3.32)

where γ is a constant which is of the order 0.04 in the sea.

If a straight log line were fitted through the data its gradient would be approximately equal to the gradient at a height $z_M = (z_L z_u)^{\frac{1}{2}}$, where z_L and z_u are the lowest and upper measuring heights. The apparent value $\tilde{u}_*$ is related to the true value u_*, and can be approximately calculated by

$$\tilde{u}_* = u_*(1 - z_M/\gamma\Lambda)$$

(3.33)

Similarly the apparent value $\tilde{z}_0$ is related to the true z_0 by

$$\tilde{z}_0 = z_0 \exp\left(\frac{\ln(z_M/z_0) - 1}{1 - \gamma\Lambda/z_M}\right)$$

(3.34)

If Equation 3.19 were used instead, both u_* and z_0 would be underestimated in accelerating currents, and overestimated in decelerating currents, typical values being ± 20 per cent for u_* and ± 60 per cent for z_0. Luckily, for most situations in the sea, the major acceleration effects occur near to slack water and they are small when velocities are high enough to cause sediment movement.

An alternative approach to the effects of acceleration using a time varying eddy viscosity has been developed by Lavelle and Mofjeld (1983).

Effect of Change in Bed Roughness

Turbulent boundary layers have a long memory and the fluid particles have a high longitudinal velocity compared with that normal to the bed. Thus a change in the velocity profile near the wall caused by an alteration in roughness will

66

gradually thicken downstream, in much the same way as a wake expands behind a boat. A change in roughness causes an internal boundary layer to form. The inner layer becomes quite rapidly adjusted to the new roughness and above it there is a zone where the profile merges into that remaining from upstream.

The transition from a smoother to rougher surface leads to a decrease in the velocity close to the bed. The near-bed velocity gradient steepens giving increased values of u_* and the value of the roughness length is increased. The form of the profiles is shown in Figure 3.9A. A decrease in surface roughness leads to the opposite effect (Figure 3.9B), the near-bed velocity increasing and the u_* and bed roughness length decreasing.

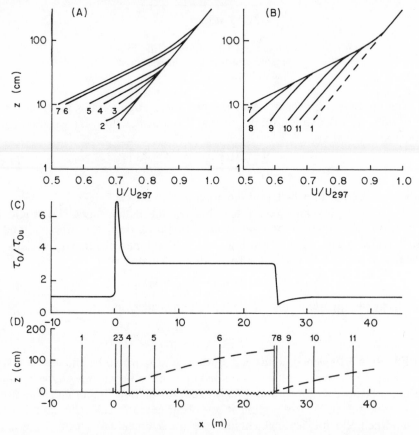

Figure 3.9 Measurements over smooth–rough and rough–smooth surface changes. *From Soulsby, 1983, in* Physical Oceanography of Coastal and Shelf Seas. *Reproduced by permission of Elsevier Science Publishers BV.* (A) Adjustment of velocity profile caused by smooth–rough change. Velocities normalized by that at a height of 297 cm. Profile positions shown in (D). (B) Adjustment of velocity profiles to a rough–smooth change. (C) Variation with position of bed shear stress relative to the upstream value. (D) Extents of the rough and smooth surfaces, the positions of the measuring stations, and the internal boundary layers (dashed lines)

The leading roughness element of the boundary, because of its exposure, will experience a higher shear stress than will occur a little way downstream. This is reflected in a slightly steeper gradient in the transition between the two profiles.

Over the change from a rough to a smoother bed there will be a sheltering effect which will decrease the shear stress initially below that at a distance. These effects are shown in Figure 3.9C.

The thickening of the internal boundary layer δ_i with distance has been shown in air to approximately follow the formula (Elliott, 1958)

$$\frac{\delta_i}{z_{02}} = a \left(\frac{x}{z_{02}} \right)^{0.8} \qquad (3.35a)$$

where a is a slowly varying function of z_{02}/z_{01}, the ratio of the downstream and the upstream roughness lengths, and which can be approximated by

$$a = 0.75 - 0.03 \ln \frac{z_{02}}{z_{01}}$$

Thus a normally takes values in the range 0.60–0.90.

An alternative formulation has been proposed by Jackson, N. A. (1976). He suggests that the thickness of the internal boundary layer δ_i increases with x according to

$$x = 3.25 \, \delta_i \left(\ln \frac{\delta_i}{z_{0m}} - 0.65 \right) \qquad (3.35b)$$

in which z_{0m} is the larger of the two roughness lengths. However, this formula is not quite as easy to use as Equation 3.35a.

As can be seen in Figure 3.9D the internal boundary layer separates two logarithmic layers with their characteristic u_* and z_0 values. Even though the upper one no longer has any contact with the bed, the u_* and z_0 values can provide an estimate of upstream conditions. A hierarchy of logarithmic profiles is formed, building up to the total flow profile. Chriss and Caldwell (1982) analysed 'kinked' logarithmic profiles, measured on the continental shelf, in terms of an upstream variation in roughness length.

Flow over Bedforms

When the roughness elements become very large, acceleration and deceleration of the near bed flow occurs over each element on the boundary. If the boundary curvature is extreme then separation can occur. Of particular interest are the characteristics of flow over wavy or triangular features that simulate the ripples and sand waves that occur in rivers and in the sea. Many experimenters have studied these effects in pipes and channels both in air and in water. On the upstream face of the wave the rising elevation of the bed causes an acceleration of the near-bed flow (Figure 3.10), and there is a decrease in pressure at the bed surface. We can explain this effect using Bernouilli's equation (Equation 3.2). The depth decreases on the upstream side of the wave and there is a decrease in pressure along the stream-line, and, as a consequence, the fluid must accelerate.

68

Type 1   Type 2   Type 3   Type 4

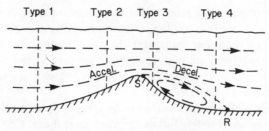

Figure 3.10 Flow over a ripple. S: Separation point.
R: Reattachment point. The streamline between S
and R defines the outer limit of the separation vortex
or lee roller. Velocity profiles corresponding to the
four types are shown in Figure 3.12

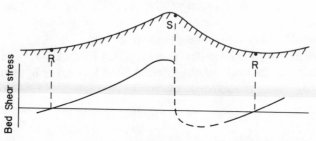

Figure 3.11 Diagrammatic representation of variation of bed
shear stress over a ripple. S: Separation point. R: Reattachment
point

The velocity gradient near the bed and also the bed shear stress increase towards
the crest (Figure 3.11), to a maximum value about equal to that for the same
flow over a flat bed (Raudkivi, 1963). At the crest the depth starts increasing, as
does the pressure, which causes an abrupt slowing of the flow. Near the bed in
the sand wave lee, the pressure increase causes the near-bed flow to be towards
the sand wave crest and the main flow detaches from the boundary, moving
upwards into the flow. This phenomenon is called boundary layer separation.
At the point of separation the zone of high shear leaves the bed and rejoins it
about six sand wave heights (Raudkivi, 1963), 4.3 wave heights (Engel and Lau,
1980), or a third of the wavelength (Karahan and Peterson, 1980) downstream
of the crest. This zone will also be a region of high turbulence intensity in both
longitudinal and vertical directions. At the re-attachment point the turbulence
intensities are greatest and surface pressures maximum (Raudkivi, 1963).
Consequently, though the shear stress is low at the re-attachment point, the
active turbulence can erode sediment. Conversely, near the crest, turbulence
intensities are at a minimum but shear stress is maximum. At a level about one
wave height above the crest the turbulence intensities are almost uniform over
crest and trough. At this level the mean flow stream-lines are approximately
sinusoidal, though not in phase with the actual sand boundary. The amplitude

of the variation is low, however, and one can consider the velocity over the wave to be constant at an elevation equal to 2–3 times the wave height over the crest. The bed features thus cause an internal boundary layer in the flow beneath which there is acceleration in the water flow over the crest of the bedforms and deceleration over the trough.

The area downstream of the point of separation and near the bed will be occupied by fluid moving down the pressure gradient, upstream against the main body of the flow. This region of reverse flow is called the separation bubble or lee roller. If the crest is at an oblique angle to the flow, a helical or spiral vortex is created as the lee roller.

Separation, however, does not necessarily occur over all features. If the pressure increase on the lee side is not too abrupt then the flow need not separate, though there will be deceleration of the flow near the boundary. Flume and wind tunnel experiments have been carried out on symmetrical, generally sinusoidal, wavelike forms, and the general conclusion appears to be that separation occurs at steepnesses (height divided by wavelength) greater than about 1/15. Thus one might expect separation to occur on features with lee slopes of avalanche steepness when the tidal velocity is at a maximum. Separation could occur at lower steepnesses, however, when the crest is sharp. On the other hand it is not necessary for separation to occur over the largest marine sand waves because of the lower slopes involved. When it occurs, the lee roller is probably an intermittent feature of variable strength which results in a reverse flow only when averaged over some time (Raudkivi, 1963). Separation is the source for most of the high frequency turbulent energy inputs into the flow, and separation can occur in the lee of grains on the boundary, as well as ripples and sand waves. The intermittency in the lee roller may be the source of turbulent boils visible at the water surface (Jackson, R. G., 1976).

Other studies have shown that the maximum shear stress also occurs slightly upstream of the crest. This feature is thought by Taylor and Dyer (1977) to be a rather critical one in determining the sediment transport across a sand wave.

The physical obstruction the bedform creates to the flow produces pressure gradients which cause an added resistance to the flow, and an internal boundary layer a few wave heights above their crests. This resistance is the form drag, and it is numerically equal to the integral of the horizontal component of the normal pressure over the length of the bedform. The form drag does not contribute to the drag causing sediment movement, but reduces the capacity of the flow to transport sediment. This is because the total drag is the sum of the skin friction and the form drag and it is only the skin friction that causes sediment movement. Near the bed the flow, though steady, is non-uniform. Nevertheless, if the bedforms are large, logarithmic velocity profiles are present and can be used as a measure of skin friction at different places on the bed wave. Above the internal boundary layer a logarithmic profile can again be measured, but the drag will then be the total drag caused by the wave bed. If the bedforms are small, such as ripples, then conventional profile measurements will only measure the total stress since the profile does not extend close enough to the bed. Consequently a true estimate of the skin friction over sands in the sea is difficult as they are generally rippled.

The interpretation of velocity profiles over bedforms as a series of logarithmic profiles was carried out by Smith and McLean (1977). They found that a total of three layers could be fitted to the data, with the maximum shear stress occurring not at the bed but at some height into the flow, because of the form drag. To provide an interpretation of the profiles an assumption was necessary about the drag of the ripples.

Velocity profiles over large gravel waves in the sea have also been studied by Dyer (1970b). He found distinctive profile shapes exemplified different parts of the wave (Figure 3.12). Type 1 showed a logarithmic profile to the water surface and was typical of the flat interwave trough areas. Type 2 occurred on the upper part of the upstream (stoss) slope of the gravel wave by a relative acceleration of the near-bed flow. Type 3 occurred in the lee of the gravel wave by sheltering of the near-bed flow by the crest. Type 4 profiles were also found over the trough areas where the near surface flow contained part of the high shear internal boundary layer visible nearer the bed in type 3 areas. Velocity profiles were also obtained over nearby sandwaves. The bed shear stress obtained from the

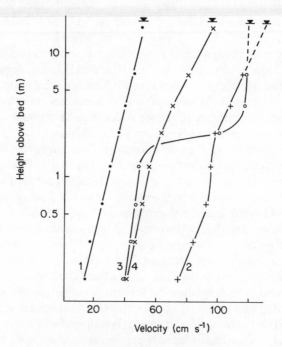

Figure 3.12 Typical velocity profiles observed in a tidal current over gravel wave bedforms. The different types occur at different positions on the bedform as shown in Figure 3.10. *From Dyer, 1970, Geophys. J. R. astr. Soc., 22, 153–161. Reproduced by permission of Blackwell Scientific Publications*

near-bed profiles varied from about 10 dynes cm^{-2} in the trough to about 100 dynes cm^{-2} at the crest (Figure 3.13). The total drag of the sand wave estimated from the upper parts of the profiles was of the order of 140 dynes cm^{-2}.

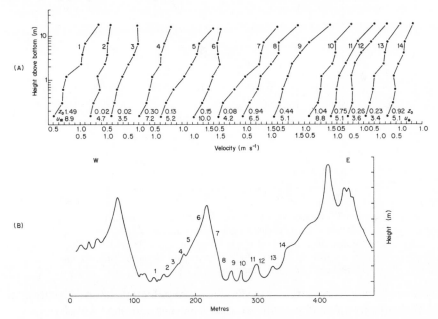

Figure 3.13 (A) Observed velocity profiles in a tidal current flowing over the bedforms shown in (B). The profile positions are shown in (B). Friction velocity u$_*$ and roughness length z values are for each profile. Separation does not seem to occur over these features. *From Dyer, 1970b, Geophys. J. R. astr. Soc.,* **22,** *153–161. Reproduced by permission of Blackwell Scientific Publications*

Many experiments have shown that the maximum resistance to flow of bedforms occurs when their steepness is about 1/10–1/15 (Davies, T. R. H., 1980). At this stage the form drag is of the same order as the skin friction. However, the form drag appears to be fairly sensitive to the steepness of the bedform and whether or not separation has taken place. Fredsoe (1982) has considered the form drag of bedforms and gives the relation in Figure 3.14, with form drag rising to about twice the skin friction at maximum bedform steepness. Smith (1977), on the other hand, has observed a form drag of 4.7 times the skin friction in a non-separating flow with suspended sediment present, but 3.1 times skin friction in a clear separating flow. At the moment prediction of form drag seems rather uncertain.

Stratification

So far we have considered a homogeneous fluid. Wherever there is a density gradient, the turbulence has to work harder to move the denser fluid upwards,

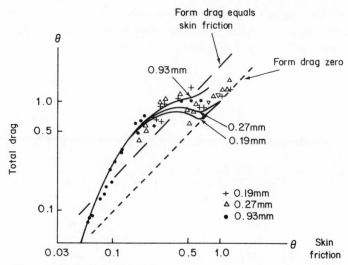

Figure 3.14 Relationship between total drag and skin friction for various grain sizes, depending on the dimensionless bed shear stress θ (see Equation 4.18). *After Fredsoe, 1982, Proc. Amer. Soc. Civ. Eng., 108, HY8, 932–947. Reproduced by permission of the American Society of Civil Engineers*

the energy being partly given to the fluid as extra potential energy. The presence of the gradient will tend to damp out the turbulence and there will obviously be a situation where the density gradient is so large that a stable interface could form with no turbulent mixing taking place across it. The relative magnitude of the stabilizing density forces and the destabilizing forces of the shear induced turbulence are measured by the Richardson number.

$$R_f = \frac{g}{\rho} \frac{\overline{w'\rho'}}{\overline{u'w'} \dfrac{d\overline{u}}{dz}} \tag{3.36}$$

R_f is the flux Richardson number. The numerator will have large values when the vertical velocity fluctuations are correlated with large fluctuations in density, i.e. when an upward velocity moves particles of high density and downwards motion lower density. The denominator is a measure of the shear induced turbulent stresses. Measurement of R_f is difficult so a gradient Richardson number is used

$$Ri = \frac{g/\rho \dfrac{d\rho}{dz}}{\left(\dfrac{d\overline{u}}{dz}\right)^2} \tag{3.37}$$

In this form the denominator is proportional to the turbulent shear stresses and the upward transfer of momentum is compared with the downward flux of mass due to gravitational forces. For $Ri > 0$ the stratification is stable, for $Ri = 0$ it is neutral and the fluid unstratified between the two depths, and for $Ri < 0$ it is unstable. When the stratification is above a certain value, turbulence will be damped out and the flow will be essentially laminar. This transition from laminar to turbulent flow under conditions of uniform flow is generally taken to occur at Ri = 0.25, though it is a gradual transition occurring over quite a wide Ri range. In the sea with unsteady, non-uniform flow the transition is likely to occur at higher Ri. It has been found that near neutral conditions occur when $0 > Ri < 0.03$, so that the effect of stratification has little effect until Ri exceeds 0.03.

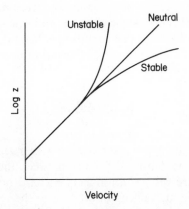

Figure 3.15 Velocity profiles under conditions of varying stratification (compare with Figure 3.8)

The presence of the density gradient will have an effect on the velocity profile as shown in Figure 3.15 and discussed by McCutcheon (1981). The presence of the stratification alters the turbulent mixing length so that it not only depends on the distance from the wall, but also on the length scale associated with the stratification. This length **L** is known as the Monin–Obukov length.

Equation 3.17 thus becomes

$$\frac{\kappa z}{u_*}\frac{d\overline{u}}{dz} = \Phi_M\left(\frac{z}{L}\right) \tag{3.38}$$

where

$$L = u_*{}^3\rho/\kappa g\overline{w'\rho'}$$

The value of z/L is dependent on Ri having the value z/L = 0 at Ri = 0 and increasing rapidly as Ri → 0.25.

Expansion of the right hand side of Equation 3.38 gives $\Phi_M = (1 + \alpha\,z/L)$. Thus

$$\frac{\overline{u}}{u_*} = \frac{1}{\kappa}\left(\ln\frac{z}{z_0} + \alpha\frac{z}{L}\right) \tag{3.39}$$

where α has a value between 4·7–5·2. This gives a log-linear velocity profile which is valid over small values of Ri and which is the same as the von Karman–

Prandtl equation for neutrally stratified flow (Equation 3.19) apart from a correction term. Because of the dependence of z/L on Ri the function φ_M is sometimes written as $\varphi_M = (1 + \alpha_1 \, Ri)^n$. A wide range of values for α_1 and n have been proposed.

On a density interface internal waves can form because of velocity shear, just as they do on the sea surface. These break like surface waves and denser fluid is mixed into the upper lighter layer. This mechanism is called entrainment and produces essentially a one-way upward mixing, unlike turbulent mixing which causes equal volumes of water to be exchanged between the two layers. Entrainment is an important process in some estuaries and where the turbulence intensity in the lower layer is less than that in the upper layer (Dyer, 1973).

An alternative way of considering the influence of density stratification is by means of an interfacial Froude number

$$F_i = \frac{u}{\sqrt{\dfrac{\Delta\rho}{\rho} \, gd}}$$

where d is the depth of the upper layer flowing with a velocity u relative to the lower layer and $\Delta\rho$ is the density difference between the layers. This has the same form as the square root of an inverse Richardson number and can be thought of as comparing the velocity of the flow to the velocity of propagation of a progressive wave along the density interface. When F_i approaches unity, interfacial waves form spontaneously and they increase in amplitude to such an extent that vigorous mixing ensues when they break at $F_i = 1$.

The effect on the velocity profiles of stratification due to the suspended sediment is dealt with in detail on p. 162

Displacement Height

If the array of current meters either digs into the sea bed or stands proud on the largest roughness elements, then the datum used for measuring the current meter heights may differ from the height of the origin of the logarithmic profile. Either will cause a slight curvature of the measured log-velocity profile. If the datum is above the bed then the profile is curved so that it is concave upwards, and if the datum is below the bed it is convex. When the height between the datum of the current meter heights and the effective bed is known, the von Karman–Prandtl equation can be modified to include the effect of this displacement height d. Thus

$$\frac{u}{u_*} = \frac{1}{\kappa} \ln \frac{z + d}{z_0} \tag{3.40}$$

If the height d is not known, the values for d and z_0 must be found by least squares adjustment to Equation 3.40, effectively altering the origin of the heights z until the velocity profile becomes straight. A method for this is given in Stearns (1970).

The example in Figure 3.16 shows a logarithmic velocity profile and profiles that would be measured if the current meter datum was 6 cm above and below the effective bed level. As can be seen, deviation from the straight log line is greatest near the bed. Generally values of *d* would be considerably smaller than this in the sea.

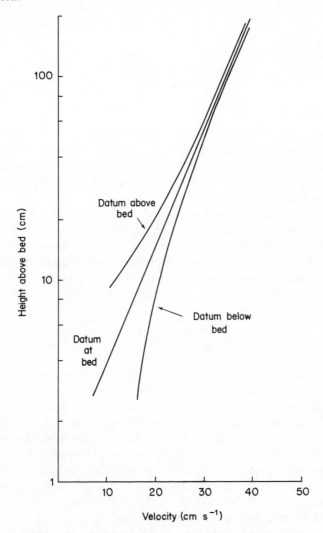

Figure 3.16 Effect of displacement height on velocity profiles. The datum of the current measurements being 6 cm above or below the true bed level

Jackson, P. S. (1981) has considered displacement height in relation to velocity profiles over rough surfaces. In this case the displacement height depends on whether the profiles are measured from the tops of the roughness elements or from the bottoms. He gives evidence that the displacement height defines the level within the roughness elements at which the mean shear acts. Other aspects of flow over surfaces with distributed roughness elements and a datum level displacement are discussed by Nowell and Church (1979) and Knight and Macdonald (1979).

Most often in the sea, however, the effect of a displacement height is small compared with random errors in the measurements and other effects, such as that of topography, and it need not be considered. However Wiberg and Smith (1983) used a displacement height of 3.5 cm to improve the fit to a log-profile of data obtained by Cacchione and Drake (1982).

Roughness Length

Obviously there is considerable interest in predicting the roughness length, or alternatively the drag coefficient, from a knowledge of the boundaries dimensions. As we have already seen, the roughness length in a rough turbulent flow was found by Nikuradse to be $z_0 = D/30$. Kamphuis (1974) in channel flow experiments found that $z_0 = D/15$.

For roughness elements larger than the individual grains, Lettau (1969) proposed the relation $z_0 = 0.5\,HS/\zeta$, where H is the average vertical extent, or obstacle height and S the cross-sectional area seen by the flow per horizontal area ζ. Consequently, for two-dimensional ripples, S/ζ approximates to the steepness, which is typically about 0.1, and for a ripple height 3 cm, z_0 would be about 0.15 cm. Wooding et al. (1973) gave an alternative form which, for ripples, is

$$z_0 = 2.0\,H\left(\frac{H}{\lambda}\right)^{1.4}$$

Again for a 3 cm ripple of steepness 0.1, the roughness length $z_0 = 0.24$ cm.

Both of the above results are in reasonable agreement with measured values. Table 3.1 shows a compilation by Heathershaw (1981) and Soulsby (1983) of many measurements of sea bed roughness length and drag coefficients that may be used for predictive purposes. The variation of results around these figures is generally encompassed by a factor of about 3, and this amongst other things, empirically takes into account the effects of biological activity, or bioturbation, in roughening the bed. Burrows, mounds, tubes and tracks all increase the hydraulic roughness, particularly of muddy sands. Nowell et al. (1981) report a doubling of the roughness caused by animal tracks.

Various studies have been carried out to define the drag coefficient, particularly C_{100}, though this can be calculated from z_0 from Equation 3.31. Sternberg (1968) examined the friction characteristics in several tidal channels and concluded that the boundary between hydraulically transitional and fully rough flow varied with the bed configuration. Using a Reynolds number based on the velocity 1 m above the sea bed, the limit was found to occur at $Re = 1.5 \times 10^5$ for rippled sand and gravel beds, whereas for complex beds of gravel and rocks the limit was at $Re = 3.6 \times 10^5$. The mean of all the data for fully rough flow gave a C_{100} of 3.1×10^{-3}.

Table 3.1   Bed roughness lengths and drag coefficients for typical sea bed types. C_{100} is the drag coefficient based on a velocity measured 100 cm above the bed (after Soulsby, 1983)

Bottom type	z_0(cm)	C_{100}
Mud	0.02	0.0022
Mud/sand	0.07	0.0030
Silt/sand	0.002	0.0014
Sand (unrippled)	0.04	0.0026
Sand (rippled)	0.6	0.0061
Sand/shell	0.03	0.0024
Sand/gravel	0.03	0.0024
Mud/sand/gravel	0.03	0.0024
Gravel	0.3	0.0047

Ludwick (1975a) carried out a similar study at the entrance to Chesapeake Bay. His values for C_{100} ranged through four orders of magnitude but with a mean value of 1.3×10^{-2}. He also pointed out that though the drag coefficient was constant for fully rough flow, for a time varying flow, a moveable bed and a size hierarchy of mobile bed forms, the value of C_{100} was not constant.

The presence of a moving layer of grains near the bed extracts energy from the flow and reduces the near-bed velocity. This alters the slope of the velocity profile and produces an effective bed roughness which is related to the thickness of the moving sediment layer, and which itself is related to the excess shear stress above the threshold of movement. This was first examined by Owen (1964) for wind blown sand and he proposed that z_0 should be related to the height of saltation. In arguments based on dimensional considerations

$$z_0 \propto \frac{u_*^2}{g} \tag{3.41}$$

However Smith (1977) has pointed out that a factor $(\rho_s - \rho)$ must be included to equate the work done on the grain by the fluid, to the height of saltation. Additionally, it is the excess shear stress above the threshold that is important rather than the total. Thus

$$z_0 = \frac{\alpha_0(\tau - \tau_c)}{(\rho_s - \rho)g} + z_N \tag{3.42}$$

z_N is the value of roughness length when the shear stress equals that at the threshold of movement and equals the Nikuradse roughness ($\sim D/30$). Comparison with measurements in the Columbia River gave $\alpha_0 = 26.3$. Dyer (1980a) found good agreement with Equation 3.42 with the same value of α_0 at low sand transport rates, but at higher rates the form drag of the ripples caused a more complex relationship.

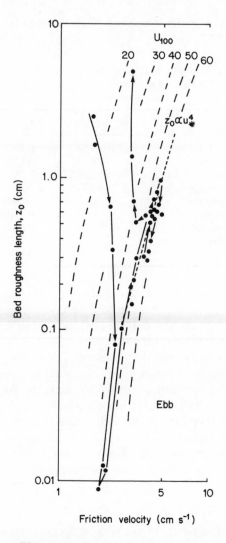

Figure 3.17 Variation of bed roughness length with friction velocity during an ebb tide. After Dyer (1980a). The data points were calculated from velocity profiles averaged over 12 min. The sequence is shown by the arrows. Curves are also shown for the values U_{100} cm s^{-1}

Dyer (1980a) investigated the variations of roughness length over a rippled sand during many tidal cycles, and found a fairly consistent pattern of change relating to the changing ripple shape and to the effects of moving sediment at spring tides. Figure 3.17 shows an example of the results. The initial decrease in apparent z_0 at the beginning of the ebb tide was due to acceleration of the flow.

The lowest value coincided with the threshold of movement and the visible commencement of separation in the lee of the ripple. The ripple then reversed its asymmetry from the preceding flood condition and gradually grew in height and moved forward. At a critical flow velocity the ripples became reduced in height, because of the intensity of the sediment movement. As the current then diminished, the roughness length increased again briefly before sediment ceased moving. There was then a rapid increase in apparent z_0 because of deceleration. During the period of ripple growth the roughness length was proportional to u_*^4, and it was estimated that about a half of the increased roughness was due to the effects of a moving grain layer, and half to form drag changes. The corresponding peak value of C_{100} was about 8×10^{-3}. At neap tides little sediment movement occurred and z_0 was approximately constant.

Heathershaw and Hammond (1979) have also found that the relation $z_0 \propto u_*^4$ matched field data in Swansea Bay.

Gust and Southard (1983) have examined the effects of weak bedload movement on the near bed velocity profiles, by carrying out comparative flume measurements on fixed and mobile beds of the same grain size. A thin viscous sublayer was present over the fixed bed, but this disappeared when up to 40 grains per cm width per second were moving. The roughness length increased by an average factor of 2 between the two situations. However, because the bed shear stress was measured in several ways, the observed increase in the velocity gradients required von Karman constant to be reduced from 0.4 to 0.32 to match with the measured bed stress. These results suggest there may be interactions between a moving grain layer and the viscous sublayer, when they are of comparable thickness, that are not present with a thicker mobile layer. Obviously there is scope for future laboratory work on these problems, though under most circumstances in the sea more than a single grain layer is likely to be mobile during most sediment transport events.

TURBULENCE

Bursting

The structure of the Reynolds stress near a boundary has been examined in the laboratory using flow visualization techniques (Nychas *et al.*, 1973; Offen and Kline, 1975). These have shown that the production of Reynolds stress is intermittent and that it is associated with a sequence of motions collectively called 'bursting'. With reference to Figure 3.18, the stress can be considered to be produced by a combination of positive and negative fluctuations in the x and z directions, i.e. four constituents.

Situations 1 and 2 represent a movement of fluid away from the boundary and 3 and 4 a movement towards the boundary. However 1 and 3 give a negative contribution to the value of $u'w'$ and consequently a positive Reynolds stress, whereas 2 and 4 give a positive contribution to $u'w'$ and a negative shearing stress. Obviously, to be realistic, motions 1 and 3 must give a higher con-

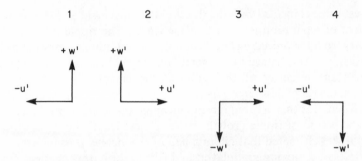

Figure 3.18 Quadrant analysis of turbulent fluctuations contributing to Reynolds stress. The origin is moving at the mean stream velocity. 1: Ejections; 3: Sweeps; 2: Outward and 4: Inward interactions

tribution to the Reynolds stress than 2 and 4. It has been found that the ejections corresponding to situation 1 and the sweeps of situation 3 comprise a surprisingly short amount of the total time. Consequently the distribution of Reynolds stress with time is very skewed. The sweeps and ejections are relatively very much more intense than the weaker outward and inward interactions of 2 and 4. It has been estimated that as much as 70 per cent of the Reynolds stress may be caused by the ejections, which seem to be about 1/3 more intense than the sweeps, but which only exist for about 20 per cent of the total time (Corino and Brodkey, 1969). Willmarth and Lu (1972) found in wind tunnel experiments that 99 per cent of the Reynolds stress was contributed in 55 per cent of the time.

Figure 3.19 shows a record of u, w and u'w' fluctuations in a tidal current. Neither the u nor w separately give a good indication of the ejections and sweeps visible in the Reynolds stress, and the large events are separated by long relatively quiescent periods. The large bursts give contributions of between 10 and 30 times the mean value of the local Reynolds stress.

Flow visualization studies very close to the boundary have suggested that the mechanism of bursting is associated with a transverse vortex rotating with the flow. The vortex is initiated from a low velocity streak in the viscous sublayer. The flow in this region consists of alternating low and higher velocity streaks. One of the streaks is lifted up in a three-dimensional disturbance and after lift up a local recirculation cell is formed beneath the streak. However, the recirculation is only apparent to an observer moving with the flow; to a stationary observer it would appear as a local velocity deficit near the bed and locally enhanced velocity at a small distance from the wall. The passage of the forward edge of the cell would produce the sweep and the trailing edge the ejection. In a rough turbulent boundary layer, when the viscous sublayer is absent, the vortex may be created by instantaneous separation in the lee of an obstacle.

As the vortices are advected downstream, they develop a 'hairpin' or 'horseshoe' shape, with the ends at the boundary and inclined at 16–20° to the horizontal (Figure 3.20). They travel outwards away from the boundary and grow larger, though less intense, as they are advected downstream. They travel downstream at about 0.8 of the mean velocity and gradually breakdown because of being stretched in the velocity gradient near the bed, and because of viscous dissipation. As the vortices are being generated intermittently at various places on the bed, at any one position different parts of the horseshoe vortices would be apparent at different times at each height above the bed, resulting in the characteristically chaotic velocity pattern. However, it is not clear how this behaviour should be scaled when applied to the sea, or even whether it applies at all.

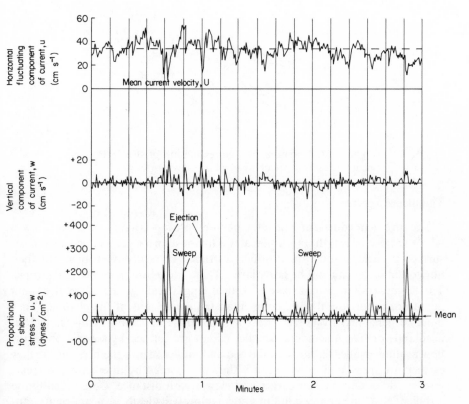

Figure 3.19 Record of the turbulent components of a tidal current and the Reynolds stress (Soulsby, personal communication)

Though the above model was obtained from laboratory observations, bursting has also been observed in the sea (Gordon, 1974; Heathershaw, 1974), though much further from the boundary. The latter author reports burst durations of 5–10 s with 20–100 s periods of lesser stress between, and 57 per cent of the stress being produced in only 7 per cent of the time. Similar intermittent turbulent events are often visible at the sea surface on calm days, especially in areas of rough topography, such as sand waves. Jackson, R. G. (1976) has described the appearance and origin of these 'boils'.

The implications of the bursting process for sediment movement in the sea are great because of the large instantaneous shears that are produced. Because most sediment movement in the sea takes place only a little above the threshold, spatial and temporal variations in burst intensity and bursting rate are likely to have significant effects on the sediment transport rates. The propagation speeds of the burst appear to be similar to the speeds of grain movement so that individual grains could be moved relatively large distances in one go. Discussion of the effects of bursting on the movement of sediment will be found on page 123.

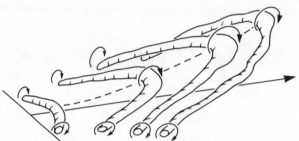

Figure 3.20 Diagrammatic representation of the development, lifting and stretching of a horseshoe vortex

Turbulence Spectra

The turbulent eddies, of which the bursting process is a manifestation, have a whole variety of length and timescales. Obviously near the sea bed or the sea surface the eddies must have rather restricted vertical dimensions, but their horizontal scales can still be fairly large. These eddies are therefore anisotropic. The largest eddies cannot have a vertical length scale greater than the water depth nor a horizontal scale larger than the dimensions of the channel. They must be generated by lateral shear against gross boundary changes, such as headlands or river meanders. Smaller eddies are formed by separation of the flow around smaller features on the bed and these rise into the body of the flow as they are translated downstream. They lose energy by interacting to create smaller eddies and become larger and slower with distance. Consequently the turbulent structure at any point in a steady flow will ideally be a stationary random function related to the boundary. The dimensions of the smallest eddies are not affected by proximity to the boundaries and become increasingly equidimensional, or isotropic with decreasing size. Eventually the interacting eddies will get broken down to scales where the energy can be dissipated as heat by viscosity. This cascade in energy from large scales to small scales has been aptly described by L. F. Richardson as:

Big whorls have little whorls,
Which feed on their velocity.
Little whorls have smaller whorls,
And so on unto viscosity,
(In the molecular sense).

The eddy sizes comprise a spectrum. The sizes of the longer period, slow eddies is limited by the channel dimensions. This is called the macroscale of the turbulence. The high frequency end is limited by viscosity and this size is the microscale of the turbulence. The spectrum is examined by analysing the proportion of the total turbulent energy which is contained in the various frequency intervals. The energy spectrum represents the distribution of energy over the various frequencies and consequently the area under the curve represents the total variance i.e. $\overline{u'^2}$. A co-spectrum can also be constructed for the cross-correlation between u and w and, in this case, the total area under the spectrum represents the Reynolds stress. However, the frequency spectrum does not necessarily give a good comparison of turbulence at different times since the eddies are being moved past on the mean current, and their apparent frequency observed at a fixed point will vary with the current speed. This can be corrected for by multiplying the frequency spectrum by $\overline{u}/2\pi$ giving the wave number spectrum.

$$\text{wave number } k = \frac{2\pi n}{\overline{u}}$$

where n is the frequency. A diagrammatic energy spectrum is shown in Figure 3.21.

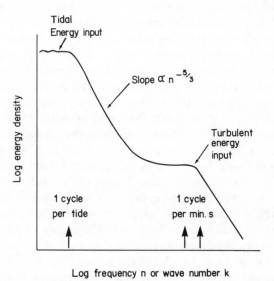

Figure 3.21 Diagrammatic energy spectrum for a tidal current

In the sea the large anisotropic eddies are the result of motions of tidal period which have very large energies. This energy is passed down to the smaller isotropic ones by the cascade process without a significant energy loss, except in areas where stratification is important. The spectrum of energy decreases proportional to $k^{-5/3}$ at frequency ranges where there is no energy input. There are major inputs of energy at tidal frequency and between a few seconds and several minutes period, as a result of the generation of turbulence by the velocity shear over the sea bed and the separation of eddies.

Averaging

Because of intermittency of the bursting process, and also longer period unsteadiness in the flow, it is apparent that one must be very careful in choosing the times over which to average the fluctuations, in order to obtain a good statistical estimate of the Reynolds stress. As far as bursts are concerned one needs a length of averaging which will include a fairly large number of bursts. Laboratory experiments have shown that the period between bursts scales with viscosity and with the bed shear stress (Blackwelder and Haritonides, 1983) according to

$$T_b = \frac{300\,\nu}{u_*^2} \tag{3.43}$$

However, this result was obtained in the laboratory close to a smooth bed in a restricted Reynolds number regime. For a u_* of $3\,\mathrm{cm\ s}^{-1}$ Equation 3.43 gives $T_b = 0.33$ s. This contrasts with measurements in the marine boundary layer (Soulsby, 1983) where for the same value of u_* a T_b of 18 s was obtained. Thus a 10 min record would sample about 30 bursts, a reasonable statistical number. The spectrum also gives us an indication of the averaging time that should be used in defining the turbulent fluctuations. The averaging time should correspond to a minimum in the spectrum of energy per unit frequency. If the chosen time is too short then some of the turbulent energy will be omitted and the values of Reynolds stress in particular will be underestimated. If the time is too long then some of the tidal energy becomes included in the spectrum and an overestimate will result. In both cases the values obtained for $\bar{u}$ will show little consistency with time, in the first case because there is still some turbulent fluctuation left to cause variation and in the other because the flow has shown significant unsteadiness, with the flow varying during the tidal cycle. Soulsby (1980) has examined the criteria necessary for selecting record length and digitization rate for near-bed turbulence measurements in tidal streams and concludes that record lengths of at least 10–15 min are required to minimize errors in turbulence parameters and Reynolds stresses. However, even with these conditions there will still be sampling errors, and Heathershaw and Simpson (1978) have outlined techniques for estimating them. With record lengths in excess of about 15 min in the tidal situation, the mean velocity and the burst period will change significantly during the record and turbulence characteristics can no longer be considered stationary. Consequently a compromise between the various factors is necessary and it is virtually impossible in most cases to achieve an estimate of shear stress to better than about ± 30 per cent.

Turbulent Properties of the Boundary Layer

There is considerable volume of laboratory data relating to turbulence in the boundary layer in both air and water, in pipes and channels, starting with the classical results of Laufer (1954), to the more recent papers by Blinco and Partheniades (1971), McQuivey and Richardson (1969) and Nalluri and Novak (1973). They cover a range of Reynolds numbers and roughnesses. Measurements in the sea have shown very large similarities between the turbulent characteristics of these different situations.

At the top of a turbulent boundary layer the turbulent eddies are not particularly restricted by the presence of the boundaries, and only the largest will be restricted in their vertical dimension, which will be limited by the height above the bottom. Towards the boundary the vertical scale becomes restricted rather faster than the horizontal. This means that the macroscale or integral length scale of the turbulence, the relative turbulence intensities and the spectral energy content all change with height. Also anisotropy moves towards increasingly shorter scales near the boundary. Of course, the absolute magnitudes of the turbulence parameters near the boundary will depend on the roughness of the boundary, the rougher boundary producing more energetic turbulence.

In a uniform channel where the boundary layer fills the whole water column, the turbulence macroscale distribution is approximately parabolic with a maximum value equal to the water depth at mid-water (McQuivey and Richardson, 1969). Near the bed the ratio of the vertical length scale to the horizontal is 0.28–0.41, increasing away from the bed (Heathershaw, 1979). Rifai and Smith (1971) in examining flow over triangular elements consider that the height of the elements determines the size of the macroscale rather than the depth of the flow, the macroscale being approximately equal to the height. The turbulent microscale is normally of the order 0.2–0.5 mm in the sea.

The laboratory results of turbulent parameters in the boundary layer are shown in Figure 3.22. Measurements in the sea have generally been obtained close to the bed, where variability tends to mask changes with height. The horizontal turbulence intensity $(\overline{u'^2})^{\frac{1}{2}}/U$ was found to be in the range 8–10 per cent (Bowden and Fairbairn, 1956) in Red Wharf Bay off the Isle of Anglesey, and the ratio of the vertical to the horizontal intensity was about 0.6.

Further results taken in Red Wharf Bay are presented by Bowden (1962). Simultaneous measurements showed that the vertical and lateral scales of the u fluctuations were similar and were about $\frac{1}{3}$ of the longitudinal scales. In the Mersey, however, both vertical and horizontal intensities were about halved compared with Red Wharf Bay (Bowden and Howe, 1963), and at the surface they were about half of the values near the bottom. The difference between the open sea and the estuarine results may have been due to stratification effects. In the absence of internal waves, stratification tends to take energy out of the vertical fluctuations and transfer some of it into the horizontal fluctuations. When internal waves are present the vertical fluctuations are relatively enhanced. Beneath the density interface, in the fairly homogeneous water close to the bottom, the major effect may be a restriction on the thickness of the boundary layer with a consequent decrease in turbulence intensities.

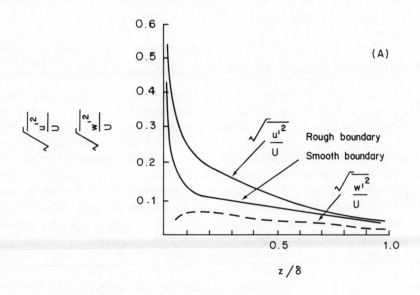

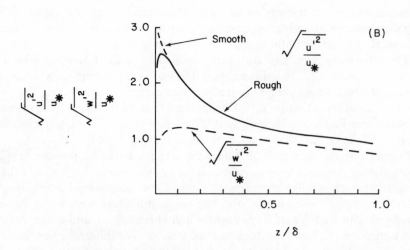

Figure 3.22 Variation of turbulence parameters through the boundary layer. (A) Horizontal and vertical turbulence intensity for rough and smooth boundaries. (B) Intensities normalized against friction velocity. Various authors

During acceleration the shear stress and the production of turbulent energy near the bed increase. These spread through the water column until at maximum current the profiles of shear stress and turbulent energy are almost linear throughout the boundary layer. As the current decelerates, the shear stress and the turbulence energy near the bed decrease, but higher in the boundary layer they decrease less rapidly, and maxima in shear stress and turbulence energy are maintained. The turbulence gradually spreads out both upwards and downwards, and decays, until it is reinforced by a new input from the bed on the next acceleration. Shear stress profiles during a tidal cycle have been measured by Bowden *et al.* (1959) (Figure 3.23). These profiles are similar to those occurring on the shorter timescale under waves and are qualitatively similar to profiles of turbulence energy.

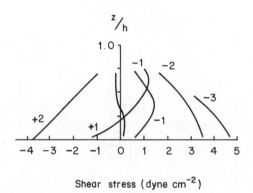

Figure 3.23 Profiles of shear stress at hourly intervals within a tidal cycle. *After Bowden* et al., *1959,* Geophys. J. R. astr. Soc., **2**, *288–305. Reproduced by permission of Blackwell Scientific Publications*

Since turbulence takes a while to decay and there is no production of turbulence near slack water, there should be, at mid-water, lower turbulent kinetic energy after slack water than before. This leads to considerable hysteresis when the turbulent kinetic energy is plotted against the Reynolds stress (Gordon, 1975). Gordon found the kinetic energy during deceleration typically twice that at the same mean flow speed during acceleration. Bowden and Ferguson (1980), however, found no such effect closer to the boundary. The persistence of turbulence over slack water should have some effect in sustaining sediment in suspension and may be a factor in hysteresis in the concentration of suspended sediment during a tidal cycle.

When waves are present the turbulence intensities at the surface will be considerably enhanced, and because of the shear stress caused by the wind blowing over the surface, the shear stress will decrease from the bottom to a finite value at the surface rather than zero. Bowden and Ferguson (1980) have considered

88

the effect of surface waves on the turbulent fluctuations measured near the sea bed. They assume a detectability level of 10 per cent of the rms fluctuations and calculate the wave-induced fluctuations from linear wave theory. Their results are summarized in Figure 3.24. This shows the mean velocities at which surface waves would become apparent in the u and w rms values in a water depth of 30 m.

Bowden and Fairbairn (1956) have examined the energy spectrum of horizontal and vertical velocities within 2 m of the sea bed in Red Wharf Bay. The vertical velocity fluctuations (w') had a peak energy at a period of 7 s and the longitudinal velocity fluctuations (u') a peak energy at a period of 70 s, but a broader spectrum. In this case the major contribution to the mean shearing stress $\overline{u'w'}$ would come from fluctuations with frequencies 0.01–0.25 Hz (4–100 s period).

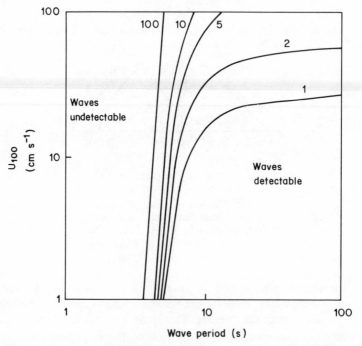

Figure 3.24 Wave amplitude (cm) at which waves at a depth of 30 m create a velocity oscillation equal to 10 per cent of the rms turbulence signal, for various wave periods and U_{100} values. After Bowden and Ferguson, 1980

Further results taken in Red Wharf Bay are presented by Bowden (1962). The spectral peak for u' was at a wave number $k = 0.25\,\mathrm{m}^{-1}$, for v' at $k = 1.5\,\mathrm{m}^{-1}$, and for w' at $k = 4.5\,\mathrm{m}^{-1}$, i.e. 50 s, 8 s and 3 s respectively for $u = 50\,\mathrm{cm\,s}^{-1}$. The peak contribution to the stress u'w' was at $k = 0.4\,\mathrm{m}^{-1}$ and most of the shearing stress was due to wave numbers k smaller than $6.0\,\mathrm{m}^{-1}$.

The limiting wave number above which the turbulence was isotropic and could not contribute to the stress was given approximately by kz = 4.0 where z is the height above the sea bed.

Soulsby (1977) has applied the interpretation of atmospheric spectra to the marine boundary layer and finds that the spectra of horizontal and vertical fluctuations, and the Reynolds stress from many positions and elevations coincide when plotted in terms of kz, where z is the height above the sea bed. This means that the peak frequency can be predicted for any situation.

TIDES

Tidal Generation

It has been known for centuries that the tidal rise and falls of the seas are related to the phases of the moon, and machines to predict tides are probably amongst the oldest of the ancestors of the modern computer. The tides can be accurately measured, predicted and modelled mathematically, and there is an extensive literature on these topics.

The earth and moon rotate around a common centre of mass that, because of its greater mass, is somewhere within the earth. The gravitational attraction between them is counteracted by the centrifugal force resulting from the orbital revolution. Since each position on the earth's surface rotates around the joint centre of mass in circles with equal radii, the centrifugal force on each point will be constant. However, on the side nearest the moon the gravitational acceleration due to the moon will be slightly more than on the side farthest away. Consequently, directly beneath the moon, the weight of an observer would be very slightly less than on the other side of the earth. On the circumferential line half way between these points there will conversely be a component tangential to the earth's surface which will be balanced by centrifugal force. The resultant tangential, or tractive forces (Figure 3.25), cause a bulge in the water surface on both sides, under the moon and on the opposite side. As the world spins, the bulge stays fixed relative to the moon and an observer at a fixed point on the earth would experience a wave going past. This is known as the Kelvin wave, and on a completely fluid covered globe it would only have an amplitude of about 30 cm. The presence of the land masses obviously causes considerable amplification, since most tides are much larger than this.

The sun has a similar effect but, despite its very much greater mass, the large distance reduces the tractive force to only about half of that of the moon. Consequently, with each daily rotation of the earth, there will be a wave component with a period of 12 hours due to the sun and one due to the moon with a slightly longer period, because the transit of the moon across the meridian is about 50 minutes later each day.

The moon also moves in an elliptical orbit round the earth. Consequently there will be a tendency for a maximum tidal range at perigee, when the moon is closest to the earth, and a minimum at apogee when it is farthest away. Additionally the tractive force will have a maximum when the sun, moon and earth are in line at new or full moon. This consequently gives the spring and neap tides that occur twice a month. Since the moon travels faster on its orbit near apogee and slower near perigee, there is a variation in the length of the basic semidiurnal tide during the lunar month between about $12\frac{1}{4}$–$12\frac{3}{4}$ hours.

Earth   Moon

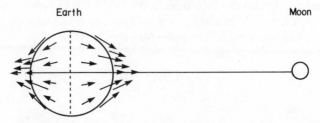

Figure 3.25 Diagrammatic representation of the tidal generating forces on the surface of the earth which leads to an accumulation of water on both sides of the earth

As the earth moves round the sun, the new or full moon occurs at intervals of 29.5 days despite the fact that the moon only takes 27.25 days to circuit the earth. Thus there will be one occasion during the year when new moon and one when full moon occur at perigee. They occur at about September and March respectively and lead to maximum spring tides, known as equinoctial spring tides, and minimum neap tides.

Further, longer term cycles occur because the planes on which the moon revolves around the earth, and that of the earth round the sun, are at an angle to each other. Consequently maximum effects will occur when eclipses occur. Additionally the earth has a perihelion and an aphelion with the sun, due to an elliptical orbit, and this gives similar effects to those of the moon.

These complex relative movements give an effectively non-repeating tidal sequence but it can be considered as resulting from a series of regular sinusoidal variations with different amplitudes and phases. The periods of oscillations can be predicted from the known astronomical movements. Some of the most important constituents are shown in Table 3.2.

Table 3.2 Principal tidal constituents

		Period (h)	Relative size %
Lunar semidiurnal	M_2	12.4	100
Solar semidiurnal	S_2	12.0	47
Lunar elliptic	N_2	12.7	19
Luni-solar semidiurnal	K_2	11.97	13
Luni-solar diurnal	K_1	23.9	58
Lunar diurnal	O_1	25.8	42
Solar diurnal	P_1	24.1	19
Lunar fortnightly	M_f	328	17

Tides in Shallow Water

Because the wavelengths of all the tidal constituents are equal to half the earth's circumference, the waves will travel as shallow water waves with a velocity of propagation which depends on the water depth, i.e.

$$c = \sqrt{gh} \qquad (3.44)$$

Because of friction, even in the deep ocean, the wave cannot keep up with the rotation of the earth so that the time of high water does not coincide with the moon crossing the meridian. Also the presence of the continents breaks up the progressive wave, reflects the energy and causes standing oscillations. The amplitude of the oscillations then depends largely on whether the fundamental period of oscillation of the ocean basin or gulf is close to $12\frac{1}{2}$ hours or not. For instance, tides in the Pacific have a higher diurnal constituent than the Atlantic because of the larger size.

The English Channel has a mean depth of about 40 m, and in $12\frac{1}{2}$ hours the tidal wave according to Equation 3.44 would travel about 900 km. The actual length is about 400 km, so that there is a half wave standing oscillation; when it is high water at the mouth, it is low water at the Dover Straits and vice versa (Figure 3.26). Thus there is an antinode at the inner end with a maximum tidal range, and a node in the middle where there is a minimum range, but maximum tidal currents. At the western end high and low water coincide with the tidal wave being generated in the Atlantic.

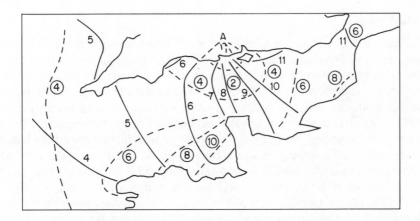

Figure 3.26 Diagrammatic representation of tides in the English Channel. Solid lines are cotidal lines joining points of equal high tide time. Numbers are hours relative to the moon passing the Greenwich Meridian. Dashed lines are corange lines. Circled numbers are mean spring tide range in metres. A: Amphidromic Point. *After Lee and Ramster, 1981,* Atlas of the Seas around the British Isles. *Reproduced by permission of the Ministry of Agriculture, Fisheries and Food*

However the tidal wave progressing into the Channel, and being reflected, will be affected by Coriolis force. This will cause the inward travelling wave to be deflected towards the right, the French coast, and the outward travelling wave to be deflected towards the English coast. Thus the wave, as shown by the co-tidal lines, progresses in an anticlockwise sense around the gulf (Figure 3.26). The node then becomes an amphidromic point, a point with minimal tidal elevation range. Also the shallow water and the consequent friction reduces the amplitude of the wave as it progresses. In the English Channel, because of the extreme frictional effect of the Cherbourg Peninsula and leakage through the Dover Straits, the tidal energy is almost entirely dissipated and the amphidromic point is forced onto Southern Britain. The tidal range at Bournemouth, being close to the 'degenerate' amphidromic point, is only about 2 m, whereas further away in the Gulf of St Malo it is in excess of 12 m.

In shallow water, friction slows down the water movement near low tide, but the crest of the tidal wave travels faster, as can be seen from Equation 3.44. The tidal wave then ceases to be a smooth sinusoid and the tidal rise becomes faster than the tidal fall, the high water becomes a rather sharply peaked event and low water is a long, flat trough. This distortion can also be described in terms of additional sinusoidal constituents as higher harmonics of the basic tidal wave. Thus there are quarter diurnal (M_4), sixth diurnal (M_6), eighth diurnal (M_8) etc. components. Every position in the gulf has a fixed relationship to the amphidromic system, and providing the basic amplitudes and phases of all the constituents can be derived from harmonic analysis of a long tidal observation, either a month or a year, then the future tides can be predicted.

Tidal Currents

Since the tidal currents are driven by the elevation changes visible at the surface, the same basic constituents can be observed in a long tidal current series. The records can also be analysed in a similar harmonic fashion, though the amplitudes and phases will differ from those of the elevations. Predictions of tidal currents can be made, but they are relatively less accurate, mainly because they are sensitive to local topography and bed friction, and these cannot be sufficiently well specified. Additionally wind effects are a further disturbing factor.

The rotary tidal wave produces a current whose direction rotates also, and whose magnitude may always be non-zero. If an enveloping ellipse is drawn to encompass all of the current vectors observed during a tidal cycle, it could be very open, almost circular tidal ellipse. In other situations, however, the tidal ellipse may become narrower and elongated. The current is then termed a rectilinear current. The degree of ellipticity can be defined by the ratio of the semi-minor to the semi-major axes. A rectilinear current thus has a small value and a rotary current approaches unity. The pattern of distribution is complex, with near circular motions, both clockwise and anticlockwise, and rectilinear motions, occurring near to and well away from coasts (see Figure 5.15 of Soulsby, 1983). A rotary analysis of the current can also be carried out by considering the ellipse as being generated by two orthogonal sine waves rotating in opposite directions, clockwise and anticlockwise (Godin, 1972).

In the vertical profile the amplitude of the tidal current reduces towards the bed, and though the directions are roughly constant within the logarithmic layer, further up in the boundary layer the instantaneous current tends to rotate in an anticlockwise direction with height above the bed in the northern hemisphere.

Time series of current meter measurements can be presented in a variety of ways; as a tidal current ellipse, a stick diagram and as a progressive vector diagram (Figure 3.27). The progressive vector diagram is constructed by drawing vectors representing the current speed and direction continuously at equal time increments. The total distance travelled after a tide and the direction then represents the residual current, the mean current averaged over the tide.

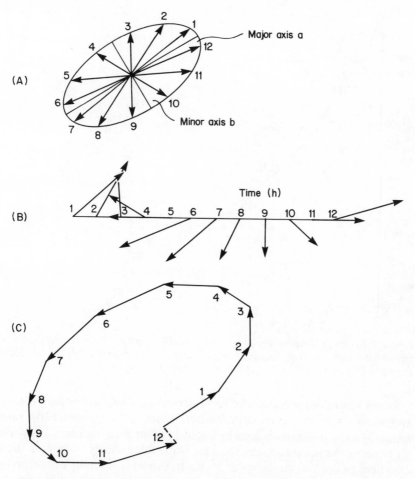

Figure 3.27 Various ways of representing tidal current vectors. (A) Tidal current ellipse. (B) Stick diagram. (C) Progressive vector diagram

The residual current can also be calculated by taking a mean over 24 hours 50 minutes of data taken at regular intervals, or applying a numerical filter to remove the tidal oscillation. Such a filter is the X_0 filter of Doodson (1928). However, there is effectively little difference between the results in most cases. Taking the mean value over only $12\frac{1}{2}$ hours gives a regular difference between consecutive tidal cycles because of the diurnal inequality.

When long periods of current measurements are obtained and analysed for residual current, the effects of storms become particularly noticeable. One example is shown by Heathershaw (1982a). On a progressive vector diagram (Figure 3.28) the residual flow changes drastically during a surge and persists for a long period afterwards in its new direction. Similar results have been obtained from many coastal situations.

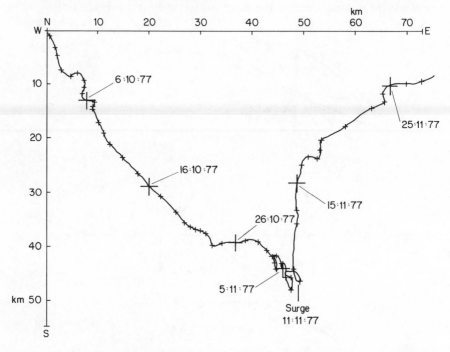

Figure 3.28 Progressive vector diagram of tidally averaged flow for a current meter record from Swansea Bay, showing effect of surge on 11 November 1977. From Heathershaw and Hammond, 1979, and Heathershaw (1982a)

Storm surges occur because of two interrelated causes; atmospheric pressure and wind. With the 'inverse barometer' effect, a low atmospheric pressure results in a rise of mean sea level by about 1 cm for every millibar. The action of the wind on the water surface causes the water additionally to move in the wind direction. In open water the speed of this transport is generally considered to be at 3 per cent of the wind speed, though more recent results suggest a figure of about 8 per cent. When the wind is blowing towards a coast there will be a downwelling of water and offshore compensating flow near the sea bed. When the wind blows offshore there will be a landward bottom flow and upwelling. Away from coasts, or on the scale of continental coastlines, the surface water flow will be at an angle to the wind, to the right in the northern hemisphere. The angle of the current to the wind increases with depth in the Ekman spiral. The result of this spiral is that the depth integrated surface water movement is at right angles to the wind direction. Thus a wind blowing along a coastline can lead to either upwelling or downwelling depending on its direction.

As far as sediment movement is concerned there is some interest in being able to describe the statistical probabilities of tides and surges interacting to lead to extreme currents. Some aspects of this combination have been considered by Pugh (1982). Figure 3.29 shows extreme current probabilities for two situations: The Inner Dowsing Light Vessel is situated in the Southern North Sea, and the probabilities are calculated from a long term current mooring (Pugh, 1982). The West Solent is a tidal channel in which the measured water surface slopes have allowed estimation of the cross-sectional mean currents (Dyer and King, 1975). Both results show the occasional occurrence of peak currents considerably in excess of the normal tidal pattern.

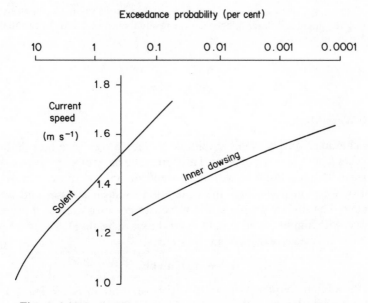

Figure 3.29 Probability of currents exceeding certain values, for Inner Dowsing (Pugh, 1982), and the Solent (Dyer and King, 1975)

Additionally for the purposes of sediment movement, progressive vector diagrams of u^3 should be considered. Because of the slight asymmetries in the tidal curve the residual water movement and the residual of u^3, which we shall see later is equivalent to the sediment transport rate, can be in different directions. Figure 3.30 shows a comparison near the Solent Bank in the West Solent where the currents are rectilinear within a straight channel. Despite this there is a 20° difference between the residual directions. In other situations where the asymmetry is particularly marked, such as some estuaries, the residual current may even be in the opposite direction to the likely sediment movement.

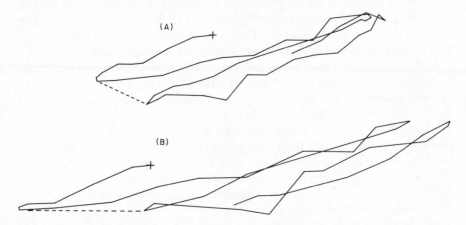

Figure 3.30 Progressive vector diagrams of (A) u, (B) u^3 for 25 hours in the West Solent, illustrating the angular difference between the residual current u and the residual power u^3. Not to the same scale

WAVES

Surface Wave Motion

The characteristics of small amplitude surface waves have been known for many years from the work of Airy. This forms the basis of extensive analysis of many wave features, but here we will concentrate only on those relevant to the motion of water near the bed, the generation of shear stresses and sediment movement. The surface profile of the wave can be considered as a progressive sine wave, with amplitude 'a' equal to half the wave height H. The wave motion is described by the wave dispersion equation

$$\sigma^2 = gk \tanh kh \qquad (3.45)$$

where the radian frequency $\sigma = 2\pi/T$, the wave number $k = 2\pi/\lambda$, T is the period, λ the wavelength of the wave, and h the water depth. An individual wave then has a velocity of propagation in the positive x direction, a phase velocity c of

$$c = \frac{\lambda}{T} = \left(\frac{g\lambda}{2\pi} \tanh \frac{2\pi h}{\lambda} \right)^{\frac{1}{2}} = \frac{gT}{2\pi} \tanh \frac{2\pi h}{\lambda} \qquad (3.46)$$

In practice it is comparatively easy to measure the period of the wave and the water depth, but the wavelength is very much more difficult. If T and h are known, λ can only be found from Equation 3.46 iteratively or from the graphical solutions shown in Figure 3.31.

In deep water when h/λ is large, tanh 2πh/λ approaches unity. Consequently

$$c = \sqrt{\frac{g\lambda}{2\pi}} = \frac{gT}{2\pi} \qquad (3.47)$$

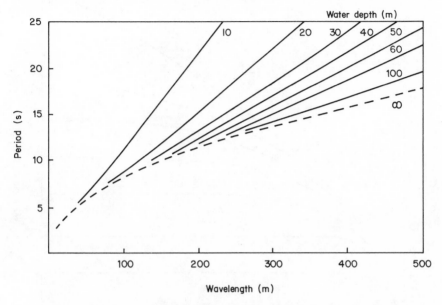

Figure 3.31 Relationship between wave length, period and water depth

In shallow water the ratio h/λ is small and tanh 2πh/λ approaches 2πh/λ. Thus

$$c = \sqrt{gh} \qquad (3.48)$$

The effective limit for deep water waves is $h > \lambda/2$ and that for shallow water is $h < \lambda/20$. For intermediate depths between these two limits the complete Equation 3.46 must be used. These relationships are shown in Figure 3.32 for a series of periods and water depths. Surface waves of period 5–15 s start to feel the bottom in approximately 20–180 m of water respectively.

In deep water the water particle orbits under the waves are circular with diameter decreasing exponentially with depth (Figure 3.33). At depths between $\lambda/2 > h > \lambda/20$ the wave feels the bottom and the vertical motion in the orbit becomes restricted. The particle orbits therefore become narrower with depth until at the bottom there is simply a to-and-fro motion. The amplitude of the horizontal water motion at the bottom is rather less than that at the surface, and is given by

$$A_b = \frac{a}{\sinh \dfrac{2\pi h}{\lambda}} \qquad (3.49)$$

and the maximum orbital velocity is

$$u_m = \frac{2\pi a}{T \sinh \frac{2\pi h}{\lambda}}$$

(3.50)

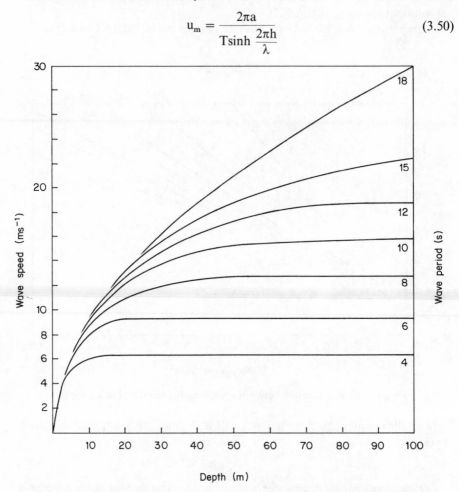

Figure 3.32 Relationship between water depth, wave speed and period

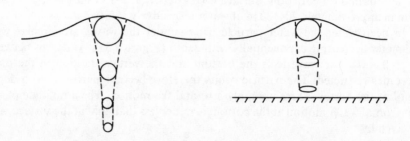

Figure 3.33 Water particle orbits under waves in deep and shallow water

The maximum velocity occurs under crest and trough of the wave.

For shallow water waves, the orbital velocity is constant with depth and equal to

$$u_m = \frac{H}{2} \left(\frac{g}{h} \right)^{\frac{1}{2}}$$

(3.51)

The deep water wave equation (Equation 3.47) shows that long wavelength waves travel faster than short wavelength ones. As the wave approaches shallow water the wave period stays constant but the wavelength decreases (Figure 3.34). The wave height decreases in the intermediate depth zone but then increases fairly abruptly in shallow water until limited by breaking. The near-bed orbital velocity also rises as the depth decreases, according to Equation 3.51.

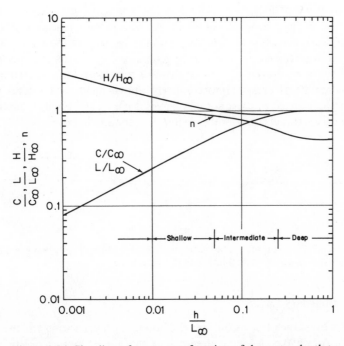

Figure 3.34 Shoaling of waves as a function of the water depth to deep water wavelength. n is the ratio of the phase to the group velocity of the wave. *From Komar, 1976,* Beach Processes and Sedimentation. *Reproduced with permission of Prentice-Hall Inc.*

The energy of the wave per unit crest length $E = \frac{1}{2} \rho g a^2$. This energy is carried forward by the wave at a velocity called the group velocity c_g. In deep water c_g is half the phase velocity, whereas in shallow water the group velocity equals the phase velocity. For deep water waves the difference can be observed in a train of waves, for the waves at the back of the group move through it to die out at the front. Thus the group progresses at less than the speed of the individual waves.

The above linear Airy wave theory was extended by Stokes for finite amplitude waves. This resulted in a wave profile that was more peaked at the crests and flatter in the troughs (Figure 3.35). For all practical purposes the group and phase velocity is unchanged from the linear theory, though the velocity of highest wave in deep water can be up to 10 per cent greater than that of the Airy wave. The one major difference between the theories is that the asymmetry of the Stokes wave is also reflected in the orbital velocities. Under the crest the horizontal velocity in the direction of wave advance is larger than that in the opposite direction under the trough. This, in particular, is likely to have an effect on the sediment movement under the waves. Because the orbits are now not closed, the particles have a mass transport, the Stokes drift, in the direction of wave propagation. This has important consequences in shallow water, as will be discussed in Chapter 11. An unfortunate feature of the mathematics of Stokes waves is that they cannot be superimposed to form an irregular sea, whereas Airy waves can be generalized by means of a spectrum.

In shallow water, waves take on an extreme form with long flat troughs and isolated symmetrical crests almost entirely above mean water level (Figure 3.35). The phase speed of these solitary waves is $c = \sqrt{(H + h)g}$. Consequently the higher waves travel considerably faster than the smaller ones.

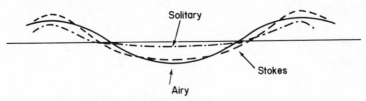

Figure 3.35 Comparison between the surface profiles of Airy wavy, solid line; Stokes wave, dashed line; and Solitary wave

The Wave Boundary Layer

In the first instance let us consider the boundary layer formed at the bottom beneath an Airy wave. We have already seen that the maximum orbital velocity just outside the boundary layer can be calculated from Equation 3.50. For a flat impermeable bed where viscosity is dominant, the velocity profile can be calculated, with the results shown in Figure 3.36, for intervals of one tenth of a wavelength for the forward stroke of the wave. As can be seen, both the phase and amplitude of the velocity oscillation vary through the boundary layer with the phase lead increasing towards the bed. For laminar flow and a smooth flat bed, the bottom stress $\tau_0 = \rho v du/dz$ leads u_m in phase by $\pi/4$. The maximum shear stress

$$\tau_m = \rho \left(v\frac{2\pi}{T} \right)^{\frac{1}{2}} u_m$$

$$(3.52)$$

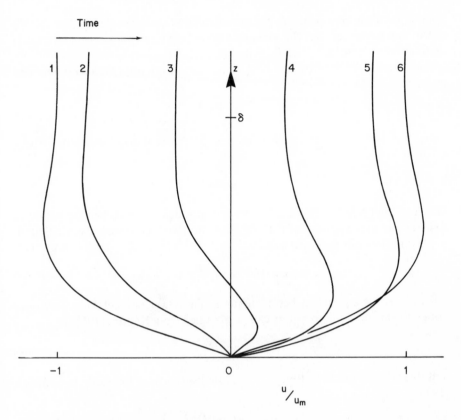

Figure 3.36 Velocity profiles in a laminar wave boundary layer at equal time intervals during half a wave cycle. δ is the boundary layer thickness. After Lamb, 1932

Consequently the maximum shear stress associated with a wave motion of a given maximum velocity at the bottom increases with decreasing wave period. The thickness of the boundary layer δ can be defined in a number of ways, but can be given by

$$\delta = (4\pi vT)^{\frac{1}{2}} \qquad (3.53)$$

This definition is rather arbitrary, in the same way as that for steady flow, since the free stream velocity is approached only slowly. The boundary layer is very thin; for a 4 s wave, δ is only about 0.7 cm. As the roughness of the bed increases, and the maximum orbital velocity increases, the laminar boundary layer thins and a transition to a turbulent boundary layer eventually occurs. The first turbulent effects that can be seen are small vortices in the lee of the individual sand grains. On reversal of the flow at the end of the stroke these vortices tend to get ejected upwards into the flow and carried back over the grain to interact with the vortex now forming on the other side. This leads to an increase in thickness of the boundary layer and when fully turbulent the boundary layer is a few centimetres thick. For small size roughness elements, but large

u_m, there is a smooth turbulent regime, analogous to that in steady flow. Sleath (1970) has suggested that the turbulent boundary layer thickness is effectively up to about double that given by Equation 3.53.

As a measure of the structure of the boundary layer, a Reynolds number can be formed

$$Re_{w1} = \frac{u_m A_b}{v} \qquad (3.54)$$

This, however, does not involve the size of the roughness elements. Alternatively

$$Re_{w2} = \frac{u_m D}{v}$$

Whether the boundary layer is laminar or turbulent depends on the Reynolds number and the roughness of the bed. Jonsson (1966) has stated that the boundary layer is laminar provided that

$$Re_{w1} < 1.26 \times 10^4 \quad \text{and} \quad \frac{A_b}{k_s} > 1.8 \, (Re_{w1})^{\frac{1}{2}} \qquad (3.55)$$

where k_s is the equivalent bed roughness. The lower limit of fully developed rough turbulent oscillatory flow can be obtained from Kajiura (1968) as

$$Re = 2000 \frac{A_b}{k_s} \qquad (3.56a)$$

and this is appropriate to $Re > 7 \times 10^3$. Below this limit the formula of Sleath (1974)

$$Re = 4130 \left(\frac{A_b}{k_s} \right)^{0.45} \qquad (3.56b)$$

can be applied. For a flat bed $k_s = 2.5 \, D_{50}$, or $2 \, D_{90}$ have been proposed, and for a rippled bed $k_s = 4 \, H_r$, or $k_s = 25 \, H_r^2/\lambda_r$, where H_r is the ripple height and λ_r the ripple wavelength.

In contrast to unidirectional flow there is a tendency for oscillating flow to remain laminar at instantaneous maximum orbital velocities which, in steady flow, would correspond to turbulent conditions.

There is some uncertainty about the form of the velocity profile in the turbulent wave boundary layer. It is thought that there is a thin logarithmic layer close to the bed, but it is inappropriate to use the von Karman–Prandtl equation to find the shear stress unless measurements are made very close to the boundary. An alternative method is by use of a wave friction factor. The maximum bed shear stress can be defined as

$$\tau = \tfrac{1}{2} f_w \rho u_m^2 \qquad (3.57)$$

where f_w is the wave friction factor, which has obvious analogies with the Darcy–Weisbach coefficient. Values for f_w can be determined analytically for a laminar boundary flow

$$f_w = 2 \, (Re_{w1})^{-\frac{1}{2}} \qquad (3.58)$$

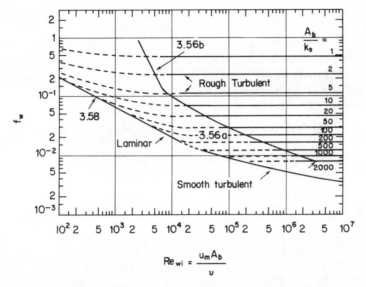

$$Re_{wi} = \frac{u_m A_b}{v}$$

Figure 3.37 Wave friction factor variation with Reynolds number and relative roughness. After Jonsson, 1966

Jonsson (1966) has shown for turbulent flows that the friction factor is also dependent on the relative roughness of the boundary. Figure 3.37 shows Jonsson's diagram for f_w against Re_{w1}. From this diagram it is possible to calculate, with use of Equations 3.49, 3.50 and 3.57, the maximum bed shear stress for a known surface wave, providing an appropriate value for the equivalent bed roughness k_s can be chosen. However, since there is no reference to phase in the wave cycle, it is not apparent when the maximum shear stress is achieved. With a turbulent boundary layer the lead of the shear stress over the maximum free stream orbital velocity is less than $\pi/4$ by a factor of about 2.

For an Airy wave and a laminar boundary layer, Longuet-Higgins (1953) has shown that as a result of the velocity in the boundary layer having a phase lead over that outside, and because the flow varies along the bed from crest to trough, a steady streaming is created just outside the boundary layer. This mass transport is in the direction of wave advance and has a magnitude given by

$$\bar{u}_b = \tfrac{5}{4} \frac{2\pi/\lambda}{\sinh 2\pi h/\lambda} \cdot u_m \qquad (3.59)$$

where $\bar{u}_b$ is the mass transport velocity and u_m is given by Equation 3.50. The magnitude of the mass transport diminishes fairly rapidly with height above the bed. When approaching a beach, there must also be a compensating return circulation higher in the flow. These two effects produce mean velocity profiles such as those shown in Figure 3.38, for two values of $2\pi h/\lambda$. Experiments have shown that this forward mass transport also occurs when the boundary layer is not strictly laminar and can even exist when ripples are present. Additionally, the mass transport is likely to be present for low amplitude Stokes waves, though the profiles shown in Figure 3.38 would be modified by Stokes drift and the additional requirements of continuity that it presents.

The forward motion near the bed is potentially of some significance for sediment transport since it can produce a residual transport of particles lifted into suspension. Linear wave motion by itself is only likely to create a stirring of the sediment and without a residual transport on a tidal current the particles will not preferentially move in any particular direction.

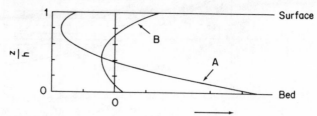

Figure 3.38 Mean drift under Stokes wave in shallow water, according to Longuet-Higgins (1953). Curve A, $2\pi h/\lambda = 0.5$. Curve B, $2\pi h/\lambda = 1.5$. *Reproduced by permission of the Royal Society*

Observed Waves in Coastal Areas

In the sea there is seldom only a single frequency of surface waves present. Generally there will be a spectrum of waves with frequency and amplitude varying with time. The longer the measurement time, the wider the spectrum, as it encompasses the larger waves in the storms as well as calm periods. Because it is obviously impractical to record waves continuously, regular short recordings are made which are assumed to be representative, and from these the longer term wave climate can be built up. A common recording sequence is 15 min every three hours.

To express the 15 min record in simple terms a characteristic wave height and period can be calculated. The significant wave height H_s is the mean of the highest one third of the waves, and the zero crossing period T_z is the average period between instants at which the surface rises above the mean water level. H_s is 0.63 of the maximum wave height for a narrow spectrum. A procedure for calculating H_s and T_z is outlined by Tucker (1963).

Using Airy wave theory (Equation 3.50) it is possible to calculate the peak near bed orbital velocity appropriate to each combination of values of H_s and T_z. These can then be plotted as a scatter diagram, as shown in Figure 3.39 which is based on one year's data, with 15 min records every three hours taken in a water depth of 20 m at Smiths Knoll in the southern North Sea. For example, a 2 m high wave with an 8 s period will cause a peak orbital velocity of 40 cm s^{-1} in 20 m of water. The contours enclose fixed percentages of the total number of observations, and show that 99.9 per cent of the records had significant wave heights less than 4 m and periods less than 9 s. Since the diagram was constructed for $\frac{1}{2}$ m increments of height and $\frac{1}{3}$ s increments of period, there is about a 0.4 per cent chance of waves with H_s of 2–2$\frac{1}{2}$ m and T_z of 6–6.5 s occurring, as can be seen by the hatched box. The equivalent orbital velocity of about 28 cm s^{-1} exceeds the threshold of movement of 150 μm sand. The overall probability of the threshold being exceeded requires integration of the areas enclosed by the contours and the threshold curve.

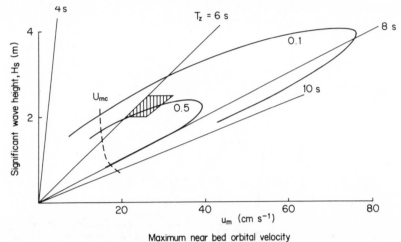

Maximum near bed orbital velocity

Figure 3.39 Scatter diagram of wave measurements over 1 year at Smith's Knoll, North Sea. See text for explanation. u_{mc}, threshold of movement for 150 μm sand

Of course the above approach takes no account of the fact that, because of attenuation, the T_z and effective H_s that would be observed at the sea bed are different from those at the surface. The effects of the short period end of the surface spectrum are lost at depth, and even in the one third highest waves some differential attenuation occurs. It is more accurate to take the observed surface spectrum, split it into narrow frequency bands, calculate the near-bed orbital velocity for each band, and then recombine them as a time series to obtain the significant wave orbital velocity and period at the bed. Figure 3.40 shows the results of such a calculation on a 15 min record taken in 100 m of water west of Uist in the Outer Hebrides. The surface wave spectrum has a peak at a period of about 13.5 s, but the T_z for the record is 8.81 s, with an H_s of 7.15 m. The spectrum of the near-bed orbital velocity peaks at about 16 s and shows a much more abrupt cut-off at shorter periods. The significant orbital velocity is 39.1 cm s^{-1}, which compares with an rms value of 9.8 cm s^{-1}. Gust and Stolte (1978) have measured the velocity spectrum 5 cm above the bed in 15 m of water in the Baltic Sea. They found the peak period was 3.3 s at the surface and 4.8 s at the bed.

Draper (1967) has used the above approach with an assumed spectral form for the surface waves to predict the bottom orbital speeds around the British Isles. Extending the analysis over a year or more of data enabled exceedance diagrams such as Figure 3.41 to be produced. Thus in 10 m water depth significant orbital speeds exceeding 75 cm s^{-1} are likely for about five days per year.

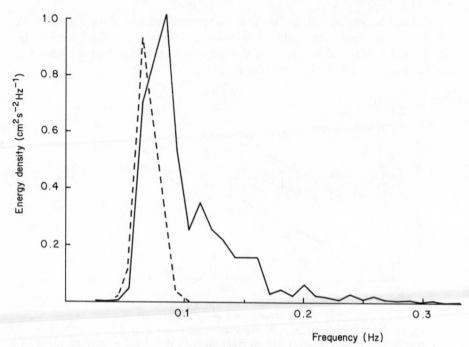

Figure 3.40 Wave spectra from South Uist, Scotland. Solid line, observed surface waves. Dashed line, calculated sea bed spectrum of orbital velocity in 100 m depth

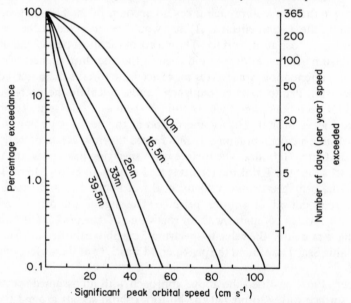

Figure 3.41 Probability of orbital speeds exceeding certain values in different water depths in the vicinity of Smith's Knoll. *From Draper, 1967, Marine Geol., 5, 133–140. Reproduced by permission of Elsevier Science Publishers BV*

Within that time, of course, considerably higher maximum speeds occur for shorter periods. The above significant orbital speeds can be multiplied by a factor derived from wave statistics to obtain the most probable maximum particle speed. According to Figure 3.42, for example, the value of $75 \, \text{cm s}^{-1}$ would have to be multiplied by the factor 1.7 to obtain the maximum speed within a half an hour interval.

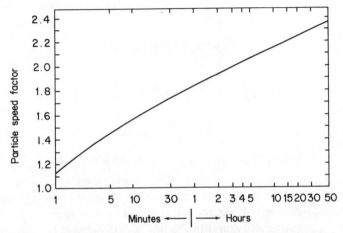

Figure 3.42 Chart for converting the significant particle obital speeds shown on Figure 3.41, to maximum orbital speeds likely to occur within the stated duration. *From Draper, 1967,* Marine Geol., *5, 133–140. Reproduced by permission of Elsevier Science Publishers BV*

In order to determine which wave is likely to be the most effective in moving sediment at the sea bed, McCave (1971a) developed an effectiveness parameter. Using the exceedance approach of Draper he plotted the product of the cube of the significant orbital speed and the probability, against the square of the speed. This represents the available fluid power compared with the bed shear stress. The curves for each depth showed a maximum, and the wave conditions giving that maximum value are taken to be those most effective in moving sediment. In the Southern Bight of the North Sea the most effective waves were those occurring between 10 and 20 per cent of the time and which had peak particle speeds of 30 to $40 \, \text{cm s}^{-1}$. Comparison of the parameter with the distribution of sand waves suggested that they were absent in areas of high wave effectiveness.

The statistical wave climate approach to near bed wave conditions has not been widely applied. Instead fairly short series of measurements have been obtained near the bed which have encompassed storms. Cacchione and Drake (1982) for instance have 'burst sampled' near-bed profiles and pressure. The burst sequence was one second samples for a minute every hour. During an intense storm maximum wave heights of 2.5 m were calculated from the bottom pressures. The maximum near-bed wave speed exceeded $25 \, \text{cm s}^{-1}$ with a mean period of 8.9 s. For sediment transport studies such direct sea bed measurements are preferred to extrapolation from sea surface data.

CHAPTER 4

Sediment Movement

SETTLING VELOCITY OF PARTICLES

In considering the interaction between the fluid and sediment particles, the simplest possible situation is that of single spherical grains settling through a motionless fluid. As we shall see, the settling velocity is an important parameter in sediment transport theory, and the results of fairly simple laboratory experiments are useful even in some complex situations.

A dense sphere released in a column of water will start to fall, and within a distance of a few grain diameters its fall velocity will be constant. This is the terminal fall velocity and is achieved when the drag forces equal the immersed weight of the sphere. Remembering Equation 3.3, the drag force on the sphere can be written in terms of a drag coefficient and thus

$$F_D = C_D \pi \frac{D^2}{4} \rho \frac{w_s^2}{2} \tag{4.1}$$

The immersed weight I is the difference between the force of gravity acting downwards and the buoyancy force acting upwards. This is expressed as

$$I = \tfrac{4}{3} \pi \frac{D^3}{8}(\rho_s - \rho)g \tag{4.2}$$

where D is the grain diameter, ρ_s and ρ are the grain and fluid densities respectively, and w_s is the settling velocity.

The flow around the sphere will differ depending on whether it falls slowly, when viscosity will be important, or fast, when the grain inertia dominates. A Reynolds number can be formed from the fall velocity and the particle diameter as a measure of these effects

$$Re_s = \frac{w_s D}{\nu} \tag{4.3}$$

Below an Re_s of about 1, laminar flow conditions exist, the spheres creep through the fluid, distorting the flow for relatively large distances from the sphere and leaving no wake. Above this number the flow separates in the lee of the particle, and vortices are shed, at first periodically and then randomly, and the boundary layer around the particle becomes turbulent.

At Reynolds number < 1 the drag can be calculated as

$$F_D = 3\pi\mu D w_s \tag{4.4}$$

where μ is the molecular viscosity. From Equations 4.4 and 4.2 we get

$$w_s = \frac{D^2}{18}\left(\frac{\rho_s - \rho}{\mu}\right)g = cD^2 \tag{4.5}$$

This is Stokes Law and it demonstrates that $w_s \propto D^2$ in the viscous regime. For quartz particles in water the value of c varies from 8975 at 20°C and 6880 at 10°C, to 5920 at 5°C. These values are decreased by about 5 per cent for a salinity of 35‰.

From Equations 4.4 and 4.1 the relationship between drag coefficient and Reynolds number is

$$C_D = \frac{24}{Re_s} \tag{4.6}$$

This has obvious analogies with laminar pipe flow (Equation 3.9).

At higher Reynolds numbers the fall velocity cannot be theoretically predicted but has been derived from experiment. The drag coefficient C_D deviates from the relationship of Equation 4.6 and eventually becomes virtually constant at a value of about 0.4 at an Re_s greater than about 10^3 (Figure 4.1). If C_D is independent of D in Equations 4.1 and 4.2, then $w_s \propto D^{\frac{1}{2}}$. This is generally known as the Impact Law.

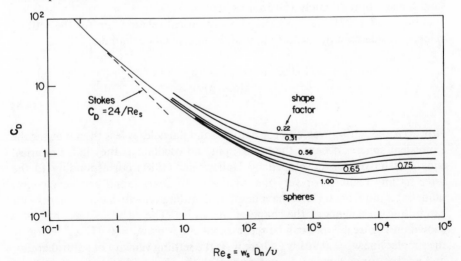

Figure 4.1 Drag coefficients of spheres and non-spherical grains with various shape factors as a function of Reynolds number, based on the nominal diameter D_n. *After Komar and Reimers, 1978,* J. Geol., *86, 193–209. Reproduced by permission of the University of Chicago Press*

The curves in Figure 4.1 are for a sphere and particles of various shape factors. A shape factor of about 0.7 can be considered appropriate for sand particles. Particles of different shapes have different curves in the range above $Re_s \sim 1$. Discs, for instance, tend to have a constant drag coefficient of about 1 at high Reynolds number. In the viscous regime the falling disc takes up an orientation with its maximum area normal to the fall direction. Clay minerals are likely to settle in this fashion.

Theoretically the fall velocity of a disc in the viscous regime is given by

$$w_s = \frac{1}{2k\mu} \frac{\overline{D}}{D_c} (\rho_s - \rho)g\overline{D}^2 \tag{4.7}$$

where $\overline{D} = \frac{1}{2}(D_a + D_b)$. The coefficient k has a value of 5.1 for broadside settling of infinitely thin particles, but the best fit for natural particles is k = 9. This gives the same result as Equation 4.5 when $\overline{D} = D_c$, i.e. for a sphere. For turbulent conditions, however, the particle oscillates while settling and even falls along a zig-zag path in extreme conditions (Stringham *et al.*, 1969).

The C_D versus Re_s plots are useful in as much as they compare the performance of grains of different sizes and densities in liquids of differing viscosity. As we shall be dealing mainly with quartz grains settling in water it is useful to replot the results in terms of grain diameter and fall velocity. Figure 4.2 is derived from Figure 4.1 and shows the variation of fall velocity with shape factor. Each curve has two sections when plotted on a double logarithmic plot. One section has a slope of 2 and corresponds to the viscous Stokes regime. The other gradually approaches a slope of $\frac{1}{2}$ for larger sizes corresponding to the Impact Law. The transition region between these two sections covers the fine to coarse sand range (approximately 150 μm to 400 μm).

Gibbs *et al.* (1971) have presented an empirical relationship for the settling velocity of spheres equivalent to that of the intermediate diameter D_b.

$$D_b = \frac{0.111608\, w_s^2\rho + 2\sqrt{[0.003114\, w_s^4\rho^2 + g(\rho_s - \rho)(4.5\mu w_s + 0.0087\, w_s^2\rho)]}}{g(\rho_s - \rho)} \tag{4.8}$$

However, the settling velocity for natural sand particles is less than that for the equivalent spheres because the grains spin and oscillate as they fall. Empirical equations have been developed by Hallermeier (1981) corresponding to the three settling regimes for very fine sand, fine to coarse sand and very coarse sand. Baba and Komar (1981) have combined settling velocity measurements with careful measurements of the shape of the grains. The curve of their results is shown on Figure 4.3 and can be represented either by $w_n = 0.977\, w_s^{0.913}$ or by the simpler line $w_n = 0.761\, w_s$, where w_n is the settling velocity of natural grains, and w_s that for spheres of diameter equal to the intermediate diameter 'D_b' of the grains. Additionally the results showed a clear dependency on the grain sphericity. The measured curve on Figure 4.3 yielded a mean shape factor for the natural grains of 0.7, as can be seen by comparison with Figure 4.2.

In nature the particles will seldom settle in motionless fluid; there will

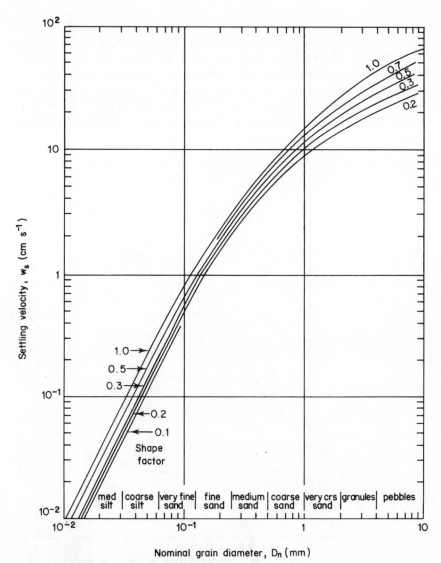

111

Figure 4.2 Settling velocities of grains in water at 20°C as a function of shape factor. *After Komar and Reimers, 1978*, J. Geol., **86**, *193–209. Reproduced by permission of the University of Chicago Press*

generally be considerable turbulence present which will affect the drag on the grains. In this case the simple Equation 4.1 has to be extended with the addition of terms which take account of acceleration, the added mass of the displaced water, and the time history of the particle. The full equation is known as the Basset–Boussinesq–Oseen equation, but most of the terms are poorly known and the effects are difficult to predict. Small particles will travel up and down with the turbulent velocity fluctuations, but in the viscous Stokes' regime they will be settling at a constant rate relative to the fluid. For coarse particles, however, the time response of the boundary layer around the particle becomes important, and any asymmetry in the up and down velocities is reflected in a varying drag on the particle. Thus the settling velocity of the particle is decreased if the fluid is oscillating sinusoidally in a vertical direction. If the oscillations are asymmetrical then the particle can move upwards if the upward acceleration is greater than the downwards acceleration and vice versa (Boyadzhiev, 1973). However this size of particle is not normally found in suspension.

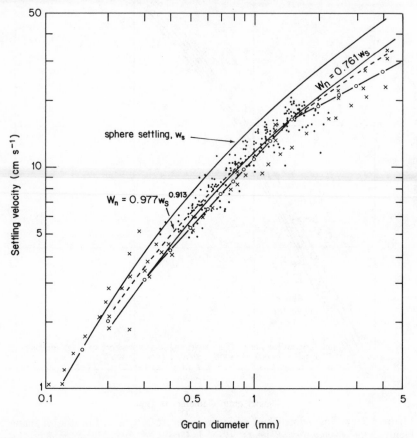

Figure 4.3 Settling velocities of natural grains of intermediate diameter D_b. *After Baba and Komar, 1981, J. Sediment. Petrol., 51, 631–640*

Ludwick and Domurat (1982) have simulated the movement of 100 and 200 µm size particles in a turbulent velocity field. They found that the dominant factor influencing particle motion is the mean vertical velocity rather than the turbulent fluctuations. When the mean velocity is upwards and exceeds the settling velocity, prolonged suspension occurs. Settling is enhanced for a mean downward flow. Though positively skewed distributions of velocity fluctuations decreased suspension time and negative skewness increased suspension time, they found settling of fine sand not significantly reduced by turbulence for flows with reasonable horizontal velocities. Thus, sand particles will settle in turbulent flow at the same velocity as in still water. However, especially for the small grains, mean vertical fluid velocities will be important in determining the rate at which suspended particles reach the sea bed. These mean velocities can occur as a result of several processes which will be considered later.

SEDIMENT MOVEMENT IN STEADY FLOW

Particles on a Boundary

Drag force

A sediment particle resting on a grain boundary will experience a drag force when exposed to a flowing fluid. This can be represented by a drag coefficient and Equation 3.3

$$F_D = \tfrac{1}{2}\rho C_D A u^2 \tag{4.9}$$

Even though the particle will not be fully exposed to the flow, the projected area A is put equal to $\pi D^2/4$. The velocity u is that at the height of the centreline of the sphere so that a relevant Reynolds number can be defined in exactly the same way as that for a settling particle (Re_s). Coleman (1972, 1977) has investigated the drag coefficient of a stationary sphere resting in rhombohedral packing on a bed of identical spheres and obtained a drag coefficient consistently about 30 per cent higher than that of a settling sphere (Figure 4.4) over the Re_s range $0.5–10^4$. These differences could be caused by the sheltering factor and that $A < \pi D^2/4$. Garde and SethuRaman (1969) have investigated the drag coefficient of a sphere rolling down a rough and a smooth slope and obtained somewhat higher values. However, in this instance, the fluid was stationary and the sphere moving relative to both the bed and the fluid. Nevertheless the values have been confirmed by Aksoy (1972) for a smooth boundary in a flume study. It is not yet clear what the drag coefficient would be of a grain in a group moving over a boundary.

The drag can also be considered in terms of the bed shear stress by replacing u in Equation 4.9 with u_*. The relevant Reynolds number is then the grain Reynolds number Re_* (p. 58).

Lift force

The grain sitting on the rough or smooth boundary will disturb the flow, the stream-lines being deflected over the top. Because of the distortion, the fluid accelerates over the top of the grain and, as shown by Bernouilli's equation, causes a lowering in pressure. Consequently there will be a difference in pressure vertically across the grain which will cause a lift force, similar to that induced by

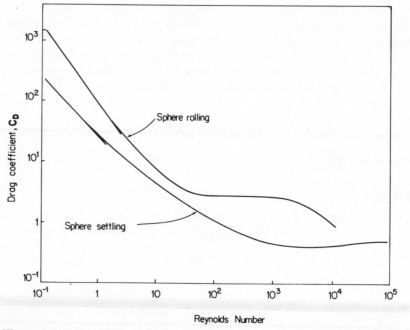

Figure 4.4 Drag coefficient of sphere rolling down an inclined boundary, in comparison with one settling. *After Coleman, 1977,* La Houille Blanche, *4, 325–328.* *Reproduced by permission of La Houille Blanche*

flow over an aerofoil, giving a tendency for the grain to rise. This lift force can be considered in a similar way to the drag force

$$F_L = \tfrac{1}{2}\rho C_L A u^2 \tag{4.10}$$

Einstein and El-Samni (1949) obtained a value for C_L of 0.178 when the velocity was measured 0.2D below the top of the bed roughness spheres. Aksoy (1972) obtained values between 0.1–0.2 for an Re_s of about 5×10^3 and also observed that though the mean lift forces were only about 1/7 of the drag force, the instantaneous lift force would reach almost three times the mean. Chepil (1961), however, has shown that the lift force was 0.75 of the drag force for a grain on a boundary of similar grains in air, and that the lift decreased rapidly as the grain was raised above the bed, disappearing one or two grain diameters above the mean surface. Bagnold (1974) also found that the lift decreased to zero when the grain was about one diameter above the bed. Additionally, he found that there were drastic differences if the particle was allowed to rotate, because of the Magnus effect. When the rotation was in the same sense as the velocity gradient near the bed (top spin) the lift force was small, but with backspin, like a well hit golf ball, the lift force was increased. With a backspin the relative velocity on the upper surface of the sphere is greater than that on the lower. The lower pressure above the sphere then causes the lift force.

For a particle sitting on the bed the lift force opposes the gravitational force acting on the particle (F_G). It is consequently interesting to compare the two. Müller *et al.* (1971) for two high Reynolds numbers (38800 and 25400) obtained ratios F_L/F_G of 0.44 and 0.18 respectively, but estimated that the lift would need to be 0.7–1.0 of the immersed weight in order for a particle to move on the bed. Coleman (1967) investigated a wide Reynolds number range and found that the lift force was 0.4 of the immersed weight of the grains for $Re_s > 200$, but at $Re_s < 100$ the lift was negative, actually pushing the grains downwards. This result was confirmed by Davies and Samad (1978) who obtained negative lift for a grain Reynolds number $Re_* < 5$. This is remarkably coincident with the limit of smooth turbulent flow, and they explained the result as due to a significant flow of water beneath the sphere at low Re_*, which would reduce the pressure difference across the grain. Consequently, at low Re_*, the grains must be relatively more difficult to erode from the bed and suspension probably relies on the velocity fluctuations that are present even in the laminar sublayer. Einstein and El-Samni (1949) have measured these fluctuations, though at high Reynolds number, by measuring the pressures developed around hemispheres on a bed. The pressure fluctuations were distributed according to the normal error law with a standard deviation of 0.364 of the mean value. This means that there will be instants when the lift is zero and also instants when the lift is several times the mean value. Cheng and Clyde (1972), however, obtained values for the standard deviation of 0.18 the mean lift.

Threshold of Grain Movement

When water is flowing over a bed of loose grains, there will be a certain velocity at which the combined drag and lift forces on the uppermost particles will be sufficient to dislodge them from their equilibrium positions. This velocity is known as the critical or threshold velocity, and, of course, there is also an equivalent critical or threshold shear stress.

If the flow velocity is increased in small increments, motion will first occur of a few particularly exposed grains, but it will die away after a time as they come to rest in new equilibrium positions. With increasing velocity, movement will become more general and prolonged. The threshold of motion is consequently a difficult thing to define, and much of the scatter of results is because observers have used different numbers of grains moving per unit area per unit time as their threshold criterion.

Let us consider a simple model of the forces on spherical grains, developed by Chepil (1959). The exposed grain sits on several others, but with their centres lying in the plane corresponding to the direction of the flow. There will be a gravitational force on the particle equal to its own immersed weight, which will act through the centre O of the sphere (Figure 4.5).

$$F_G = \frac{\pi}{6}(\rho_s - \rho)g\,D^3 \tag{4.11}$$

The drag and lift forces, however, need not necessarily act through O because the flow round the sphere will be influenced by the configuration of the surrounding grain surface. When the grain is on the point of movement the resultant of the three forces must act through the point of contact P between the grains. With a further increase in the drag there will be a couple which will rotate the grain about P and dislodge it.

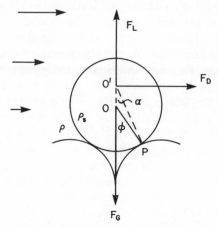

Figure 4.5 Forces acting on a static
grain resting on a grain boundary

If the angle between OP and the vertical is small then the grain will tend to roll over the other, however, but if it is larger then the grain will be forced to hop out of its resting place. The critical values of lift will obviously be different in the two cases. The former case will be more likely to occur if the uppermost grain is larger than those beneath, and the latter, if the top grain is smaller than the bed. Let us consider equisized spheres, and that the vertical lift force also passes through the centre of the sphere.

The fluid drag will consist of the drag on the area occupied by the grain and, because it protrudes above the general bed level, a large proportion of that on the area covered by its wake. If the drag force per unit area of the bed, the shear stress, is τ, then the equivalent drag on the grain is τ/N, where N is the ratio of the mean drag and lift forces over the whole bed to the drag and lift on the protruding grains. The value of N ranges between 0.20 and 0.30, implying that the stress on an area between five and three times the grain's projected area in the grain's lee, is imposed on the grain. Chepil has measured a value of N of 0.21 for hemispherical elements placed three diameters apart in a hexagonal pattern.

Thus the drag on the grain is

$$F_D = \pi \frac{D^2}{4} \frac{\tau}{N} \tag{4.12}$$

Chepil has also shown that the drag acts some way above the centre of the sphere, 0.29 of a grain diameter below the top of the grain. This gives a value for α of 24°. The angle α is the angle of dynamic friction, and contrasts with the angle of static friction φ which is equal to the angle of repose of the grains.

When on the point of movement, the resultant of the forces will act along the line O′P. Then

$$F_D = (F_G - F_L) \tan \alpha$$

Substituting for F_D and F_G from Equations 4.12 and 4.11 gives

$$\frac{\pi D^2}{4} \frac{\tau_c}{N} = (\frac{\pi}{6} D^3 g(\rho_s - \rho) - F_L) \tan \alpha \qquad (4.13)$$

where τ_c is the threshold value of the shear stress. For spherical particles Chepil (1958) has shown that $F_L = 0.85 F_D$, so that Equation 4.13 becomes

$$\tau_c = \rho u_{*c}^2 = \frac{0.66 \, Dg(\rho_s - \rho)N \tan \alpha}{1 + 0.85 \tan \alpha} \qquad (4.14)$$

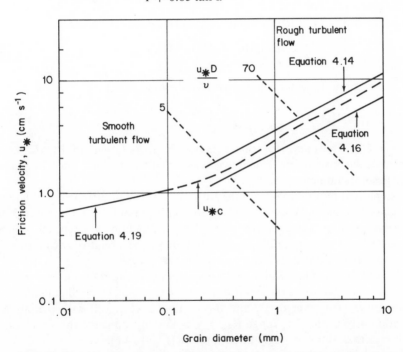

Figure 4.6 Threshold friction velocity for grain movement. Dashed line, observed values according to Miller *et al.* (1977), *Sedimentology*, **24**, 507–527. Line of Equation 4.19, observations of White (1970). Also shown are the grain Reynolds number limits for smooth and rough turbulent flow

The curve for u_{*c} versus D for quartz grains in water is shown in Figure 4.6 for $N = 0.3$ and $\alpha = 24°$. As u_{*c} is proportional to $D^{\frac{1}{2}}$, the threshold curve is parallel to that for the Impact Law.

Coleman (1967) has shown that for $Re_s > 200$ the lift force $F_L = 0.4 F_G$. This leads to effectively the same result as Equation 4.14.

The instantaneous bed shear stress can be much higher than the mean values and it is the maximum that will cause the sediment to move. It is easier to measure the mean fluctuating pressures close to the particles on the bed rather than the actual shear stresses on them, but it can be assumed that the fluctuating pressures will have the same relationship to the mean pressure as would the fluctuating to the mean shear stresses. In statistical terms the maximum fluctuation in pressure will be about three standard deviations in excess of the mean. Thus the maximum shear stress can be calculated as $\tau\, T = \tau \left(\dfrac{\overline{p} + 3\sigma_{p'}}{\overline{p}} \right)$ where p is the mean pressure of drag and lift and σ_p is the standard deviation of the fluctuations. Chepil (1959) measured values of $\tau = 0.49\,\overline{p}$. This gives a value for the turbulence factor T of 2.5. There are several other studies (e.g. Grass, 1970) which show that the standard deviation of the shear stress $\sigma_\tau = 0.4\,\overline{\tau}$, which would give a value of T of 2.2.

To take account of turbulent fluctuations Equation 4.14 becomes

$$\tau_c = \rho u_{*c}^{2} = \frac{0.66\, Dg(\rho_s - \rho)N \tan \alpha}{(1 + 0.85 \tan \alpha)\, T} \tag{4.15}$$

For quartz grains in water Equation 4.15 reduces to

$$\tau_c = 41.4\, D \tag{4.16}$$

for T = 2.5. The curve for this equation is also shown on Figure 4.6. It represents the onset of movement, whereas the curve for Equation 4.14 presents more general grain movement.

Also plotted on Figure 4.6 are the Re_* limits of rough turbulent and laminar boundary layer flow. When a laminar sublayer is present it can be expected that the value of N will change, as it is probably a function of Reynolds number. Below $Re_* = 5$ when negative lift occurs, there will be a relative rise in the value of the threshold shear stress with decreasing grain size.

There have been many studies in flumes of the threshold of uniform particles on a plane bed. Many of these results have been compiled by Miller *et al.* (1977) into a curve of u_{*c} versus D, and this compares very well with the curve of Equation 4.16 in the Re_* range $Re_* > 5$. Below this limit the threshold values are higher than those given by Equation 4.16 (Figure 4.6).

If the grain is resting on a slope, as is often the case when bedforms are present, the force balance becomes rather complicated unless the centre of drag is assumed to coincide with the centre of gravity, i.e. it is assumed that $\alpha = \varphi$. In this case the forces will be as shown on Figure 4.7. There will now be a component of the weight of the grain acting down the slope and assisting the fluid drag. Similarly to the previous analysis,

the force down the slope = $F_D + F_G \sin \beta$

The restoring force resisting movement = $(F_G \cos \beta - F_L) \tan \varphi$

Thus

$$\tan \varphi = \frac{F_G \sin \beta + F_D}{F_G \cos \beta - F_L}$$

Consequently the equation for the threshold on a slope will be equal to Equations 4.14 or 4.15, but multiplied by a factor

$$\frac{(\tan \varphi - \tan \beta)}{\tan \alpha} \cos \beta$$

where $\varphi \sim 30°$. When $\beta = \varphi$, the shear stress for movement will obviously be zero and the slope will naturally avalanche.

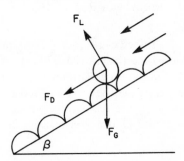

Figure 4.7 Forces acting on a particle resting on a sloping grain bed

Shields Threshold Curve

For the high Reynolds number regime where the drag and lift coefficients, and probably N and T also, become independent of Re_*, then both Equations 4.14 and 4.15 can be written as

$$\frac{\tau_c}{(\rho_s - \rho)g\,D} = \text{constant} \tag{4.17}$$

As can be seen from the fact that the relationship between u and u_* is a function of Re_*, the general form of Equation 4.17 is

$$\theta_c = \frac{\tau_c}{(\rho_s - \rho)g\,D} = f(Re_*) = f(\frac{u_* D}{\nu}) \tag{4.18}$$

The dimensionless group on the left hand side effectively compares the threshold shear stress with the immersed weight of a unit grain thickness layer of the bed. It is called the Shields entrainment function θ, with a threshold value θ_c. The relationship shown in Equation 4.18 was first investigated by Shields in 1936 and he produced the threshold curve shown in Figure 4.8. This has three distinct zones whose limits correspond to the three boundary layer flow regimes.

1. Up to $Re_* \sim 3$, smooth boundary flow exists and the particles are embedded in the viscous sublayer. Shields assumed that the threshold shear in this zone was independent of grain diameter and in that case the curve should have a slope of $-45°$ and $\theta_c \propto Re_*^{-1}$. Comparison with experiments suggested that $\theta_c = 0.1\,Re_*^{-1}$. Bagnold (1956) has suggested that the maximum value

of θ_c will occur when the whole top layer of grains yields simultaneously. This would be equivalent to the layer moving on an avalanche slope, and the $\theta_c = C \tan \varphi$, where C is the bed grain concentration. Since both C and $\tan \varphi$ have a value of about 0.63, then the maximum value of θ_c may be 0.4.

2. Between $3 < Re_* < 200$ there is a transitional region where the grain size is of the same order as the thickness of the viscous sublayer. There is a minimum value of θ_c of about 0.03.

3. In the rough turbulent regime with $Re_* > 200$, θ_c has a constant value of between 0.03 and 0.06.

The similarity to the drag coefficient versus Reynolds number curves for spheres, pipes, etc. is obvious, and this leads one to expect that the Shields curve is only one of a series that could be constructed. For non-uniform and non-spherical particles parallel curves may be expected, and at the larger grain sizes in shallow flows a dependency on the relative roughness h/D may be possible, which would compare with the Darcy–Weisbach coefficient for pipe flow of various roughnesses.

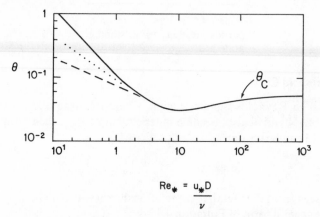

Figure 4.8 Shields threshold curve, solid line, compared with the results of White (1970), dashed line, and those of Grass (1970), dotted line

There have been few studies of the threshold characteristics in the low Reynolds number regime and Shields original curve was not confirmed at the time through lack of experimental results. White (1970) has investigated the threshold of natural quartz grains and of glass ballotini down to an $Re_* \sim 0.05$, corresponding to a grain size of about 20 μm. The results fit a curve $\theta_c = 0.06 \, Re_*^{-\frac{1}{2}}$, or, in terms of the friction velocity

$$u_{*c}^{\frac{5}{2}} = 0.06 \, g(\frac{\rho_s - \rho}{\rho}) \, v^{\frac{1}{2}} D^{\frac{1}{2}}$$

(4.19

where v is the kinematic viscosity.

This result is shown in Figures 4.6 and 4.8. Though there is a dependence on viscosity, the threshold is not independent of the grain diameter as Shields proposed. A possible reason for this is that the bed cannot be made completely smooth and the small piles of grains are sufficient to compare with the viscous sublayer thickness. Also, as discussed in Chapter 3, there are significant 'bursting' events occurring in the viscous sublayer which are likely to cause movement of these exposed grains.

The effect of turbulent fluctuations on the viscous sublayer and on grain motion have been studied by Grass (1970). He considered both the distribution of available shear stress and the distribution of the threshold stresses required by the potentially mobile grains in statistical terms, each with a mean and a standard deviation. When the mean shear stress is low, it is only the occasional large fluctuations that will move the most unstable grains. At higher stresses grain movement is more general. Consequently the degree of grain movement will be a measure of the overlap of the two distributions. Thus

$$\bar{\tau} + n\sigma_\tau = \bar{\tau}_t - n\sigma_t \tag{4.20}$$

where n determines the degree of overlap of the distributions of ambient shear stress $\bar{\tau}$ and of the stress required to move the topmost grains $\bar{\tau}_t$, whose standard deviations are σ_τ and σ_t respectively. From measurements $\frac{\sigma_t}{\bar{\tau}_t} \sim 0.3$ and $\frac{\sigma_\tau}{\bar{\tau}} \sim 0.4$. These values in Equation 4.20 give

$$\bar{\tau} = \frac{\bar{\tau}_t(1 - 0.3n)}{1 + 0.4n} \tag{4.21}$$

Matching a line through experimental results to the Shields curve at $Re_* = 2$ gave a value of $n = 0.625$. This means that the criterion for the threshold of movement is when the mean fluid shear stress is 0.65 the mean threshold shear stress required to move the topmost grains. As can be seen in Figure 4.8 the results of Grass (1970) are slightly higher than those of White (1970).

There have been many studies of the threshold of sand and, despite the scatter, the general dip to about $\theta_c = 0.03$ at an Re_* of about 10 is fairly well established. However, the threshold shear stress increases smoothly with grain size.

There have been fewer studies of the threshold at Re_* greater than about 1000, and there are fairly large discrepancies between the results. Some of the discrepancies are due to the relative protrusion of the grains. Fenton and Abbott (1977) investigated this by pushing a grain gradually above other particles in a level bed and they found that the value of θ_c could be reduced by almost an order of magnitude since the grain sheltered an increasingly large area. This is equivalent to decreasing the value of N in Equation 4.15.

Bathurst et al. (1983) have measured the threshold of gravel in rivers in the range $h/D = 1$ to 11. They found that there was a dependency of the threshold θ_c on the ratio h/D and that $\theta_c \propto Re_*^2$. This latter result suggests that the grain size D was more or less constant.

122

Carling (1983), on the other hand, found that $\theta_c \propto Re_*^{\frac{1}{2}}$, a trend at right angles to that of Bathurst. These results were explained as due to the increasing relative protrusion of the larger grains, and are equivalent to a simultaneous decrease of tan φ and of N in Equation 4.15, with increasing grain size. Both Bathurst *et al.*'s and Carling's results intersected the Shields curve.

An alternative explanation of the deviations from the Shields curve at large grain sizes results from consideration of the movement of mixed size spheres (Komar and Li, 1985), or non-spherical particles. Using Equation 2.18 in Equation 4.15 and assuming $\alpha = \varphi$ suggests that, for large particles on a boundary of smaller particles, the threshold curve would be parallel to the Shields' curve at a lower value of θ. For fine particles the value of φ would be greater than for equisize particles and consequently θ_c would be higher. This would give results similar to those obtained by Fenton and Abbott (1977), but caused by a different mechanism. Ippen and Verma (1953) studied the movement of glass and plastic spheres over a fixed boundary of smaller size natural sand. When the sand diameter was used in determining the value of the entrainment function, and the diameter of the moving spheres that of the Reynolds number, the θ_c versus Re_* curves over the range $20 < Re_* < 150$ were of the same form as the Shields curve. However the plastic spheres showed higher values, and the glass spheres lower values than the Shields curve at the same Re_*. Mantz (1973) considered the threshold of flaky particles in the range $Re_* = 0.1$ to 1.0 and obtained different curves depending on which measure of particle size was taken.

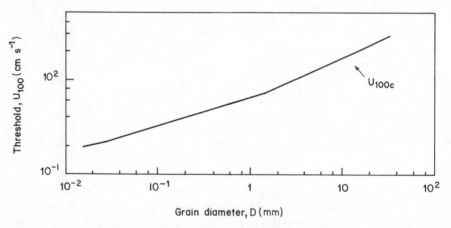

Figure 4.9 Curve of observed values of the threshold of movement on a flat bed for a current measured at a height of 100 cm. *From Miller* et al., *(1977),* Sedimentology, **24,** *507–528. Reproduced by permission of International Association of Sedimentologists*

Much of the scatter of the observed threshold values for any particular grain size may be the result of using different criteria for the relative intensity of transport. Several criteria have been developed. The most widely used is that of Neill and Yalin (1969). They proposed that the number of grains N moving in an area A during a time t should be defined as

$$\frac{N}{At} = \xi \left(\frac{\rho D^5}{(\rho_s - \rho)g} \right)^{-\frac{1}{2}}$$

ξ is an unknown function of the threshold θ_c and Re_*, but experiments could be compared using an agreed value of ξ. A value of 1×10^{-6} was proposed. Thus for experiments with larger grains the experimenter must observe either for a longer duration, or over a larger area than for small ones, and the apparent threshold θ_c will decrease as the number of grains in motion decreases.

The Shields curve is useful in that it allows direct comparison of results obtained using different densities and viscosities. However since u_* and D appear in both ordinate and abscissa, the relationship between them is not obvious. In particular when u_{*c} or D are the only variables, all others being kept constant, slopes of -1 or 2 imply that u_{*c} or D respectively are also constant.

Curves such as Figure 4.6 and 4.9 are of more direct use. These are mainly derived from Miller *et al.* (1977), who have compiled most of the available data for flat beds and produced useful threshold curves of θ_c versus Re_*, U_* versus D, and τ_c and U_{100c} versus D. These can be summarized by best fit lines represented by (Bridge, 1981a)

$$D = 0.0078 \, \tau_c^{1.69} \text{ for } \tau_c < 4.0 \text{ dynes cm}^{-2} \text{ and } D < 0.08 \text{ cm}$$
$$\text{and } D = 0.023 \, \tau_c^{0.9} \quad \text{ for } \tau_c > 4.0 \text{ dynes cm}^{-2} \text{ and } D > 0.08 \text{ cm}$$

The threshold of movement has also been considered in terms of the ratio of the friction velocity to the fall velocity u_*/w_s. Figure 4.10 shows the Shields curve

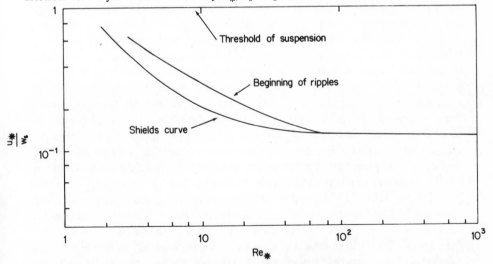

Figure 4.10 Shields curve transposed onto a graph of movability number u_*/w_s versus grain Reynolds number, including the lower limit of the occurrence of ripples. After Liu, 1957, *Proc. Amer. Soc. Civil Eng.*, **83**, *HY2, Paper 1197*, and the threshold of suspension $w_s/u_* = 1$ according to Middleton (1976)

transposed onto a u_*/w_s versus Re_* plot. The dimensionless number u_*/w_s is called the movability number, and in the rough turbulent regime it attains a constant value of about 0.13.

Turbulent Bursting and Sediment Movement

Sutherland (1967) appears to be the first to have reported that an essential feature of sediment transport is the tendency of grains to move in intermittent bursts. He found that near the threshold on a flat bed, motion was restricted to isolated spots with a small number of grains moving briefly a few centimetres. Bursts of motion occurred from positions all over the bed with frequencies that appeared to be constant from one location to another. Near the threshold on a rippled bed the motion occurred first in troughs and in the lee of the ripples the grains could move upstream as well as downstream. Sutherland connected the intermittent movement with penetration of eddies into the viscous sublayer. He illustrated the mechanism by experiments with vortex rings impingeing on the bed, and concluded that the particle is lifted into suspension by the velocity vectors inclined at an angle to the horizontal in the centre of the vortex.

Müller et al. (1971) carried out similar experiments, but measured the pressure gradients developed on the surface of a spherical grain within the vortex. This led to an estimate of the lift at the threshold of motion of 0.7 times the immersed weight.

The vortex model of Sutherland (1967) has most of the essential features of the bursting sequence. The sweep is an inrush of faster fluid towards the boundary and, consequently, has a high shear stress associated with a high velocity at the level of the tops of the particles. On the other hand, the ejection, being an upward movement of lower velocity fluid, has a high stress coupled with a lower local velocity. One would anticipate, therefore, that the sweeps would be relatively more effective at moving bedload, and the ejections in promotion suspension. Outward interactions, the part of the fluid motion having a high longitudinal fluctuation coupled with a vertical one, may be even more effective in causing suspension, but they are relatively fewer in number than ejections. If the above suppositions are correct, there must be a level, presumably near the mean level of the top of the trajectory of saltating particles, where sweeps give way to ejections as the prime sediment moving events. Studies on particle motions were carried out by Sumer and Oguz (1978) and Sumer and Deigaard (1981), and interpreted in terms of the turbulent bursting process as visualized by Offen and Kline (1975). Sumer and Oguz measured particle trajectories over a smooth bed and proposed that particle lift up from the bottom is the result of the temporary local adverse pressure gradient created by the burst passing over the particle. The lifted particle is then carried into the body of the flow by the ejection and, as it gradually breaks up, the particle settles towards the wall, only to be caught up in the next burst. The process is illustrated in Figure 4.11. Sumer and Deigaard (1981) showed that the same mechanism held for experiments with a rough boundary also.

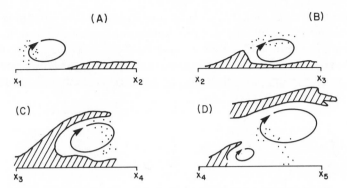

Figure 4.11 Diagrammatic sequence of suspension caused by a burst. Shaded area is zone of low velocity fluid. After Sumer and Oguz (1978). (A) Grain suspended in local adverse pressure gradient near the bed. (B) Grains held in suspension in the vortex. (C) Grains start settling. (D) Some grains return towards the bed, the rest remain with the vortex

From measurements of gravel movement in the sea, using acoustic sensing of particle movement coupled with Reynolds stress measurements, Heathershaw and Thorne (1985) have shown that bedload movement is predominantly related to sweep events.

Within each burst some differentiation of grains will be possible. The burst may be able to suspend the finer grains, but only start the coarser grains moving as bedload. As the burst dies away the suspended grains will settle and those with the largest settling velocity will reach the bed first. Bridge (1978) has used these processes to explain the formation of laminae in a depositing sediment. Net deposition will occur when the sediment transport rate decreases in the direction of transport, or with time. Following the occurrence of maximum bed shear stress during a sweep, the coarsest sediment will be deposited. The local deceleration then ensures deposition of sediment with progressively smaller settling velocities, until the next ejection occurs. The ejection disperses suspended particles and some of the saltating grains upwards. These particles then gradually settle to the bed before the next burst sequence.

Thus hydraulically-equivalent grains will be deposited together in a sequence most likely to have coarsest grains at the bottom and fining upwards. A single laminae is therefore presumed to be developed by a single bursting cycle, with individual laminae reaching up to a millimetre or so thick. Ungraded or reversely-graded laminae may be produced by a combination of varying burst duration and intensity, and settling lag.

Threshold of Motion in the Sea

In the sea, the bed is almost invariably rippled when it is composed of sand. Consequently it may be more appropriate to consider the threshold for grains on a rippled bed rather than that for a flat bed. Bagnold (1963) has proposed a threshold Shields curve for a rippled bed, but without experimental verification. This is shown in Figure 4.12. As a consequence of the form drag of the ripples, which must be partly offset by the acceleration of the water flow towards the ripple crests, it is virtually parallel to the flat bed threshold curve, but with a u_* value about double the flat bed value.

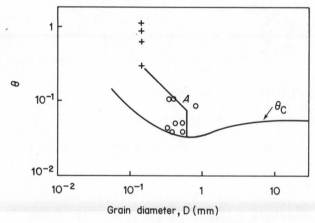

Figure 4.12 Shields threshold curve plotted against grain diameter. Curve A, threshold of movement on a rippled bed according to Bagnold (1963). Observed threshold on rippled beds in the sea; +, Dyer (1980a), ○, Sternberg (1971)

The first measurements of the threshold of sediment movement in the sea were by Sternberg (1966). Analysis of measurement from several sites, with grain size varying from 330–1090 μm show reasonable agreement with the threshold curves as shown in Figure 4.12 (Sternberg, 1971). Further measurements at a grain size of 125 μm are reported by Dyer (1980a). Both sets of measurements were obtained using underwater television and motion inferred from a cloudy appearance near the bed.

The addition of only a small percentage of clay to a sandy bed will cause the threshold of movement to rise. A minimum of between 5–9 per cent will start to bind the grains together because of the cohesive nature of the clay particles. As the percentage of clay rises then so does the threshold. Much of this material can settle into the pore spaces of the sand from suspension, but will only do so when the sand is not mobile. Consequently muddy sediments will initially only form where the currents have low values for long enough for the clay to settle. Once formed, however, the muddy bed will survive currents that would be strong enough to move the sand by itself. Muddy sediments will be discussed further in Chapter 9.

There appears to be only one study of the threshold of gravel in the sea and this confirmed the importance of protrusion and sheltering. Hammond *et al.* (1984) investigated the threshold of gravel in the range 0.2–5 cm using underwater TV recordings. They summarized the gravel movement as:

1. $u_* < 5\,cm\,s^{-1}$, rocking of granules (0.2–0.4 cm) in the interstices between the pebbles.
2. $u_* = 5\,cm\,s^{-1}$, movement of granules between pebbles, accompanied by rocking of pebbles 0.5–1 cm in diameter.
3. $u_* > 5\,cm\,s^{-1}$, sudden and intense sheet movement of granules, accompanied by movement of material 0.5–2 cm through distances of several centimetres by sliding and rolling.
4. For peak values $u_* \sim 10\,cm\,s^{-1}$, pebbles 2–4 cm and even up to 6 cm depending on exposure moved by sliding, and bursts of granule movement lasting 1–5 s occurred.

The combined threshold data was represented by the line $\theta_c = 0.51\,Re_*^{-0.45}$ which is subparallel to the results of Carling (1983). In terms of u_* and D the results conformed to $u_{*c} = 7.0\,D^{0.2}$. These curves intersected the Shields curve at $Re_* \sim 180$ and $D \sim 3\,mm$, but gave values below that curve at larger sizes. Hammond *et al.* (1984) explained the results in terms of the relative protrusion of the larger grains and obtained the relationship $\Lambda = 0.22\,D^{0.55}$, where Λ is the relative protrusion.

The above relationship $u_{*c} \propto D^{0.2}$ is the same as that found by White (1970) for very fine sand and coarse silt (Equation 4.19). This raises the speculative observation that the form of the threshold curves at small grain Reynolds numbers may be related to grain protrusion. In a flume it is easy to obtain a flat bed with coarse grains, but not with fine ones. If a flat bed could be made then perhaps, apart from the effect of ionic forces at small sizes, the threshold would be at a constant value of $\theta_c \sim 0.06$ throughout.

Effects of Biological Activity on the Threshold of Movement

Biological activity can either bind or destabilize sediments. Jumars and Nowell (1984) have defined four mechanisms which may affect the sediment stability.

1. Alteration of fluid momentum impingeing on the bed, by changing the near bed flow or bed roughness.
2. Alteration of particle exposure to the flow, by burrowing or bioturbation.
3. Adhesion between particles produced by mucus films.
4. Alteration of particle momentum by filter feeding or ejection of pseudo-faeces.

In many ways it is difficult to separate the different effects and examine them independently. However, Jumars and Nowell indicate how the various factors may be incorporated into sediment transport formulae, though lack of quantitative data prevents testing the concepts.

Organisms can alter the entrainment of sediment by influencing the properties of the boundary layer flow. Isolated polychaëte tubes can cause local enhancement of the velocity as the flow diverts round the tubes, and this can cause scour. Higher densities of tubes have been supposed to protect the substrate. To investigate this, Eckman *et al.* (1981) carried out experiments in the laboratory with the polychaete *Owenia fusiformis* at various densities. They found that the substrate was destabilized at all densities, rather than the expected stabilization at high densities. They explained this as being due to the lack of mucus binding in their experiments, and suggested that stable beds persist despite the destabilizing effect of the animal tubes because of mucus films.

The organisms make mounds, pits and tracks, all of which are likely to increase the hydraulic roughness of the bed. Nowell *et al.* (1981) show that tracking doubled the boundary roughness, and decreased the critical entrainment velocity by 20 per cent. They also reported that faecal mounds were not easily entrained because of mucus adhesion, but faecal pellets were readily transported as bedload.

Microbial growth and the grazing of infauna produces mucus films that can bind the sediment. For example, Grant *et al.* (1982) examined the seasonal erosion of fine sands taken from Barnstable Harbour, Massachusetts. During the winter they found that threshold conditions agreed with the Shields value, but during the fall when adhesion reached a maximum they were about double the predicted values. De Flaun and Mayer (1983) have shown that there are seasonal changes in total bacterial numbers and their associated mucus coatings in sediments, related to the temperature variations. The bacteria inhabited shallow depressions on grains larger than about $10\,\mu m$, but their mucus secretions accumulated the clay size grains. Rhoads *et al.* (1978) reported that ten days of microbial growth increased the threshold velocity of coarse silts by 45 per cent.

On the other hand biological reworking, bioturbation, reduces the expected enhancement in the threshold. Grant *et al.* (1982) explain this as being due to physical breakup of the bacterial mucus films, but as all of the binding materials are still present there is still a net stabilization of the interface. Rhoads and Young (1970) found that by increasing the pore fluid content burrowing activity decreased the threshold velocity of fine grained sediments by 25 per cent, though again break-up of mucal films may have been important.

The presence of filter feeding organisms can affect the concentration of suspended material in transport by taking particles out of the near bed flow. The animals then deposit the material as faecal pellets or eject pseudo-faeces back into the water. Some interesting statistics on biodeposition have been presented by Haven and Morales-Alamo (1972). They point out that faecal pellets in the Clyde area can be deposited at a rate of $33\,\mathrm{mgm\,cm^{-2}\,week^{-1}}$. Oysters in a laboratory each deposited a maximum of 3 to 9 gm week^{-1}, and these deposits were 50 to 80 per cent clay, and 9 to 23 per cent organic. This material will degrade slowly, but can be mixed by burrowing activity to depths of 10–15 cm in less than three weeks, which is similar to the degradation time. Thus pelletized sediment can be formed which eventually reduces the mean grain size of the sediment, and being clay increases its cohesiveness and the threshold of motion.

On balance it is impossible at present to relate changes in sediment stability to any single biological process because several different ones are probably operating simultaneously. Nevertheless since most biological activities involve the production of extensive mucus films, they are likely to enhance the stability of the bed.

MODES OF PARTICLE TRANSPORT

Once the grains are moved by the fluid drag they do so in three different ways: by rolling, saltation and suspension. These have been examined by a nunber of researchers who have photographed the movement in flumes of single particles over flat beds of fixed particles of the same size. Features of the movement such as speed of particle movement and height and length of their trajectories, have been related to the transport stage, defined as the ratio of the ambient friction velocity to the threshold friction velocity, i.e. u_*/u_{*c}. Of course, though the three modes are fairly distinct, they all occur together throughout a very wide range of transport stages, just the proportions of grains moving in the different modes varies. This is illustrated in Figure 4.13. Near the threshold rolling is dominant, but the proportion of grains rolling decreases very rapidly as the transport stage rises and suspension increases.

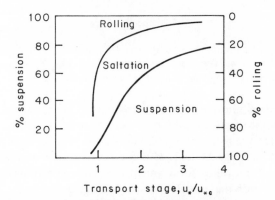

Figure 4.13 Relative percentages of rolling, saltation and suspension as a function of transport stage. *From Abbott and Francis, 1977*, Phil. Trans. R. Soc. London, *A284, 225–254. Reproduced by permission of The Royal Society*

Rolling

At stresses very close to the threshold the exposed grains will be forced to roll out of their positions and roll downstream until they find a new stable position. Bagnold (1973) regards rolling as incipient saltation.

Saltation

Momentary initial impulses of fluid drag and lift cause particles to jump into

130

the flow at an angle of about 50°. They then perform a ballistic trajectory which is continuously concave downwards. The maximum height of rise is about two to four grain diameters (Francis, 1973). The height of rise and the length of the hop increase to a maximum at a transport stage of about 2.5 (Figure 4.14) and the landing angle decreases from 17° at $u_*/u_{*c} = 1.04$, to 8° at a transport stage of 2.52 (Abbott and Francis, 1977). On landing the grain may bounce, but normally there is insufficient inertia in the grain to cause other grains to saltate after the impact. In air, however, the 'splashing' effect is much more important.

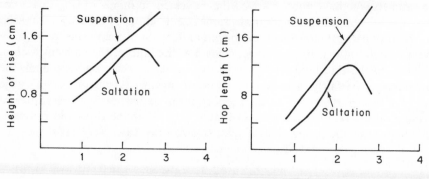

Figure 4.14 Mean maximum height and length of trajectories of particles undergoing saltation and suspension. *From Abbot and Francis, 1977,* Phil. Trans. R. Soc. London, *A284, 225–254. Reproduced by permission of The Royal Society*

From analysis of photographs of the saltating grains, Francis (1973) calculated the dynamic friction angle α of the grains striking the bed. Though there was a large scatter, the mean values of α decreased from 27° at a transport stage of 1.43 to 22° at a stage of 2.25. However the method was only considered approximate and these values for α were probably underestimates.

During the saltation the particle spins and this gives rise to the Magnus force. When the grain is accelerating and rising from the bed, the sense of spin and the relative velocity between the grain and fluid gives an upward Magnus force. At this time the particle is travelling slower than the local fluid. A downwards force occurs when the particle is decelerating and falling. Its forward velocity is then greater than that of the local fluid. White and Schultz (1977), have shown that the particle trajectory reaches a height 50 per cent greater than a theoretical one because of this lift. They also found that the impact angles and velocities were approximately normally distributed. However the lift off angles and velocities were considerably skewed. This means that when there are a number of saltating grains there are likely to be a few high trajectories and many lower ones, which are probably interspersed with periods of rolling.

The mean forward speed of the grains has been examined by Francis (1973) and compared with the depth mean current. The grain speed increases rapidly above the threshold and approaches the depth mean velocity asymptotically

(Figure 4.15). At a value of u_*/w_s of about unity the grains and the fluid are travelling at about the same velocity and suspension is likely to be active. As can be seen from Figure 4.15 this limit is equivalent to a transport stage of about 5. Angular grains travel more slowly than the rounded ones, and they spin more quickly. Experiments with a cloud of grains gave velocities only 5 per cent less. For large values of Re_* Francis (1973) shows that saltation occurred for transport stages between 1.0–2.2 for angular grains, and 1.0–5.5 for rounded.

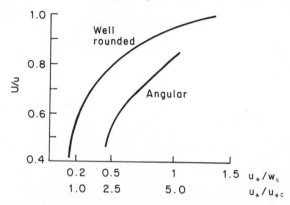

Figure 4.15 Mean forward speed of grains U relative to the fluid speed u, as a function of transport stage, for well rounded and angular grains. *From Francis, 1973,* Proc. R. Soc. London, *A332, 443–471. Reproduced by permission of The Royal Society*

Suspension

When the vertical components of the turbulent velocity were approximately equal to the settling velocity of the grain, i.e. when $w_s/u_* \sim 1$, the form of the grain trajectories change (Francis, 1973). They become wavy in their upper part, with convex downwards sections, and the jumps become much longer (Figure 4.16). These jumps are also higher, the grains seldom approaching less than two to three diameters from the bed. If they do, they normally return to the bed, where they can saltate or be taken again into suspension.

It is commonly assumed, following Bagnold (1956), that suspension occurs when the upward directed components of the turbulent velocity fluctuations w'_{up} exceeds the fall velocity of the grains. Thus $w'_{up} = w_s$. Bagnold estimated that $w'_{up} = 1.56 (w'^2)^{\frac{1}{2}}$, where $(w'^2)^{\frac{1}{2}}$ is the rms vertical velocity fluctuations. Bagnold used the relationship $(w'^2)^{\frac{1}{2}} = 0.8 u_*$, and full suspension thus occurs when $w_s = 1.25 u_*$, i.e. at a transport stage

$$\frac{u_*}{u_{*c}} = 0.8 \frac{w_s}{u_{*c}} \qquad (4.22)$$

Bagnold's hypothesis has recently been validated from a consideration of turbulence data by Leeder (1983). For large grain sizes the fall velocity is proportional to $D^{\frac{1}{2}}$ and consequently the suspension criterion of Equation 4.22 bears a constant relationship to the threshold for large Re_*. It can be seen from Figures 4.6 and 4.3 that above about $D = 700\,\mu m$, $w_s/u_{*c} = 4.5$. Thus from Equation 4.22 suspension effectively occurs when $u_*/u_{*c} \sim 3.6$. Additionally it is apparent that when $D < 150\,\mu m$, suspension of sand particles is likely at the threshold of movement.

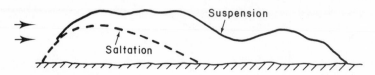

Figure 4.16 Diagrammatic comparison between the trajectories of a saltating and suspended particle

The above criterion can also be transferred onto the Shields curve. Substituting $w_s = 1.25\,u_*$ into the Shields Entrainment function gives

$$\theta_s = 0.4\,\frac{w_s^{\,2}}{gD} \tag{4.23}$$

At high Re_*, $w_s = 4.5\,u_{*c}$, and $\theta_c = 0.06$, gives $\theta_s = 0.78$. The curve of Equation 4.23 is shown together with the Shields threshold curve on Figure 4.17.

McCave (1971b) has argued, using the same concept as Bagnold, that the near-bed value of $(w'^2)^{\frac{1}{2}} \sim 1.20\,u_*$ should be used, rather than the average boundary layer value (see Figure 3.22). This gives a value for the coefficient in Equation 4.23 of 0.19.

Inman (1949) and Middleton (1976) have considered suspension to occur when $(w'^2)^{\frac{1}{2}} = w_s$. For a value of $(w'^2)^{\frac{1}{2}} = 1.2\,u_*$, this gives the same result as Equation 4.23.

Bagnold (1973) has argued that the major distinction in grain motion is between suspended and unsuspended transport, the latter comprising rolling, saltation and sliding. This is because particles in unsuspended transport receive no upward impulses other than those due to successive contacts between the solid and the bed, the fluid impulses on the grains being essentially horizontal. Unsuspended transport is equivalent to the term bedload.

During transport the water must be expending some energy in supporting the grains in suspension. Also, within the bedload layer the collisions between grains and between the grains and the bed provides a resistance to the flow which decreases the shear stress towards the bottom of the mobile layer. At the base of the mobile layer the shear stress must be equal to threshold shear stress, or otherwise more grains could be eroded and put into motion. The excess shear stress above the threshold is consequently absorbed in maintaining movement, and a balance is achieved at the bed, with as many grains being deposited as are being eroded.

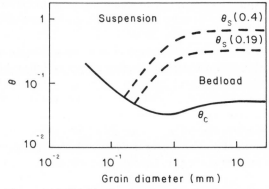

Figure 4.17 Shields diagram showing the threshold of suspension θ_s according to Bagnold (1956) with a coefficient of 0.4 and McCave (1971) with a coefficient of 0.19. *After McCave, 1971b, Marine Geol., **10**, 119–225. Reproduced by permission of Elsevier Science Publishers BV*

The distinction between suspension, saltation and creep or rolling is not as apparent in nature. Saltation is particularly important in air, but in water this mode is restricted to only a few grain diameters in height and has been thought to be relatively unimportant. Because of the intermittency in the shear stresses there is considerable variability almost from second-to-second in the sediment movement above the threshold in a tidal current. Dyer (1980a) has described the sequence of events occurring with increasing flow velocity above the threshold of motion on a rippled bed.

1. In the lee of the ripples organic debris and shell fragments begin intermittent slow, random movements. This phase may well indicate the onset of separation in the ripple lee.
2. Occasionally small numbers of sand grains are pushed off the ripple crests and cascade down the lee slope. These cascades often occur at preferred locations. No movement up the stoss slope is observed, the result being a change in the location of the crest.
3. Bursts of movement that carry a stream of sand grains over several ripple wavelengths as a fairly thin carpet 1–2 cm thick. The width of these patches of movement is of the order of 10 cm and the duration about a second. The movements are intermittent in space and time, though they often seem to originate at bifurcation points on the ripple crest.
4. Bursts of movement become longer in duration, more frequent and wider.
5. Swirls, during which sand grains are picked up in a whirling turbulent fashion to a height of at least 30 cm. Their duration is several seconds.
6. Suspension carpet. The whole bed becomes obscured by a zone of the order of 10 cm thickness of grains moving in suspension. The upper surface of the carpet is fairly distinct though it is a gradual rather than an abrupt change in concentration, but with a billowing form not unlike clouds seen from above.

The reverse sequence of events is visible during the decelerating tide but it is completed more quickly. At neap tides only the lower parts of the sequence are covered.

The term saltation would cover stages 3 to 5, since the grains are only intermittently in motion and are deposited after a brief period of movement. However, when averaged through time, at any one position there is likely to be a concentration gradient with concentration decreasing with height above the bed and extending a few tens of centimetres into the flow. This will have the appearance of suspension and will grade into true suspension when the grains are maintained in the flow for longer periods with increasing flow velocity. The apparent lower limit of stage 6, in the above sequence, gave a value of w_s/u_* of about 2.

SEQUENCE OF BEDFORMS

Once sediment starts to move the boundary deforms into a series of bedforms. The acceleration of the water flow over the resulting undulations causes form drag which decreases the shear stress available at the bed to create grain movement. Additionally the sediment has to be pushed up the slope on the upstream face of the bedforms, increasing the shear stress required to move the sediment, and the work done in moving the grains a unit distance along the bed. Consequently the sediment transport over a bed with bedforms is less than it would be on a flat bed with the same surface water slope. The acceleration of the flow towards the crest obviously provides additional potential for movement to partly offset the component of the gravitational resistance, but as the shape, height and form drag of the bed features varies with shear stress or stream power, the total effect on the sediment transport will be complicated. This is one of the main reasons why the prediction of sediment transport is still a topic with uncertainties and conflicting results.

There is considerable confusion in the nomenclature of bedforms, with different authors using different terms for the same feature, and parallels being drawn between desert bedforms, laboratory flumes and the sea. Nevertheless there is a sequence in bedforms with increasing stress above the threshold. In laboratory flumes the sequence is:

Ripples → Dunes → High Stage Plane Bed → Antidunes.

Some of the major bedform characteristics are summarized in Figure 4.18, and equivalent nomenclature is shown in Table 4.1. In the sea there is a particularly marked hierarchy of bedforms with smaller features being superimposed on the larger ones. The ripples respond to the flow most readily, change their shape and orientation with the oscillating tidal current and exhibit the fastest sediment transport rates. The largest sand waves, which can reach 15 m in height, as far as can be determined scarcely move at all. The superimposition of bedforms in the sea may be related to the varying oscillatory nature of the flow, and size relationships may not be as clear as in the laboratory because of possible paucity of material available for movement.

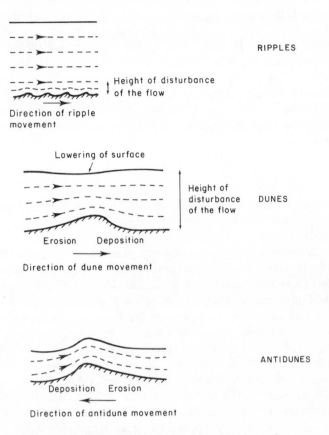

Figure 4.18 Diagrammatic representation of the flow over ripples, dunes and antidunes, and their movement

Table 4.1   Nomenclature of bedforms

(This book) Oscillatory tidal flow	Ripples	Megaripples	Sand waves
	Ripples	Dunes	Sand waves
	Ripples	Small sand waves	Large sand waves
Unidirectional flow	Ripples		Dunes

There have been many flume experiments carried out in order to try and define limiting conditions for the various bedforms. These have shown that the grain size of the sediment, the depth of water and the fluid velocity are particularly important. In engineering terms, the Froude number has been used (e.g. Simons *et al.*, 1961) since it includes the water depth and velocity. Other studies have considered the depth mean velocity, but rather fewer studies have considered the limits in terms of the Shields entrainment function. These laboratory measurements are generally taken as the starting point for developing relationships applicable to the sea. Only the first two bedforms are found in the sea as the flow velocities are not normally strong enough for the high stage forms. However they can often be observed temporarily occurring in the backwash of waves on a sandy beach.

Ripples

When the grains start moving at the threshold, the movement is intermittent initially and the grains find new equilibrium positions, forming a lower flat bed regime. When all the most exposed grains have moved to fill the recesses, further movement causes grains to pile up against stable groups of grains. With a further slight increase of velocity, separation of flow occurs around these groups, a vortex is created in their lee and the bed quickly forms itself into a series of regular ripples. The grains are pushed up the stoss, or upstream slope and on reaching the crest cascade down the avalanche lee slope. At the crest there is flow separation and a 'lee roller', a regular vortex, is formed in the lee. This causes a certain amount of erosion in the next trough and the amplitude of the ripple will grow. The interaction of the vortex with the bed downstream causes erosion and leads to a supply of grains available to form the next ripple, and the train of ripples grows in a downstream direction (Southard and Dingler, 1971).

Bagnold (1956) called the first features 'rolling grain' or primary ripples and the second features secondary ripples. The secondary ripples have appreciable form drag and initially the transport of grains decreases significantly when they are formed. In one example Bagnold quotes values of $\theta_c = 0.043$ for the threshold of movement and $\theta = 0.09$ for the lower limit of the secondary ripples. These values are within the range of normally quoted thresholds, and because of the impossibility of maintaining an even distribution of shear stress the two thresholds are likely to be considered as one. However, Liu (1957) also considers a separate threshold for the beginning of ripples and this contention is supported by the results of Williams and Kemp (1971) (Figure 4.19).

Nevertheless other authors consider that ripples will eventually form providing movement near the threshold can be maintained for long enough. Consequently the primary ripple phase, or the lower flat bed regime, may simply be a metastable one. Some additional support for the stability of the initial flat bed stage is found in the results of Southard and Dingler (1971). They found that in principle ripple formation could take place behind mounds higher than a critical height, at stresses insufficient to cause movement on a flat bed.

Many authors (e.g. Southard and Boguchwal, 1973) have found that ripples only form in grain sizes less than 0.6–0.7 mm (600–700 μm), or below about $Re_* = 13$ (Jackson, R. G., 1976). These limits coincide with the beginning of the upper limit on the presence of the viscous sublayer and suggests that the appearance of ripples may be related to an instability in the boundary layer. It is possible that the piles of grains being of the same thickness as the viscous sublayer perturb the layer causing a regular wave-like thickening and thinning. This would produce regular strips of increased and decreased bed shear stress promoting alternate erosion and deposition.

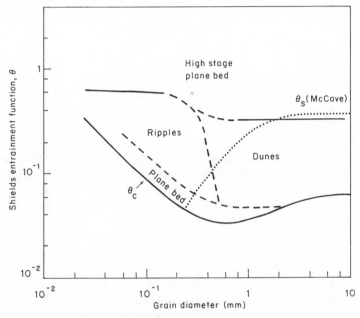

Figure 4.19 Limits of occurrence of bedforms (various authors)

Williams and Kemp (1971) consider ripples to result from separation in the lee of small deformations. These were caused by the action of turbulent bursts on the bed and were of two or three grain diameters in height, and of height to length ratio of about 1/100. The initiation of ripples occurred when the bed shear stress distribution overlapped the threshold shear stress distribution with a factor $n = 0.2$. This contrasts with Grass's (1970) plane bed n value of 0.625 (see p. 121). They found no evidence to support Bagnold's explanation for primary and secondary ripples.

It is interesting to note that the rippled bed regime coincides largely with the predicted presence of significant suspension at the threshold (Figure 4.19). Rees (1966) concluded, from experiments with a 10 μm silt, that the stability of the ripple configuration is in some way dependent on the presence of load in suspension, and that in the absence of sufficient load the bed may become plane. He explains the result by considering that there may be effectively two thresholds on the bed, one of the stable bed grains, and the other of grains involved in the suspension, the latter would be the lower threshold.

The ripples grow with increasing shear stress and eventually reach an equilibrium size. They commonly have a wavelength λ_r of 20–30 cm and a height H_r of a few centimetres. Yalin (1964), by consideration of experimental results from flume studies, has shown that

$$\lambda_r = 1000D \tag{4.24}$$

A similar result was found by Richards (1980) from a theoretical stability analysis of the bed. Thus the wavelength of the ripples is dependent on grain size. Though there is a limiting grain size for ripples of about 600 μm from laboratory measurements, in the sea bedforms in gravel of 10–25 mm diameter have been observed in the tidal channel of the West Solent (Dyer, 1971). These bedforms had wavelengths up to 18 m and could be termed ripples according to Equation 4.24. In flumes, the upper grain size limit on the ripple regime may not be well defined since the water depth necessary to move the coarse grains ensures that dune features grow at the expense of ripples. Consequently in the deeper water of the sea ripples may be present with larger grain sizes.

Ripples have a maximum steepness, H_r/λ_r, of about 1/10, an upstream slope of about 5° and a downstream avalanche slope of about 32°. At low stresses the ripples have relatively continuous, but somewhat sinuous crests, fairly uniform height and wavelength, and well oriented avalanche faces with a sharp crest line. At higher stresses the ripples are less regular in pattern, the crests being less continuous and more strongly curved. Although the average height and wavelength are relatively constant, the variability of height and wavelength increases with the flow (Harms, 1969).

The ripples move downstream, but not in a regular fashion, because they migrate through a process of localized scour and erosion. Both advance and retreat can be seen at various times at different places on the ripple crest. Sternberg (1967) has determined average rates of movement in the sea of 1 cm in 5 minutes and found that the rate of ripple advance U_r could be represented by $U_r = 2.02 \times 10^{-10} U_{10}^5$, where U_{10} is the velocity 10 cm above the bed.

Dunes

These bedforms have a very similar cross-sectional form to ripples, but are much larger and their size appears to be controlled by the flow depth rather than the grain size. There is considerable evidence that the dune wavelength is related to the water depth in steady uniform flow, and Yalin (1964) has shown that

$$\lambda_D \sim 2\pi h \tag{4.25}$$

This relationship is also supported by the theoretical analysis of Richards (1980). Nevertheless some slight dependency of wavelength on grain size is reported by some authors. The height of dunes is limited by the constriction of the flow over the crest which causes increased flow velocity and enhanced sediment movement. Yalin (1964) considers from flume data that there is a critical dune height to water depth ratio of 1/6, above which two-dimensional dunes do not occur. This, together with Equation 4.25 suggests that the maximum dune steepness is likely to be about 1/30. However, Yalin (1977) plotted many observations and showed that the maximum steepness of dunes is approximately 1/17 and this occurred at a transport stage of 3.5 (Figure 4.20). At this steepness separation of the flow is likely. Dunes eventually disappeared at a transport stage of about 8.

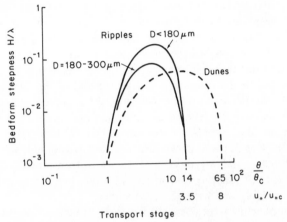

Figure 4.20 Steepness of ripples and dunes as a function of transport stage. *After Yalin, 1977,* Mechanics of Sediment Transport, *Pergamon Press, Oxford, 290 pp. Reproduced by permission of Pergamon Press*

Unlike flow over ripples where the water surface is smooth, over dunes the surface is uneven and turbulent. Over the crest the water surface is lower than over the trough. This is the result of acceleration towards the crest and deceleration towards the trough, and can be explained by examination of Bernouilli's equation (Equation 3.2). The water surface wave is thus out of phase with the bed wave, 'Boils' are also often visible at the water surface. These are energetic uprushes of water, in which the water spills outwards on the surface within a generally circular upwelling. Boils, together with the underwater 'Kolks' are caused by intermittent separation effects in the lee of the dune, and have been related to turbulent bursting events by Jackson, R. G. (1976).

One of the major features distinguishing dunes from ripples is that on the upstream stoss slope of dunes, which has an angle of about 5°, ripples can form. When present these ripples progress downstream at a faster rate than the dune. At the dune crest the finer fraction of the sediment is lifted into suspension, and the coarser fraction avalanches into the trough. As the velocity rises so does the amount of material in suspension. The internal structure of dunes created by the avalanching has been extensively described by Allen (1982).

In laboratory flumes dunes form in sediments coarser than about 600 µm directly following the lower stage flat bed. Simons *et al.* (1961) have also proposed that dunes do not form in sediments of mean size coarser than 8 mm. At grain sizes less than 600 µm ripples give way to dunes with increasing flow in an ill-defined and complicated transition range. However as can be seen by comparison of Equations 4.24 and 4.25, $\lambda_r \approx \lambda_D$ for these grain sizes at normal flume water depths. Simons *et al.* (1961), for a 450 µm sand, found ripples giving way to dunes at a Froude Number of about 0.3. The generalized limit is shown in Figure 4.19.

Theoretical analysis of the growth and stability of bedforms under unidirectional flow has been carried out by, amongst others, Engelund (1970), Smith (1970), Fredsoe (1974) and Richards (1980). These contributions have gradually developed an understanding of the interaction of the turbulent flow with the bed. It has been found that there are two modes of instability of the bed, with wavelengths related to the roughness of the bed and to the depth of the flow. These obviously compare with the formation of ripples and dunes. Both instabilities can occur separately, or together, depending on the values of friction angle in the bedload transport equation and the ratio of water depth to bed roughness length. For realistic flow variables Richards (1980) found development times of 600 s and 10 days and forward translational velocities of 3×10^{-3} cm s^{-1} and 1 m day^{-1} for ripples and dunes respectively. These values agree reasonably well with observed values.

An important conclusion of these studies is that there is a phase lag between the bedform and the fluid velocity. Thus the maximum bed shear stress occurs on the order of 20° upstream of the crest, and unless this lag is present, growth of bedforms does not seem to be possible. Additionally separation of the flow appears to be necessary for the propagation of a bedform without change of shape.

Sand Waves and Megaripples

The relationship of Equation 4.25 appears to hold for sandwaves in the sea, although there is quite a large scatter. Additionally marine sandwaves occur in areas where the Froude number is less than 0.1. However, as pointed out by Jackson, R. G. (1976) it is the boundary layer thickness that is important in the scaling, and for some of the marine data the boundary layer may not occupy the entire water column. Also variation in the dynamic friction angle tan α may be important (Richards, 1980). On the other hand Allen (1980) has suggested that large sand waves may be related to the oscillatory tidal wave in the same way as wave-induced ripples are to surface waves, but he admitted that there is no theoretical or experimental evidence for this concept.

Sand waves, though they commonly have steepnesses of 1/10–1/15, more often than not have lee slopes with angles much less than the angle of rest. Even in the asymmetric form the lee slopes seldom exceed 10°. Ludwick (1972), for example, found lee slopes averaging 1.5°, and only occasionally reaching 6°.

Consequently the degree of asymmetry is very much less than bedforms in unidirectional flows, and the crest of the sand wave is much nearer midway between the troughs. The degree of asymmetry can be quantified by the ratio of horizontal length of the lee slope divided by the horizontal length of the stoss slope. The values then are $\leqslant 1$, with symmetrical sand waves having a value of unity. In the extreme, completely symmetrical features are found with slopes of equal angle, and they are often of greater steepness than asymmetric ones. It is commonly assumed that the degree of asymmetry is a measure of the magnitude of transport of sediment across the wave. The low angle of the lee slope means that, providing the flow does not separate, the sediment will move over the crest and will be driven down the lee slope by the fluid stress, rather than cascading freely down. On the reverse phase of the tidal cycle a smaller amount will be driven back up this slope. The net result will be a gradual migration of the wave. For symmetrical features an eventual balance is possible with equal amounts of sediment carried backwards and forwards over the crest on each phase of the tide. These processes have been examined by modelling by Taylor and Dyer (1977) and Richards and Taylor (1981).

The smaller megaripple features found on the backs of the sand waves are rather more problematical. They invariably have lee slopes at avalanche angles and the direction of their asymmetry is frequently observed to reorientate with the reversal of the tide. Typical dimensions are height 0.5–1 m and wavelength 5–10 m. Thus their steepness is in accord with dunes in flumes, but their height does not appear to be related to the water depth. Instead a relationship to the roughness caused by the presence of normal ripples has been proposed, leading to $\lambda_m \sim 1000 \, H_r$. Though no field proof of this is available, it supports usage of the term megaripple rather than alternatives. Variation of megaripple wavelength at different positions on a sand wave has been demonstrated (Langhorne, 1977) with wavelength greatest at the crest. This suggests a relationship with flow velocity, and hence with ripple size, rather than with water depth.

Further discussion of the characteristics of megaripple and sand waves will be found on page 272.

High Stage Plane Bed

At fairly high velocities throughout the grain size range the dunes eventually become washed-out by the erosion occurring at the dune crests. The concentration of moving material near bed becomes very high and eventually the bed becomes planar. In this situation the stress at the level of the non-mobile bed must be about equal to the threshold stress, and the excess stress above threshold must be equal to the frictional forces within the moving grains. Bagnold (1956) has proposed that the critical stage should occur when the stress is sufficient to erode a unit grain layer of the bed. The immersed weight of the layer is $(\rho_s - \rho) \, g \, D \, C_0$, where C_0 is the volume concentration of the bed. The shear stress required to maintain the motion of the load is $(\rho_s - \rho) \, g \, D \, C_0 \tan \alpha$

where α is the dynamic friction angle. Thus the critical stage occurs when

$$\theta = C_0 \tan \alpha \qquad (4.26)$$

For natural beds $C_0 \sim 0.63$.

As we have seen at the boundary, for large grains Chepil (1959) found that $\tan \alpha = 0.445$ ($\alpha = 24°$), and Francis (1973) calculated $\tan \alpha = 0.5$–0.4. However, grain interaction is likely within the flow, as well as at the boundary. As grain concentration is increased grains cannot avoid hitting each other. On impact there is a force acting along the line joining the centres of the two particles. The mean angle of impact will equal the angle of dynamic friction, and the force will have normal and longitudinal components. Bagnold (1954) carried out a well-designed experiment to examine the value of $\tan \alpha$ in a shearing flow for a wide range of particle sizes, concentrations and shear rates. A series of neutrally buoyant spheres was sheared in the angular space between two coaxial drums. The normal stress was measured as the inward radial pressure (the dispersive pressure) on the wall of the inner drum, and the tangential shear stress as the torque upon it. The ratio between these two stresses equals $\tan \alpha$. For concentrations less than about $C_0 = 0.57$ there were limiting values of $\tan \alpha$ of 0.32 for inertial conditions, and 0.75 for viscous conditions. For C_0 in excess of 0.57, $\tan \alpha$ was equal to 0.4. The grain interaction appeared to be negligible at concentrations lower than about 0.08 (i.e. 8 per cent by volume).

The two limiting cases for inertial and viscous conditions are effectively equivalent to relatively large grains in a high shear field, and relatively small grains in a lower shear field. It is likely that the large grains physically hit each other during transport, and this would abrade the grain surfaces. On the other hand, the smaller grains do not physically come into contact since the viscosity of the thin layer of water is sufficient to overcome the grain inertia and cushions the impact. The transition between the two will be governed by grain size, concentration and shear stress. In nature, grains larger than about 125 μm are reasonably well rounded, whilst those less than this value become increasingly angular. Consequently the transition zone is likely to be centred on this grain size.

Putting the values of 0.75 and 0.32 in Equation 4.26 and taking $C_0 = 0.63$, gives the effective lower limit of the plane bed stage as $\theta = 0.49$ and 0.20 respectively for small and large D. Allen and Leeder (1980) have considered Bagnold's results together with those of others, and obtained values of $\theta = 0.62$ and 0.34. These limits are shown in Figure 4.19, but do not agree too well with the limits of ripples and dunes suggested by Yalin (1977).

Experiments in flumes have shown that the upper limits of the dune bedform regime and the transition to the upper stage flat bed can be related to Froude number, with a value of $Fr = 1$ being the limit. Generally in coastal areas the Froude number never approaches this limit and dunes are the highest stage bedform encountered. Nevertheless in the backwash of waves on a sandy beach it is often possible to see the whole range of bedform and flow stages within a few seconds. In the upper reaches of estuaries with high tidal ranges, high stage flat bed conditions can occur over the tidal flats at maximum flow.

Antidunes

When the Froude number exceeds unity antidunes can form. These are symmetrical features and there is a corresponding wave of about equal amplitude on the water surface, the water depth being more or less constant. The surface wave is unstable and can grow and break in an upstream direction at times. Both the surface wave and the antidune move upstream against the flow, in contrast to ripples and dunes. The amount of sediment in motion is high, and it is trapped on the upstream face of the antidune, and eroded from the downstream face. A good description of antidunes is given in Simons *et al.* (1961).

CHAPTER 5

Sediment Movement Under Waves

THRESHOLD OF GRAIN MOVEMENT

The movement of individual spherical particles on a flat bed can be treated in the same way as the simple model derived for steady unidirectional flows. However, because of the oscillatory nature of the flow, the changing surface water slopes cause instantaneous horizontal pressure gradients and acceleration of the water and sediment particles. The resulting force can be divided into two components, one relating to the force required to accelerate a volume of fluid equivalent to that of the particle and the other that required to accelerate an additional mass of fluid around the particle.

The force can be written in terms of the fluid acceleration du/dt, and is equal to the inertia force of the displaced fluid. Thus

$$F_p = \pi \frac{D^3}{6} \rho \frac{du}{dt} \qquad (5.1)$$

Similarly the virtual mass force is given by

$$F_{VM} = C_M \pi \frac{D^3}{6} \rho \frac{du}{dt} \qquad (5.2)$$

where C_M is the coefficient of virtual mass. For a sphere in an unbounded flow C_M has a value of 0.5. This means that there is the mass of a volume of fluid equivalent to about half the volume of the sphere helping to resist the motion. In the present case the value of C_M will alter because of the presence of the boundary.

In comparison with Equation 4.13, the stability equation at the phase in the wave cycle of maximum near bed orbital velocity u_m is

144

$$C_D \pi \frac{D^2}{4} \rho \frac{u_m{}^2}{2} + (1 + C_M) \frac{\pi D^3}{6} \rho \frac{du}{dt} =$$
$$\left(\frac{\pi D^3}{6} g (\rho_s - \rho) - C_L \frac{\pi D^2}{4} \rho \frac{u_m{}^2}{2} \right) \tan \alpha \qquad (5.3)$$

This can be arranged with equivalent assumptions to Equation 4.4 to yield

$$\frac{\rho u_m{}^2}{g(\rho_s - \rho)D} = \frac{0.66 \left(\tan \alpha - (1 + C_M) \dfrac{\rho}{(\rho_s - \rho)g} \cdot \dfrac{du}{dt} \right)}{C_D + C_L \tan \alpha} \qquad (5.4)$$

The group on the left hand side of Equation 5.4, which has similarities to the Shields Entrainment Function, is called the mobility number **M**,

$$\mathbf{M} = \frac{\rho u_m{}^2}{(\rho_s - \rho)gD} = \frac{u_m{}^2}{\gamma gD} \qquad (5.5)$$

For sinusoidal waves the maximum velocity u_m occurs when the acceleration is zero, in which case Equation 5.4 is very similar to Equation 4.14. In practice, for irregular sea waves there can be a large range of accelerations for any value of u_m, because the wave period is involved. Also the value of du/dt may not be zero at maximum u_m. Additionally the values for the three coefficients are not well defined in an unsteady boundary flow. Consequently the approach to the threshold of movement under waves has been largely an empirical one.

There have been many laboratory studies of the threshold of movement of sand under waves. Bagnold (1946) proposed that threshold conditions were given by

$$\frac{\sigma^{\frac{1}{4}} u_m{}^{\frac{3}{4}}}{\gamma^{\frac{1}{2}} D^{0.325}} = 21.5 \text{ (cgs units)} \qquad (5.6)$$

where $\sigma = 2\pi/T$, T being the wave period.

Manohar (1955) proposed

$$\frac{u_m}{(\gamma g)^{0.4}(\nu D)^{0.2}} = \begin{cases} 7.45 & \text{for initial movement} \\ 8.20 & \text{for general movement} \end{cases} \qquad (5.7)$$

Komar and Miller (1973, 1975a,b) have shown that for grain sizes less than 500 μm, the boundary layer was still laminar at the threshold of motion and it could be represented by a critical value $\mathbf{M_c}$ of the mobility number,

$$\mathbf{M_c} = 0.21 \left(\frac{d_0}{D} \right)^{\frac{1}{2}} \qquad (5.8)$$

where d_0 is the orbital diameter ($= 2A_b$)

For grains coarser than 500 μm the relation becomes

$$\mathbf{M_c} = 0.46 \pi \left(\frac{d_0}{D} \right)^{\frac{1}{4}} \qquad (5.9)$$

Since the maximum orbital velocity and the orbital diameter d_0 are related by $u_m = \pi d_0/T$, the threshold can be written in terms of two out of the three variables. Komar and Miller (1975a) have produced the curves shown in Figure 5.1 which gives a simple means of relating the threshold orbital velocity to the grain size of quartz particles and to wave period. The transition between Equations

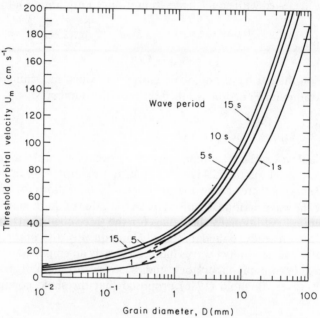

Figure 5.1 Near bed maximum orbital velocity u_m for the threshold of sediment movement under waves of different period. *From Komar and Miller 1975a*, J. Sediment. Petrol., **45**, 362–367. *Reproduced by permission of the Society for Economic Paleontologists and Mineralogists*

5.8 and 5.9 is dashed at around 500 μm. There are of course many combinations of water depth, wave period and wave height which could produce the necessary orbital velocity. Using the linear wave equations (Equation 3.50), it is possible to calculate the wave height which, for a particular wave period, would move sediment of various sizes in different water depths. Komar and Miller (1975a) provided a computer program for carrying out this calculation. An example of the resulting relationship is shown in Figure 5.2 for a period of 15 s.

Lofquist (1978) on the basis of field observations and laboratory data by many authors proposed the threshold curve shown in Figure 5.3, which relates M_c to $d_0/2D$ for a flat bed. The latter parameter is equivalent to the relative roughness of the bed, which together with the Reynolds number defines laminar, transitional and rough turbulent oscillatory flow. He also proposed a lower curve given by $M = M_c/5$ as the lower bound for equilibrium ripples. This effectively defines the threshold of sand movement on a rippled bed, which is lower than the flat bed value because the bed shear stresses at the crests are enhanced above the average value .

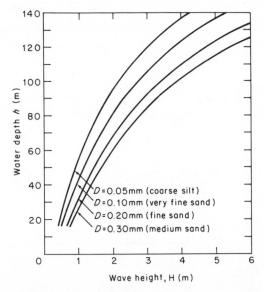

Figure 5.2 The height of a wave of period 15 s which causes movement of various size particles on a flat bed in different water depths. *From Komar and Miller 1975a, J. Sediment. Petrol., 45, 362–367. Reproduced by permission of the Society for Economic Paleontologists and Mineralogists*

An alternative approach to that is to write the threshold conditions in terms of the bed shear stress by use of the wave friction factor, f_w. Thus

$$\theta_c = \frac{\tau_c}{(\rho_s - \rho)gD} = \frac{\frac{1}{2}f_w\rho u_m^2}{(\rho_s - \rho)gD} \tag{5.10}$$

This was carried out by Madsen and Grant (1975) and Komar and Miller (1975a) and gives a good comparison with the unidirectional Shields curve. The result is shown in Figure 5.4. For the fine grain sizes in laminar boundary layer conditions the curve may be obtained analytically for a smooth bed as

$$\theta_c = 0.21\sqrt{2}(Re_{w2})^{-\frac{1}{2}} \tag{5.11}$$

The implication from the good comparison between the unidirectional and oscillatory results is that, providing the wave friction factor is used, the acceleration during the orbital oscillation is of minor importance.

However, this view is not supported by the results of Chan *et al.* (1972). They have compared the threshold of particle motion in steady and unsteady flows.

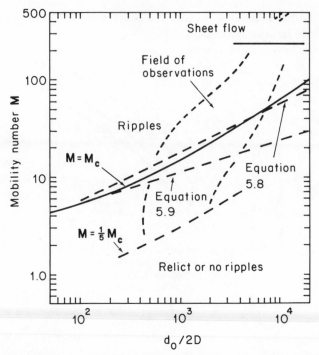

Figure 5.3 Threshold of movement of sediment in terms of mobility number. M_c is the flatbed threshold according to Lofquist (1978). Equations 5.8 and 5.9 are the curves of Komar and Miller (1975), largely based on the same data as that of Lofquist. The curve $M = 1/5\ M_c$ is the lower bound for equilibrium ripples. The dashed lines enclose the majority of quoted field observations. After Lofquist, 1978

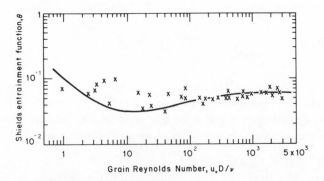

Figure 5.4 Comparison between Shields threshold curve for unidirectional flow (solid line), and the observed threshold under waves (crosses) calculated using the friction factor. *From Komar and Miller, 1974,* Proc. 14th Coastal Eng. Conf., *756–775. Reproduced by permission of American Society of Civil Engineers*

Their results indicate that in a wave cycle, the peak oscillatory shear stress needed to cause motion is about twice the threshold shear stress in unidirectional flow. They explained the difference as being due to the response time for a particle to be moved out of its place on the bed surface. In steady flow this response time would be unimportant, but with an oscillatory stress it would result in a high apparent value at the threshold of motion. However, a factor of two in the stress is within the scatter of observations for steady flows, and they took no account of the possible influence of acceleration.

The laboratory data used for evaluation of the threshold curves are all for regular sinusoidal waves. In the sea the waves are generally anything but regular and within the spectrum of motion there is considerable variation of both wave height and period. Obtaining and interpreting field data to compare with the above equations has been very difficult, as we shall see later.

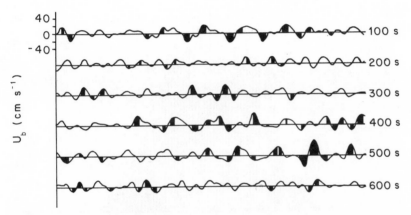

Figure 5.5 An example of a 10 m record of the near bed velocity observed over a rippled bed, and the periods of observed sediment movement near the ripple crests, shown in black. This illustrates that orbital velocity may not be a good determinant of the threshold. *From Davies, 1980,* Coastal Eng., *4, 23–46. Reproduced by permission of Elsevier Science Publishers BV*

It is widely assumed that wave boundary layers in the field are always turbulent at and above the threshold of movement. However, in some cases the flow is transitional at the threshold. Figure 5.5 shows a 10 minute record of the velocity recorded above a rippled bed together with the periods of bed movement determined from an underwater television. Though some of the largest velocities caused movements, many of the instances of movement occurred with relatively small velocities. However consideration of velocities may be unsuitable and it may be more fruitful to try and compare the instances of motion with the bed shear stress. Davies, A. G. (1980) outlines a method whereby bed stresses are calculated from the measured velocities assuming that the boundary layer is almost laminar. He found the calculated shear stress to be very sensitive to the level at which it is assumed to act on the surface grains. By an optimiza-

150

tion technique the best match was achieved between the shear stresses and the instants of movement, and the spread of threshold values was minimized by choosing an 'effective bed level'. This optimum level was determined as 0.093 cm (above a notional smooth flat bed) when the grain size was 0.14 cm (1400 μm). Figure 5.6 shows a histogram of the deduced shear stress for all waves, as well as those causing movement. There is obviously a transition range when only some waves caused movement, but above about 10 dynes cm^{-2} all waves moved sediment. The values in the transition range span those expected from the modified Shields curve.

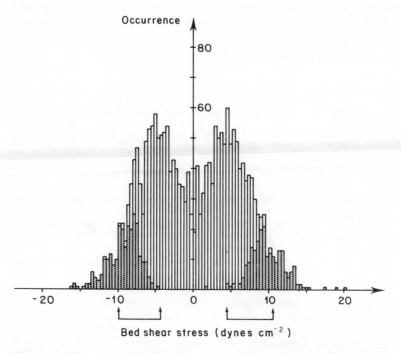

Figure 5.6 Histogram of the bed shear stress calculated at the effective bed level. The shaded portions are those waves causing sediment movement. The transition ranges are indicated by the arrowed limits. *From Davies, 1980,* Coastal Eng., *4, 23–46. Reproduced by permission of Elsevier Science Publishers BV*

The above type of analysis was used by Davies (1985) to compare the observed thresholds with the formulae of Bagnold, Manohar, and Komar and Miller (Equations 5.6–5.9). Very good comparison was achieved for the critical velocity amplitude; values of u_m from the three formulae being within a few cms^{-1} of each other. When the threshold stress was calculated from the results using Jonsson's friction factor, good comparison for a flat bed was achieved with Shields curve provided the roughness k_s was taken as $k_s = 2D_{90}$. When the bed was rippled and the roughness defined in terms of the ripple height, there was a very large difference which suggested that the form drag of the ripples was an order of magnitude greater than the skin·friction. This is an important contrast with form drag in steady flow, which is about equal to the skin friction.

There have been very few measurements of the threshold of sediment motion under the combined action of waves and currents. Hammond and Collins (1979) found that for combined flows, fine and medium sand moved at lower critical velocities under longer period waves than under short period waves. This is the reverse of the situation under waves alone, in which the critical velocity is lower for waves of short period than for long waves. They propose that the probability of turbulent bursts caused by the steady flow, coinciding with orbital velocities near the maximum becomes larger with increasing wave period. Consequently the maximum orbital velocities which are responsible for movement under waves alone become superseded by statistically fluctuating events.

In the laboratory, of course, only the codirectional situation can be investigated. I am not aware of any published field measurements of the threshold of movement under combined waves and currents.

SEQUENCE OF BEDFORMS

As in steady flow, a sequence of bedforms occurs above the threshold of movement as the magnitude of orbital diameter is increased:

Flat bed → rolling grain ripples → vortex ripples → flat bed

Ripples

Wave-induced ripples can be distinguished from ripples in steady flow by their shape (Harms, 1969). Wave ripples have symmetrical and rounded profiles and do not slope at the angle of repose. A comparison between the cross-sectional form of current and wave ripples is shown in Figure 5.7. Additionally the crests are long and straight, and ripple height and spacing is very uniform. Some crests divide in a simple Y-shaped junction while other crests end by abrupt tapering, and adjacent ripple crests bend to achieve the characteristic ripple spacing.

Figure 5.7 Diagrammatic comparison of the cross section of (A) unidirectional, and (B) wave, ripples. Not to scale

Bagnold (1946) has described two types of wave ripples: rolling-grain ripples and vortex ripples.

Rolling grain ripples

At the threshold of motion, grains start to be rolled to and fro over the surface but are not lifted off it. Since the grain movement only occurs at around the maximum orbital velocity the length of the grain path is short. It lengthens as the oscillation velocity increases.

Though motion is initially distributed randomly, the moving grains become organized eventually and tend to accumulate in parallel zones, which form wavy ridges a few grains high, whose crests sway from side to side during the fluid oscillation. As the ridge grows it shelters a wider and wider strip of the flat troughs from the water action. When this sheltered region extends the full distance between the ridges, movement is restricted to the crests. Since the ridges cannot collect any further grains, the arrangement becomes stable according to Bagnold.

The rolling grain ripples occurred on all sands. With fine sands the troughs remained flat, but with larger grains they became a nearly circular arc of large radius. The distinguishing feature of this type of ripple was the entire absence of grain movement in the troughs. However, other authors believe that rolling grain ripples are only a transient stage and are not permanently stable. Sleath (1976) has developed relations for rolling grain ripple wavelength based on an analytical approach.

Vortex ripples

Bagnold (1946) found that once the orbital velocity exceeds twice the critical speed for the threshold of movement, the character of the flow abruptly changes. A vortex forms in the lee of the crest in each wave half cycle because of separation. The vortex sweeps the grains towards the crests until the flow reverses. The vortex is then swept towards the crest and is lifted up into the flow, the bulk of the grains coming to rest at the crest of the ripple. However with the finer grains, or with more energetic waves, some grains are carried up into suspension. A diagrammatic sequence of the generation of vortices during the wave cycle is shown in Figure 5.8. Towards the end of the return stroke a second vortex is formed which, in turn, is projected into the flow. As the orbital velocity increases, the number of vortices present in the flow increases, but they gradually decay. At the beginning and in the middle of the stroke, while there is no vortex in the ripple trough, the velocity of the water flow over the crest tends to flatten it by dislodging the grains at the crest and rolling them into the trough.

This velocity effect at mid-stroke becomes more pronounced as the stroke lengthens, so that the ripple height decreases. When the stroke is very long and the velocity high, the crests becomes flattened to the extent that the end of stroke vortex cannot develop. Consequently the ripple steepness decreases as the orbital velocity increases. Generally the steepness is 0.1 to 0.2. Sleath (1975) has investigated the development of vortex ripples based on a numerical solution of the equation of vorticity.

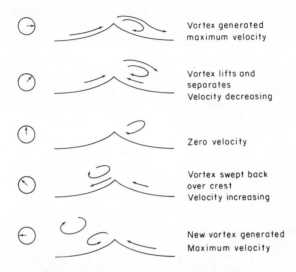

Figure 5.8 Diagrammatic sequence of events during a wave oscillation showing the generation, separation and advection of vortices. The wave phase is shown by the rotating vector on the left

Tunstall and Inman (1975) have shown that the maximum circulation velocities in the oscillatory flow bed vortices are large, being two or three times the amplitude of the wave induced velocities near and parallel to the bed. Thus vertical velocities of the same order as the horizontal velocities occur very close to the boundary (see Davies, 1984, 1985). The vortex formation and ejection process is an extremely efficient mechanism for entraining sediment on a rippled bed and placing it into suspension. The turbulence produced by these vortices is usually confined to a layer less than about 2 ripple heights above the bed.

As a consequence of the sediment suspension in the vortices, there are considerable temporal changes in concentration at any point. The sediment cloud entrained near each crest will be swept over a distance of a fraction of a ripple wavelength, to more than a wavelength, depending on the ratio of the fluid excursion to the ripple wavelength, all the while dispersing and settling towards the bed. When the velocity reverses, the clouds entrained during the preceding stroke may be swept back in the opposite direction. Meanwhile another burst of sediment is being entrained from the other side of the crest. Thus at any point an observer would see several concentration peaks, depending on the excursion of the bursts, and the position relative to the crest. The same cloud may be observed on both forward and backward strokes. The rate at which sediment is

154

entrained is governed by the ripple spacing, the strength of the vortices and the shear stress (Kennedy and Locher, 1972).

There have been many studies of the ratio of ripple wavelength to orbital diameter. Bagnold (1946) pointed out that the ripple wavelength λ_r was about equal to the orbital diameter d_0 up to a certain diameter, the 'break-off' point, above which the wavelength remained constant. The terminal wavelength above the break-off point was independent of the grain density and varied nearly as the square root of the grain diameter. This pattern of wavelength change has been confirmed by several laboratory and field studies. Many of the results have been compared by Komar (1974) and Miller and Komar (1980a), and are shown in Figure 5.9. The best empirical fit to the observed data gives $\lambda_r = 0.8 \, d_0$. However Miller and Komar (1980b) quote $\lambda_r = 0.65 \, d_0$, and this agrees with Lofquist (1978) who proposed that $\lambda_r/d_0 = \frac{2}{3}$ for $10^2 < d_0/2D < 5 \times 10^3$ and $0 < M < 30$.

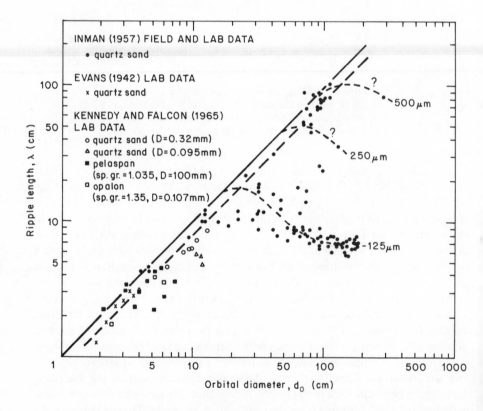

Figure 5.9 Wavelength of wave induced ripples versus orbital diameter. *From Komar, 1974, J. Sediment. Petrol., 44, 169–180. Reproduced by permission of the Society for Economic Paleontologists and Mineralogists*

Nielson (1981) has combined many laboratory results into the overall relationship

$$\frac{2\lambda_r}{d_0} = 2.2 - 0.345\,\mathbf{M}^{0.34} \tag{5.12}$$

As can be seen in Figure 5.10, this curve gives a virtually constant λ_r/d_0 of between 0.6–0.8 for $\mathbf{M} < 40$.

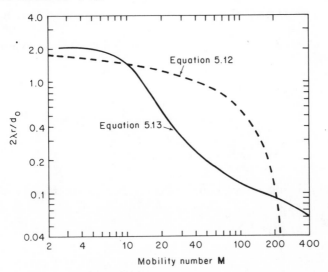

Figure 5.10 Relationship between ripple wavelength/orbital diameter ratio and mobility number $\mathbf{M}$. Dashed line Equation 5.12, laboratory results; solid line Equation 5.13. After Davies (1984) and Nielsen, (1981)

However wave induced ripples in the field are considerably shorter than those in the laboratory due to the spectrum of velocities. It is difficult to obtain reliable field data since the ripples may not be in equilibrium with the waves; the waves have an irregularity that ensures the ripples may be continually changing. For field measurements Nielson (1981) proposed the relationship

$$2\frac{\lambda_r}{d_0} = \exp\left(\frac{693 - 0.37\ \ln^8\mathbf{M}}{1000 + 0.75\ \ln^7\mathbf{M}}\right) \tag{5.13}$$

The curve of Equation 5.13 is shown in Figure 5.10.

These results probably incorporate ripples formed both above and below the break-off point. The observed ripples are dominated by the largest waves present, and the best agreement with laboratory results is obtained if values of d_0 based on the significant wave height are used. Miller and Komar (1980a) concluded from their own field data that longer ripple wavelengths are associated with a polymodal wave spectrum, and smaller wavelengths with single peaked spectra.

The wavelength of the ripples at the break-off point can be represented by (Miller and Komar, 1980a)

$$\lambda_r = 14700 \, D^{1.68} \tag{5.14}$$

where D and λ_r are in cm. The data can also be approximately represented by $\lambda_r = 1000 \, D$, the unidirectional flow relationship, which indicates that for large oscillations the ripple dimensions may be controlled more by the sediment movement over the crest at mid-stroke rather than by the vortices in the ripple lee.

Grant and Madsen (1982) have proposed that the break-off point can be predicted from

$$\frac{\theta'}{\theta_c} = 1.8 \, S_*^{\,0.6} \tag{5.15}$$

where

$$S_* = \left(\frac{D}{4v}\right)\left(\frac{\rho_s - \rho}{\rho} \, gD\right)^{\frac{1}{2}}$$

and θ' is the skin friction component of stress. For quartz grains in water Equation 5.15 reduces to

$$\frac{\theta'}{\theta_c} = 114 \, D^{0.9} \tag{5.16}$$

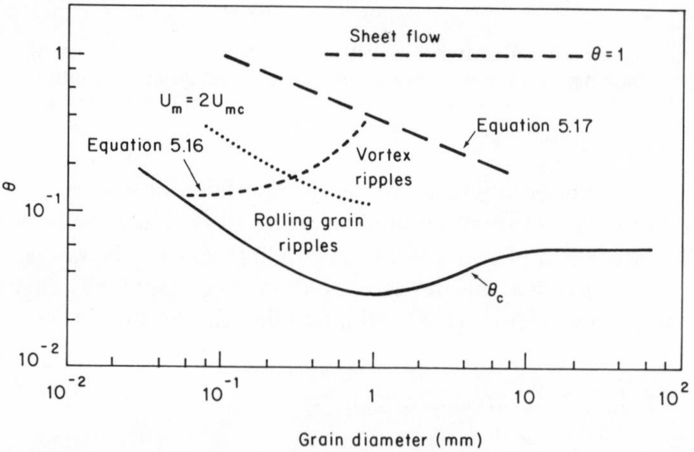

Figure 5.11 Shields diagram showing the limits of wave induced bedforms. The line separating rolling grain and vortex ripples is from Bagnold (1946). Equation 5.16 is the break-off point according to Grant and Madsen (1982). Equation 5.17 is the upper limit of ripples according to Komar and Miller (1975b) and $\theta = 1$ that according to Nielsen (1981)

The curve of Equation 5.16, shown on Figure 5.11, suggests that for grain sizes larger than about 1 mm, ripples with wavelengths equal to the near bed orbital diameter are present throughout the ripple stage. Komar (1974) argues that for λ_r to be equal to d_0 the wave period would have to be short, and consequently ripples with $\lambda_r \sim d_0$ would be restricted to very shallow water and areas of limited fetch, for the finer grain sizes. For coarse particles the condition $\lambda_r \sim d_0$ corresponds to the higher periods found in deeper water.

Flat Bed

With increasing orbital velocity the height of the ripples decreases and a 'sheet flow' of suspended sediment occurs at the bed as a layer a few centimetres thick. Some workers consider that rolling grain ripples are also found in the transition state between vortex ripples and the flat bed, but if this stage occurs it is distinguished from the lower stage of rolling grain ripples by the intense suspended sediment movement.

Komar and Miller (1975b), using laboratory data of Manohar, proposed that the transition from vortex ripples to flat bed can be expressed by

$$\theta = 0.413\, D^{-2/5} \tag{5.17}$$

when D is in centimetres, or

$$\theta = 4.4\,(Re_{w2})^{-1/3} \tag{5.18}$$

The former expression bears a very close correspondence to the limit of dunes in steady flow shown in Figure 4.19. The results are shown in Figure 5.11.

Dingler and Inman (1976) show from field observation that the transition from vortex ripples to sheet flow commences at $\mathbf{M} = 40$ and ends at $\mathbf{M} \sim 240$ (Figure 5.3). Nielson (1981) proposes the criteria that sheet flow occurs when $\theta \geqslant 1$. This is also shown in Figure 5.11. The lower limit of the flat bed and the transition to the underlying ripple field has been proposed as the field in which hummocky cross-stratification is developed. These low undulating features have heights of 0.1 to 0.5 m and spacing of several metres. Their mode of formation is still under intense discussion, but generation probably involves simultaneous steady flows, since their wavelengths are too large to be produced by surface waves alone (Allen, P. A., 1985).

CHAPTER 6

Sediment Suspension

We have already seen that, above certain critical conditions, sediment particles are maintained in suspension by the exchange of momentum from the fluid to the particles. To calculate the suspended load transport rate it is necessary to be able to predict the profile of concentration throughout the depth either based on a measurement at one level, or on knowledge of the power of the stream and the bed sediment size. The profile of concentration can be multiplied by that of the velocity and then numerically integrated to obtain the transport rate. This has been done empirically in some of the total load sediment transport formulae described in the following chapter. To carry the process out more rigorously reveals that there are difficulties created by the fact that the suspended sediment takes energy from the flow and this affects the shape of the velocity profile and the shear stress.

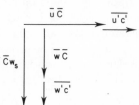

Figure 6.1 Mean and fluctuating components of the suspended sediment flux

In a turbulent tidal stream the fluxes of suspended sediment will have mean and turbulent components in both the horizontal and vertical axes (Figure 6.1). In the downstream direction the mean flux will be $\overline{u}\,\overline{C}$, determined from the product of the mean velocity and mean concentration. In addition there is a turbulent component $\overline{u'c'}$, which could be either positive or negative. The assumption that it is a small positive quantity is supported by laboratory and field measurements. Nevertheless it may be negative under some conditions.

In the vertical direction there is an equivalent mean vertical flux $\overline{w}\,\overline{C}$, which can be either positive or negative depending on secondary flow circulations. The particle settling flux $\overline{C}\,w_s$ will always be towards the bed. The vertical turbulent flux $\overline{w'c'}$ may also be upwards or downwards. However, if $\overline{C}\,w_s + \overline{w}\,\overline{C} + \overline{w'c'} > 0$, then deposition at the bed is required, and the reverse for erosion. The turbulent fluxes are known as the Reynolds fluxes, by analogy with the Reynolds stresses.

SUSPENSION PROFILES

Under steady state conditions and in the absence of a mean vertical velocity, the flux of settling particles $\overline{C}\,w_s$ is equal to the vertical turbulent flux $\overline{w'c'}$. Thus

$$\overline{C}\,w_s = -\overline{w'c'}$$

The mean concentration profile therefore depends on the variation with height above the bed of the vertical turbulent flux.

Making a gradient diffusion assumption, the turbulent flux is assumed to be equal to the concentration gradient times an eddy diffusion coefficient for sediment particles. This gives

$$\overline{C}\,w_s = -\mathbf{K}_s\frac{d\overline{C}}{dz} \tag{6.1}$$

Integration gives (for convenience ommitting the overbar denoting mean values)

$$\ln\frac{C}{C_a} = w_s\int_a^z \frac{dz}{\mathbf{K}_s} \tag{6.2}$$

A variety of formulations is possible for the variation of $\mathbf{K}_s$ with height. Assuming a linear shear stress distribution from surface to bottom, $\mathbf{K}_s$ can be expressed as

$$\mathbf{K}_s = \beta\kappa u_*z(h - z)/h \tag{6.3}$$

where β is a constant of proportionality between the diffusion coefficients for suspended sediment and fluid mass, $\mathbf{K}_s = \beta\,\mathbf{K}_m$. Normally $\mathbf{K}_m$ is considered to equal the eddy viscosity, which is valid provided the stratification is near neutral. Equation 6.3 can be used with Equation 6.2 to give the relative concentration profile

$$\frac{C_z}{C_a} = \left(\frac{h - z}{z}\cdot\frac{a}{h - a}\right)^{w_s/\beta\kappa u_*} \tag{6.4}$$

where C_z is the concentration at a height z relative to that at a reference height a. This equation was first derived by Rouse in 1936 and is called the Rouse Equation.

Assuming that $\beta = 1$ in Equation 6.4 gives the concentration profiles shown diagrammatically in Figure 6.2. For large w_s/u_* i.e. large grains in a relatively

160

slow flow, the upper flow is clear. For small grains in a faster flow (w_s/u_* small) there will be relatively high concentration throughout the flow.

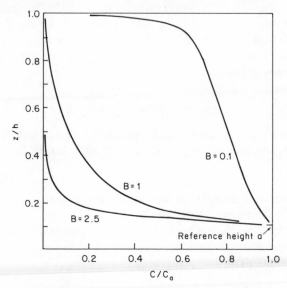

Figure 6.2 Relative concentration profiles calculated from Equation 6.4 for varying values of $B = w_s/\beta\kappa u_*$

Close to the boundary it is a reasonable approximation to assume that $K_s = \beta\kappa u_* z$, which together with Equation 6.2 gives

$$\frac{C_z}{C_a} = \left(\frac{a}{z}\right) w_s/\beta\kappa u_* = \left(\frac{a}{z}\right)^B \tag{6.5}$$

This is sufficiently accurate provided $(h - z) \sim (h - a)$, and would obviously be most useful for large grains where the bulk of the suspension is close to the boundary.

As can be seen from Equations 6.4 and 6.5, the concentration has to be measured or predicted at one level in order to derive the complete profile. The reference concentration C_a is a particular problem which is treated in more detail on page 167.

There is no general agreement on the value of β. Values both greater and less than unity have been obtained. However, it is normally assumed that $\beta = 1$. An evaluation of the value of β was made from velocity and concentration measurements in the sea by Lees (1981a). Von Karman's constant was considered to be equal to 0.4 and the concentration profiles of separate fractions of the suspension were measured. Values for β for each grain size were then determined using β as the free parameter in the von Karman–Prandtl velocity profile and the Rouse concentration profile. The values for β increased with grain size, being approximately 1 for the very fine sand fraction, but reaching almost 10 for medium sand. Additionally there was a significant inverse correlation between the value of β and the concentration at 100 cm above the bed. The values of β approached unity at a concentration of about 180 mg l^{-1}. It is difficult to explain why the values of β should be so high at low concentrations; however there is likely to be an inverse correlation between grain size and concentration at a height above the bed.

The Rouse equation matches well with profile measurements made in flume experiments, when fitted using the exponent as a free parameter. The results however, deviate from the Rouse profile near the bed in high concentrations. This has been explained by Hunt (1954) by considering the volume occupied by the particles. Equation 6.1 then becomes $(1 - C)Cw_s = K_s \, dC/dz$. The resulting concentration profile is more complex than the Rouse profile and, consequently, is not often used. Additionally, Lavelle and Thacker (1978) have included the hindered settling modification of settling velocity, written as $w_s = w_0 (1 - C)^5$ to account for the effects of high concentration near the bed (see p. 211).

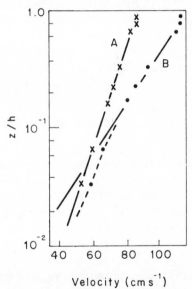

Figure 6.3 Comparative velocity profile obtained with the same flow depth and surface slope, but with different suspended sediment concentrations. (A) Clear water, κ = 0.4. (B) With 15.8 gm l^{-1} of 100 μm sediment, κ = 0.21. *After Vanoni and Nomicos, 1959,* Proc. Amer. Soc. Civil Eng., **85,** *HY5, 77–107. Reproduced by permission of American Society of Civil Engineers*

Measurements of the near bed velocity profiles during experiments with suspensions have shown drastic differences from clear water (Vanoni and Nomicos, 1959). For flows of the same depth and surface slope, the flow with suspended solids has a higher mean velocity than clear flows (Figure 6.3), and this implies a lower friction factor, or a drag reduction. The velocity gradient in the body of the flow is increased, suggesting a higher u_* and z_0. In order to equate this to the shear stress calculated from the surface water slope the value of κ needs to be reduced. Near the bed, however, the data points deviate from the log profile and indicate an actual value of τ at the bed and z_0 not greatly different from the clear water value. Most studies only have velocity profiles measured in the outer part of the flow and as a consequence they give an overestimate of the actual bed shear stress and using these values in the Rouse equation does not give a match to concentration profile data. To get around this discrepancy the value of the von Karman constant is reduced to obtain agreement. Thus as the concentration increases, the apparent value of κ decreases, ultimately reaching values of ~ 0.2. However it is apparent that this is really a misrepresentation of the physics (see Coleman, 1981), though it has been convenient in many respects. An example of the interpretation of velocity and concentration measurements from the sea in terms of a variable κ is shown in McCave (1973).

In order to understand how this arises we need to consider the effect that the suspended sediment has on the turbulent mixing and the velocity profiles near the boundary.

STRATIFICATION BY SUSPENDED SEDIMENT

The suspended sediment produces a vertical gradient of density, and this must have a stabilizing effect on the flow, inhibiting vertical exchange by turbulence. This suggests using a modification of the mixing length theory of turbulent mixing. This was done by Barenblatt in 1953 in a Russian publication, and it has been restated by Taylor and Dyer (1977). A somewhat similar and apparently independent analysis has been presented recently by Itakura and Kishi (1980). Various aspects of the approach have also been considered by Soulsby and Wainwright (1985).

The essence of this theory is that the mixing length or the eddy sizes involved in the turbulent exchanges of mass and momentum become modified by the presence of the concentration gradient. There is a restoring buoyancy force in the vertical direction that requires the turbulence to work harder to create any exchange. The length scale involved is known as the Monin–Obukhov length L and this can be written as a ratio of the buoyancy flux to a velocity of energy dissipation. It can be expressed in terms of u_* and the concentration gradient.

The eddy diffusivity

$$K_s = \frac{\kappa u_* z}{\Phi_M} \tag{6.6}$$

The non-dimensional shear Φ_M is a function of z/L (see page 73).

When Equation 6.6 is put into Equation 6.1 the resulting equation can be integrated to give the concentration profile

$$\ln \frac{C}{C_0} = -B \left[\ln \frac{z}{z_0} + \frac{1}{B} \ln \left\{ 1 + \frac{AB}{1-B} \left(\left(\frac{z}{z_0} \right)^{1-B} - 1 \right) \right\} \right] \quad (6.7)$$

This equation is slightly different from that derived by Taylor and Dyer (1977) since their boundary condition of $z + z_0$ has been replaced by z in order to give $u = 0$ at $z = z_0$.

In Equation 6.7 the constants A and B are defined as

$$A = \frac{\alpha w_s \kappa g (\rho_s - \rho) C_0 z_0}{\rho_s \rho \, u_*^3} \qquad \text{and} \qquad B = \frac{w_s}{\kappa \, u_*},$$

The value of α is normally taken between 4.7 and 5.2. As we will see later it is convenient in many respects to define the concentration C_0 at the height of the roughness length z_0, where C_0 is the mass of sediment per unit volume.

The corresponding velocity profile is

$$u = \frac{u_*}{\kappa} \left[\ln \frac{z}{z_0} + \frac{1}{B} \ln \left\{ 1 + \frac{AB}{1-B} \left(\left(\frac{z}{z_0} \right)^{1-B} - 1 \right) \right\} \right] \quad (6.8)$$

Because the terms in the squared brackets are the same in both Equations 6.7 and 6.8, the concentration and velocity profiles bear a very close relationship to each other. Additionally, since the second term in the squared brackets is always positive, the value of u is always greater than the sediment free value at any particular height. Thus u_*^2/u^2 is smaller in sediment-laden flow, explaining the drag-reduction effect mentioned above.

For very small suspended concentrations C_0 approaches zero, and Equation 6.8 reduces to the von Karman–Prandtl equation. Additionally for that situation Equation 6.7 approaches Equation 6.5.

It is illustrative to consider some examples of velocity and concentration profiles derived from the above theory. The profiles are shown in Figure 6.4 for four cases.

Case 1: For a moderate flow of fine sand (200 μm) over a rippled sand or sandy gravel bed we can set $u_* = 4 \, \text{cm s}^{-1}$, $w_s = 2.24 \, \text{cm s}^{-1}$, $z_0 = 1.0 \, \text{cm}$ and $C_0 = 0.10$. Then $B = 1.4$ and $A = 5.0$. In this case with $B > 1$, the sediment concentration falls off rapidly with height. The velocity profile is modified close to the bed, but further up has the same slope as the sediment free situation. Thus u_* from the profile would be the same as for clear water, but z_0 considerably reduced. The concentration profile similarly is concave upwards.

Case 2: For the same parameters as in Case 1, but with a much higher $u_* = 10 \, \text{cm s}^{-1}$, then $B = 0.56$ and $A = 0.32$. In this case the effect of the stratification influences a thicker zone of the flow and the velocity profile is now modified at all heights. Although the velocity profile is convex upwards, it is almost a straight log line. If a straight line were forced to fit it, an overestimate of both u_* and z_0 would result.

Case 3: For a coarse silt (60 μm) with $w_s = 0.32 \, \text{cm s}^{-1}$, $u_* = 4 \, \text{cm s}^{-1}$, $C_0 = 0.14$ and $z_0 = 0.1 \, \text{cm}$, $B = 0.2$ and $A = 0.1$. The sediment is distributed through an even thicker layer. The effect on the velocity profile is more drastic than Case 2, with a convex upwards profile occurring well away from the bed.

Case 4: For values the same as Case 3, apart from $z_0 = 0.001$ cm, B = 0.2 and A = 0.001. When the change in z_0 is taken into account, the curves of Cases 3 and 4 are similar, but the velocities are higher at every dimensional height z above the sea bed for case 4. Additionally the relative concentration in Case 4 is less than that for Case 3 at all heights.

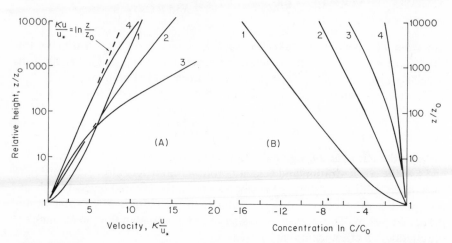

Figure 6.4 Velocity and concentration profiles for the four cases discussed in the text. (A) Velocity profiles calculated from Equation 6.8. (B) Concentration profiles calculated from Equation 6.7. After Taylor and Dyer, 1977

Somewhere between Cases 1 and 2 the profile shape must change from concave upwards to convex upwards. This occurs when A = 1–B/B. For this situation Equation 6.8 reduces to

$$u = \frac{u_*}{B\kappa} \ln \frac{z}{z_0} \qquad (6.9)$$

Equation 6.9 defines a logarithmic profile giving an unmodified value of z_0, but having a slope which is simply w_s/u_*^2. Similarly Equation 6.7 reduces to

$$\ln \frac{C}{C_0} = - \ln \frac{z}{z_0}$$

which gives a concentration which varies inversely with height.

Additionally, for reasonably large heights above the bed, and provided B < 1, Equation 6.8 becomes

$$u = \frac{u_*}{B\kappa} \left[\ln \frac{z}{z_0} + \ln \frac{AB}{1-B} \right]$$

Consequently for Cases 2–4 the velocity profiles away from the bed can be interpreted as a logarithmic profile with modified values of z_0 and κ. The value for the apparent von Karman constant is w_s/u_*, and this should conform with the interpreted reduction in von Karman's constant that has been observed in sediment laden flows.

The condition $A = 1-B/B$ has been evaluated by Soulsby and Wainwright (1985), by defining the concentration at a reference height equal to z_0, and expressing both in terms of the friction velocity. Thus

$$C_0 = \rho_s \gamma_1 \left(\frac{u_*^2 - u_{*c}^2}{u_{*c}^2} \right) \qquad \text{(see page 168).}$$

and

$$z_0 = \frac{26.3\rho(u_*^2 - u_{*c}^2)}{g(\rho_s - \rho)} \qquad \text{(see Equation 3.42)}$$

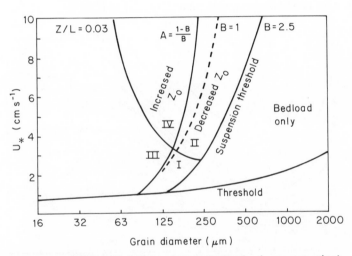

Figure 6.5 Regimes of the various velocity (and concentration) profiles in a suspension. For explanation see text. From Soulsby and Wainwright, in preparation

Using a value of $\gamma_1 = 1.56 \times 10^{-3}$ gives the curve for the criterion $A = 1-B/B$ shown in Figure 6.5. Between this curve and the limit $w_s/u_* = 1$ (i.e. $B = 2.5$), which approximately defines the threshold of suspension, the velocity profiles would give a reduced z_0, but the correct shear stress. For smaller values of B, the velocity profiles would lead to apparent values of both z_0 and u_* higher than actual. In practice it may be sufficiently accurate for most purposes to take the simpler criterion $B = 1$.

Soulsby and Wainwright (1985) have also considered the field of sediment suspension in terms of the conditions for near neutral stability. When $z/L \leqslant 0.03$ at some height above the bed, the profile will show minimal effects due to stratification. z/L is a function of u_*, z and D, and if $B < 1$ then z/L increases with height, whereas if $B > 1$, z/L decreases with height. It is con-

166

venient to define the value of the stability parameter z/L at the roughness height z_0. When combined with the criterion $A = 1-B/B$, the parameter z_0/L divides the field of suspended sediment into four regimes; as shown in figure 6.5.

I. When $A > 1-B/B$ and $z_0/L < 0.03$. Stratification decreases with height above the bed, but the suspension is not significantly stratified, and departures from the clear water logarithmic profile are small.

II. When $A > 1-B/B$ and $z_0/L > 0.03$. The flow is significantly stratified at the bed, but the effect decreases with height. This results in velocity and suspended sediment profiles similar to Case 2.

III. When $A < 1-B/B$ and $z_0/L < 0.03$. The flow is near neutral at the bed, but z/L increases with height. This gives a velocity profile deviating from the log profile with height similar to Case 4.

IV. When $A < 1-B/B$ and $z_0/L > 0.003$. The flow is stratified throughout, but becomes increasingly so with height, similar to Case 3.

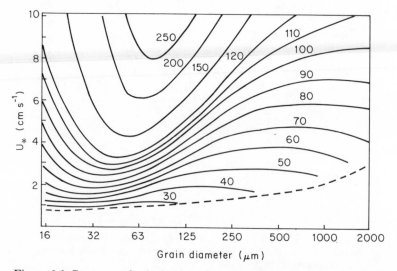

Figure 6.6 Contours of velocity 1 m above a mobile sand cm s^{-1} against friction velocity u$_*$ cm s^{-1} and grain size in microns. From Soulsby and Wainwright, in preparation

In order to calculate the bed shear stress in regimes III and IV, the complete Equation 6.8 needs to be considered. However Soulsby and Wainwright (1985) have calculated a useful diagram (Figure 6.6) which relates the grain size of a mobile sediment, and the velocity measured a metre above the bed, to the friction velocity. Despite the assumptions and approximations involved in its derivation, this diagram should provide an excellent means of obtaining a reasonable estimate of the shear stress in a sandy suspension from the velocity measured 1 m above the bed. Using the friction velocity estimated from Figure 6.6, and the measured velocity at 100 cm allows calculation of the terms in the squared brackets in Equation 6.8, and this, in Equation 6.7 gives the relative concentration profile. In order to obtain the actual concentration profile the concentration at some level must be known.

However, to fit Equations 6.7 and 6.8 to profiles of measured velocity and concentration requires an iterative procedure.

It should be apparent from the above examples that the concentration and velocity profiles undergo a continuous transformation, and are particularly sensitive to the value of w_s/u_*. In practice there are likely to be deviations from these curves since the particles will not all have the same settling velocity. A graded suspension will be formed with the coarser grains occurring nearer the bed and B would decrease with height.

Even when suspension is not developed, but there is a thin layer of grains saltating near the bed, modifications to the velocity profile and to the roughness length are apparent (page 77). Consequently in the zone of bedload movement shown in Figure 6.5 there may be roughness length greater than the Nikuradse or grain roughness. Since saltation and suspension are both active at intermediate transport stages, the initial decrease in z_0 apparent in case 1 above may not be particularly obvious. However, this effect may partially account for some of the reduction in z_0 at the beginning of a tidal cycle described by Dyer (1980a) (see p. 78).

The effect of suspended sediment on the boundary layer has been modelled by Adams and Weatherly (1981). Results from the model suggest that turbulent kinetic energy and bottom stress are reduced by about 40 per cent and 45 per cent respectively. The slope of the velocity profile is consequently decreased, and the thickness of the boundary layer is reduced by as much as 50 per cent. They also point out that the effect of suspended sediment on the velocity profile can be considered in terms of a modified von Karman constant. The modified value $\kappa' = \kappa (1 + 5.5 R_f)$, where R_f is the flux Richardson number.

Further modelling of suspended sediment profiles has been carried out by Smith (1977), and Smith and McLean (1977).

In summary, the presence of suspended sediment reduces the turbulent kinetic energy and modifies the velocity profiles. If the velocity profiles are interpreted as being taken in clear water then the shear stress and bed roughness length will in many cases be overestimated.

REFERENCE CONCENTRATION

In most of the above concentration profiles it is necessary to specify the concentration C_a at a reference level. Since the concentration increases rapidly towards the bed, the reference level should be at the junction where suspension merges with bedload. However, it is difficult to define precisely. If the reference height is taken at a = 0, then Equation 6.4 predicts an infinite concentration. However it is commonly assumed that the concentration equals the static bed concentration (~ 0.63) at a level slightly above the bed level, either some small multiple of the grain size, or at the height of the bed roughness length.

168

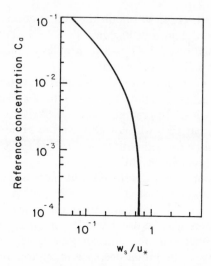

Figure 6.7 Relationship between reference concentration at the level 0.05 times the water depth against w_s/u_*. *After Itakura and Kishi, 1980*, Proc. Amer. Soc. Civil Eng., **106**, *HY8, 1325–1343. Reproduced by permission of American Society of Civil Engineers*

Itakura and Kishi (1980) have proposed a relationship which provides a reasonable fit to a large scatter of experimental data by many authors (Figure 6.7). They have taken the reference level a at 0.05 h and the reference concentration is a function of w_s/u_*, decreasing rapidly as w_s/u_* approaches 0.7.

Smith (1977) and Smith and McLean (1977), however, have proposed that the reference level should be taken at the height of the roughness length z_0. They further suggest that the volume concentration C_a can be written in terms of the bed concentration C_0 and the normalized excess shear stress $s = (\tau - \tau_c)/\tau_c$ in the form

$$C_a = C_0 \gamma_0 s/(1 + \gamma_0 s) \tag{6.10}$$

where γ_0 is an empirical constant which they evaluate as 2.4×10^{-3}. For low values of s, i.e. low suspended loads, Equation 6.10 reduces to $C_a = \gamma_1 s$. Based on suspended sediment measurements 10 cm above the bed of the Columbia River, a value of γ_1 of 1.24×10^{-3} was obtained when the concentration was defined at the level of z_0 as given by Equation 3.42. In a tidal flow over a rippled sand bed Dyer (1980a) found a value of γ_1 of 0.78×10^{-4}. This was in terms of mass concentration of sediment calculated at the level of the observed bed roughness. Wilkinson (personal communication) found a value of $\gamma_1 = 0.84 \times 10^{-4}$ for similar data.

Because of its crucial role in the scaling of the concentration profiles, some agreement on the level for the reference concentration and its magnitude is essential. Smith (1977) concluded that the specification of the reference concentration was one of the weakest points in sediment transport prediction, and this is still so.

SUSPENDED SEDIMENT MEASUREMENTS IN THE SEA

Within a turbulent tidal current the statistical frequency distribution ensures that the suspension of sediment will be dominated by the occasional high stress events (Lesht, 1979). It is only recently that methods have been developed by which these events can be measured. Previously the most commonly used method was sampling the suspension by pumping through a filter. This provided an average concentration at any one height, and sequential sampling at a number of heights to obtain a profile took quite a long time. Obviously this technique was incapable of resolving high frequency events, or measuring concentration profiles when the current was changing rapidly.

Measurements within a tidal flow by many people have shown that there is a lag of the suspended sediment concentration behind the velocity, or the bed shear stress. On the decreasing current the concentrations at any one level are higher than those on the increasing current (Figure 6.8). This effect is known as settling lag, and may be due to a combination of several processes.

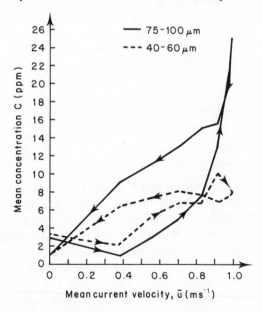

Figure 6.8 Hysteresis of concentration with current velocity during an ebb tide of two grain size fractions measured at a height of 4 m above the bed. From Davies, 1977, *Geophys. J. R. astr. Soc.*, **51**, 503–529. *Reproduced by permission of Blackwell Scientific Publications*

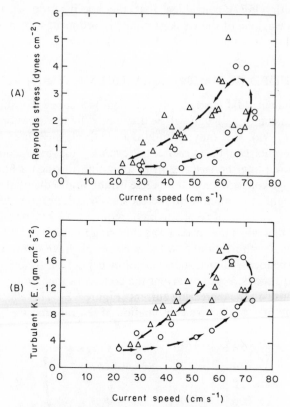

Figure 6.9 The hysteresis of (A) Reynolds stress, and (B) Turbulent kinetic energy, as a function of flow speed. ○ Accelerating flood tide; △ Decelerating flood tide. *From Gordon 1975, Marine Geol., 18, M57–M64. Reproduced by permission of Elsevier Science Publishers BV*

Direct measurements of the turbulent Reynolds stresses and kinetic energy have shown that both are lower on the accelerating current than on the decelerating one (Gordon, 1975; McLean, 1983). The results shown in Figure 6.9 illustrate this hysteresis. Since concentration is related to the boundary shear stress and the turbulence intensity influences particle diffusivity, a similar hysteresis in concentration is to be expected. However, when one considers bed shear stress, rather than the stress about 2 m above the bed as Gordon has done, a different hysteresis pattern would be expected (Lavelle and Mofjeld, 1983).

Additionally, a certain amount of lag will be generated because of the time taken for the sediment to be entrained from the bed and lifted to a measurement level on accelerating flow, and enhancement of concentrations because of settling of grains from above on deceleration. Settling lag is likely to be small close to the bed and increase with height. This is shown by the fact that near the bed the concentration can be reasonably well represented as a simple function of velocity or shear stress. For instance Dyer (1980a) found that the concentration at a height of 10 cm was proportional to u_*^5. An idea of the timescale of the settling lag can be obtained by consideration of turbulence diffusion. The mean height of suspension $\bar{z}$ increases with time according to

$$\bar{z} = 0.77\,\kappa u_* t \qquad \text{(Monin and Yaglom, 1971)}$$

Consequently for 1 m height the time delay would be $2\frac{3}{4}$ min for a u_* of $2\,\text{cm s}^{-1}$. During this time the sediment cloud would have travelled about 50 m downstream. Settling from this height would take in the region of 30 mins. Analytical modelling of settling lag has been carried out by Davies (1977).

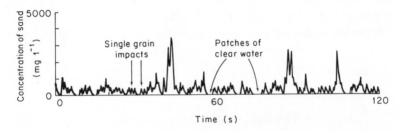

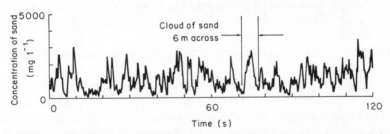

Figure 6.10 High frequency fluctuations in the concentration of suspended sand measured by the impact sensor at a height of 18 cm above the bed in the Taw Estuary. The tidal current at the same height was $83\,\text{cm s}^{-1}$ for the upper trace, and $96\,\text{cm s}^{-1}$ for the lower trace. The site is sheltered from wave activity. Soulsby, personal communication

Direct measurements of the instantaneous sand concentration have recently been made using a fast-response impact sensor, used in conjunction with electromagnetic flow meters (Soulsby *et al.*, 1984). Measurements were made at a height of 18 cm above a rippled sand bed of median diameter 200 μm. The suspended sand moved in clouds, with almost clear water between them, the mean streamwise length of the clouds being about 3 m (Figure 6.10). The presence of the sediment significantly lowered the turbulence intensity from about 29 per cent to about 12 per cent. In contrast, the intensity of fluctuations in the concentration was very high, being of the same order as the mean concentration at times of low concentration, but decreasing to a minimum value of 67 per cent of the peak mean concentration. Despite this high variability, the downstream turbulent flux of sediment never exceeded 1.6 per cent of the mean flux.

Unfortunately, because of vibration effects, the vertical turbulent fluxes were not clear. However subsequent measurements near the crest of a sandwave (Soulsby, personal communication) have shown that the depositional flux $\overline{C}\, w_s$, was about five times the erosional flux $\overline{w'c'}$. This could have resulted in accumulation of sediment at the bed, but the downwards flux could also have been balanced by a gradient in the upwards advective flux. However, in that instance, the values of mean vertical velocity $\overline{w}$ were too small to be accurately measured.

SUSPENDED SEDIMENT TRANSPORT RATE

The suspended sediment transport rate will be equal to the integral through depth of the suspended sediment flux, the product of the velocity and concentration.

$$q_s = \int_0^h uCdz \qquad (6.11)$$

One must remember that

$$\int_0^h uCdz \neq \int_0^h udz . \int_0^h Cdz,$$

since the depth mean velocity is near mid depth, well above the level of the mean concentration. Consequently the velocity/concentration product must be considered at each increment of height in the integration. For most situations the level of mean suspended sediment transport is within the bottom few per cent of the flow.

If the concentration is specified in terms of the Rouse profile, and the velocity specified by the von Karman–Prandtl equation, then Equation 6.11 cannot be integrated analytically. Alternatively numerical integration using stated values of the variables has provided useful nomograms enabling calculation of the suspended load transport rate (Einstein, 1950). For a deep flow and for relatively coarse grains some realistic assumptions are possible (Soulsby, personal communication). Using Equation 6.5 with $\beta = 1$, Equation 6.11 becomes

$$q_s = \rho \int_{z_0}^h C_{z0} \left(\frac{z_0}{z} \right)^B \frac{u_*}{\kappa} \ln \left(\frac{z}{z_0} \right) dz$$

with the reference height and the lowest level of interest as z_0. This gives

$$q_s = \frac{\rho C_{z0} u_* z_0}{\kappa (1 - B)^2} \left\{ \left(\frac{h}{z_0} \right)^{1-B} \left[\ln \left(\frac{h}{z_0} \right) - B \ln \left(\frac{h}{z_0} \right) - 1 \right] + 1 \right\}$$

which, if B is large, reduces to

$$q_s = \rho \frac{C_{z0} u_* z_0}{\kappa (1 - B)^2}$$

If

$$C_{z0} = \frac{\rho_s}{\rho} \gamma_1 (\tau - \tau_c)/\tau_c$$

where C_{z0} is now a mass concentration, then

$$q_s = \frac{\kappa \gamma_1 \rho_s u_*^3 z_0}{(\kappa u_* - w_s)^2} \frac{(\rho u_*^2 - \tau_c)}{\tau_c} \tag{6.12}$$

Equation 6.12 can thus be used with velocity profile measurements and concentration measured at any one height to calculate q_s. The inherent assumption, of course, is that the modified Rouse profile is a reasonable representation, which may not be correct in a graded suspension, or when concentrations are anything other than rather low. To obtain the suspended sediment transport rate using Equations 6.7 and 6.8 requires numerical integration of the product of these equations.

Dyer (1980a) has used Equation 6.12 in measurements over an ebb tidal cycle to calculate q_s and then obtained the relationship $q_s \propto u^7$. Since most sands in the sea are rippled this result may be a useful first estimate in other areas where complete measurements are not available.

Measurements in the Outer Thames Estuary by sampling throughout the water column has shown (Owen and Thorn, 1978) that the suspended sand transport is proportional to the fourth power of the depth–mean tidal velocity.

SUSPENSION PROFILES UNDER WAVES

The most intense suspension occurs above a flat bed in the sheet flow regime. However in the ripple regime suspension is almost entirely the result of the ejection of vortices into the flow from the lee of ripples on the reversal of flow. Consequently there is considerable variation of concentration within each wave oscillation with phase, and with height above the bed, as well as with the wave parameters governing the maximum oscillation velocity. However most people have concentrated on the mean concentration over the wave cycle. By assuming the mass of suspended sediment in the vortex is proportional to the vortex volume, Sleath (1982) has shown that the mean concentration outside the immediate vicinity of the bed is given by

$$\overline{C} = A \exp(-\beta z/X) \tag{6.13}$$

where A is a constant, $\beta = (\pi/vT)^{\frac{1}{2}}$ and X is another empirical coefficient. It is postulated that X is given by

$$X = 4.9 \left(\frac{u_m H_r}{v} \right)^{\frac{1}{6}} \tag{6.14}$$

where u_m is the amplitude of the oscillation outside the boundary layer and H_r is the ripple height. However, Sleath states that it is possible that fall velocity and ripple steepness variations may also be important, and that Equation 6.13 should be treated with caution. Of course Equation 6.13 can give a concentration profile relative to the mean concentration ($\overline{C}_a$) at a reference height ($z = a$), i.e.

$$\frac{\overline{C}}{\overline{C}_a} = \exp\left(-\frac{\beta}{X}(z-a)\right) \tag{6.15}$$

A rather different approach has been taken by Wang and Liang (1975) in treating suspension throughout the flow depth. Using the diffusion equation, they assumed that the diffusion coefficients for both turbulence and the sediment particles was proportional to the amplitude of the vertical velocity component of the wave field. The coefficient of proportionality σ had the dimensions of a length. For shallow water they derived the solution

$$\frac{\overline{C}}{\overline{C}_a} = \left(\frac{z}{a}\right)^{-R} \tag{6.16}$$

where $R = w_s/\sigma k u_m$, where $k = 2\pi/\lambda$. Comparison with laboratory measurements gave a value of $\sigma = 5.15\,\text{m}$, and a reasonable agreement was achieved with a set of field measurements using this value. Comparison between Equations 6.16 and 6.13 indicates that the effect of grain size on the concentrations appears in Equation 6.13 either in terms of the ripple height, or in the empirical constant.

Nielson (1984) performed a number of field experiments and found that the concentration decreased exponentially away from the bed, as follows

$$\overline{C} = \overline{C}_0 \exp(-z/l_s) \tag{6.17}$$

The decay length scale l_s was approximately equal to the ripple height, so that concentrations decay by a factor of 100 over four or five ripple heights. The characteristic value of l_s was 3-5 cm. Comparative profiles relative to a reference height of 1 cm are shown in Figure 6.11. Though the conditions are not the same between curves, they are broadly similar.

From long term measurement on the New York Shelf, Lavelle et al. (1978) have shown that concentration fluctuations 100 cm above the bed occurred at the same frequencies as the surface waves, and were also correlated with groups of higher waves. During these events sand sized particles are carried to a greater height than can be explained by turbulent boundary layer theory, suggesting that the wave groups produce bursts of turbulence that can suspend the sand (Clarke et al., 1982).

Using averages over about 4.25 minutes, Vincent et al. (1982) reported that the simultaneous concentrations measured at 10 elevations near the bed off Long Island could not be matched to the Rouse profile. They matched much better the profile

$$C = C_1(1 - A \ln z/z_1) \tag{6.18}$$

where C_1 is the concentration measured at a height of 1 cm. The constant A is proportional to the height at which the sediment concentration approaches zero and had a value A = 0.22. This is equivalent to negligible concentrations at a height of 1 m.

The concentration C_1 correlated reasonably well with the bedload concentration C_* calculated using Equation 7.27, giving the result $C_1 = 330\,C_* + 3.8$. The intercept probably represents the background turbidity of the shelf water. Consequently from the known wave parameters the value of C_* and C_1 can be calculated, and this could be used with Equation 6.18, or any of the other profile equations. However this result may well be a site specific one.

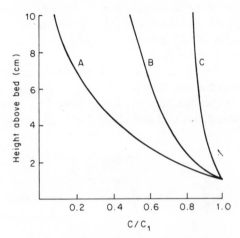

Figure 6.11 Comparative concentration profiles beneath waves, relative to concentration at a reference level a = 1 cm. (A) Equation 6.13 (Sleath, 1982) for T = 8 s, $u_m = 30\,\text{cm s}^{-1}$, $H_r = 3\,\text{cm}$. (B) Equation 6.18 (Vincent et al., 1982). (C) Equation 6.16 (Wang and Liang, 1975) for $w_s = 0.8\,\text{cm s}^{-1}$, $\lambda = 100\,\text{m}$, $u_m = 30\,\text{cm}$ s^{-1}

Profiles of suspended sediment concentration derived from acoustic backscatter measurements in several areas have been analysed by Clarke et al. (1982). They found a very good correlation between concentration 1 m above the bed and rms wave velocity, and used the predictive relationship

$$\overline{C}_{100} = C_B + \beta\,(u_m - u_{mc}) \tag{6.19}$$

where C_B is the background concentration, and β is the resuspension rate. At two of the sites the threshold orbital velocity u_{mc} was $10.9\,\text{cm s}^{-1}$ and best-fit values of β were 0.9 and $1.0\,\text{mg l}^{-1}\,\text{cm}^{-1}\,\text{s}^{-1}$. At the third site the values $u_{mc} = 6\,\text{cm s}^{-1}$ and $\beta = 3.3$ were quite reasonable. They also examined two mechanisms for the resuspension of the silty material from the silty sand. One process resulted from the pressure-induced pore water motion through the sediment carrying the silt particles out of the bed. The other mechanism was based on the release of the silt by the progressive stripping off of layers of sand in bedload motion. A combination of the two models gave values of β of 0.63 and 0.76 for the comparison with the first two sites.

In shallow water the waves may break and cause enhanced suspension. Nielson (1984) examined the concentration profiles under breaking waves and found that they were unaffected in the lowermost 2 cm or so of the flow, but higher up the increased turbulence and mixing of the breaking increased the concentration to an almost uniform value. However, the results were not comprehensive enough to quantify these effects. Similarly Kana (1978) found an exponential increase of concentration 10 cm above the bed with height of the waves at the break point. Spilling waves gave mean concentrations of the order of 0.2 gm l^{-1} and plunging waves up to 1.8 gm l^{-1}.

It is apparent that one of the major problems in suspended sediment is in specification of the reference concentration. Since it is the finer fraction of the bed sediment which is suspended, this problem is bound up with determination of the relationship between bedload and suspended load.

CHAPTER 7

Sediment Transport Rate

PREDICTIVE FORMULAE FOR STEADY FLOW

There is, of course, great interest in quantifying sediment movement so that predictive formulae can be developed. Essentially, the sediment transport rate is the total weight of grains passing through a section per unit time, and is the product of the weight of moving grains present in the water over a unit area times the velocity at which they move. As we have seen, grain movement gradually develops from saltation-dominated to suspension-dominated with increasing transport stage. Both the concentration of moving sediment and the velocity of movement are functions of the excess shear stress, and bedforms are generated and produce form drag. Consequently the relationship between the sediment transport rate and the flow is a complex one.

Many sediment transport formulae have been proposed. It is impossible to describe all of them here, or even some of them completely, because of the intricacies of their derivations. However a few of the most popular ones are chosen in order to describe their basic principles, in particular those that have been applied in the sea. It will be necessary to go to the original publications for enough detail to be able to use them in specific examples. Further background on them is available in some of the more complete text books on sediment transport, such as Raudkivi (1967), Yalin (1977) and Graf (1971).

The formulae fall into three basic groups:

1. Experimental. Using a large number of flume results and effectively obtaining completely empirical relationships between the sediment transport rate and a characteristic flow variable. The equation of Meyer-Peter and Müller (1948) is an example of this approach.
2. Theoretical. Starting from the basic physics of the movement of individual grains relations are developed. However, averaging these movements in space and in time inevitably involves constants which have to be determined from a set of calibration data. Examples of this approach are Einstein (1950) and Bagnold (1956, 1966).

3. Dimensionless analysis. The sediment and flow variables are grouped together in dimensionless numbers. The constants and coefficients are then determined by comparison with flume results, e.g. Ackers and White (1973), Yalin (1963).

As can be seen all of the formulae depend on calibration with data. Therein lies both their strength and their weakness. The formulae will only be as good as the data used to prove them, and though they will be restricted in range by the data, an appropriate one can be found for most situations. All were developed for steady unidirectional flow, but have been applied in the sea, with some modifications, due to the lack of anything better. These applications will be considered later.

Some of the formulae apply to bedload only, others to total load, by including suspension. The effect of bedforms is included in some, but others only apply to plane beds. Nevertheless there is a degree of conformity between most of them, in that the sediment transport rate q is proportional to u_*^3 over some part of the range, and this has proved valuable in some simplified approaches.

There are also differences between formulations derived by the various authors since the sediment transport rate can be expressed either as a dry weight transport rate, the immersed weight transport rate or the volumetric transport rate. The dry weight transport rate will be larger than the immersed weight transport rate by the factor $\rho_s/\rho_s - \rho$, and the volumetric transport rate will be less than the immersed weight transport rate by the factor $(\rho_s - \rho)g$.

The transport rates are generally non-dimensionalized, with the dimensionless transport being

$$\varphi \equiv \frac{q\rho^{\frac{1}{2}}}{((\rho_s - \rho)gD)^{\frac{3}{2}}}$$

for the dry weight transport rate

$$\equiv \frac{q}{\rho_s g}\left(\frac{\rho}{\rho_s - \rho}\right)^{\frac{1}{2}}(gD^3)^{-\frac{1}{2}}$$

for the immersed weight transport rate,

or alternatively $\equiv q/w_s D$

for the volumetric transport rate

Hereafter the dry weight transport rate will be used unless otherwise stated.

Meyer-Peter and Müller

Extensive experimental results led to the dimensionless transport rate

$$\varphi = \frac{q\rho^{\frac{1}{2}}}{((\rho_s - \rho)gD)^{\frac{3}{2}}} = 8(\theta - 0.047)^{\frac{3}{2}} \tag{7.1}$$

where q is the dry weight transport rate.

This formula is valid for bedload only and is restricted to $D > 400\,\mu m$. However it should cover the presence of dunes. Clearly the constant 0.047 is the threshold Shields function for the coarse grains and as the ambient shear stress becomes very large, $q \propto u_*^3$.

Einstein

Einstein considers that the grains move by saltation as a result of the fluctuating lift force and then perform a jump which, on average, has a length L of about 100 D. A function is then developed to describe the probability both in space and time of jumps occurring.

The number of grains passing through a unit width in time interval T

$$ N = \frac{L}{\pi/4\ D^2} \sum_{n=1}^{\infty} P_n $$

where P is the probability of a jump occurring upstream of the point of interest on any increment n of the bed within the particle trajectory. The jump length is considered to be only a function of D, whereas, as we have seen in Chapter 3, it is also a function of θ. The sediment transport rate will be given by the number of grains passing per unit time N, multiplied by the weight of a single grain. Therefore

$$ q = N \cdot \frac{\pi}{16}(\rho_s - \rho)gD^3 $$

$$ = \frac{(\rho_s - \rho)g}{4D} \cdot \frac{L}{T} \sum_{n=1}^{\infty} P_n $$

The time interval T is considered proportional to D/w_s, where w_s is the settling velocity.

Combining these various relationships gives

$$ \varphi = \frac{qp^{\frac{1}{2}}}{((\rho_s - \rho)gD)^{\frac{3}{2}}} = \frac{1}{A_*} \sum_{n=1}^{\infty} P_n $$

where A_* is a constant, and P_n is assumed to be given by the normal error law. Evaluating the summation term gives

$$ \varphi = \frac{1}{A_*} \cdot \frac{P}{1-P} \tag{7.2} $$

where

$$ P = 1 - \frac{1}{\pi^{\frac{1}{2}}} \int_{-\frac{B_*}{\theta} - \frac{1}{n_0}}^{\frac{B_*}{\theta} - \frac{1}{n_0}} e^{-t^2} dt $$

and t is a variable of integration. The constants A_*, B_* and n_0 were calculated from flume data obtained for low transport rates using 27 mm gravel, and for high rates using 795 µm sand. The values were $A_* = 43.5$, $B_* = 0.143$ and $n_0 = 0.5$. For this procedure it is necessary to separate the form drag and friction drag and use only the latter in Equation 7.2.

Very small transport rates are approached when $\theta \sim 0.04$ implying that the threshold of movement is included by use of calibration data. For low transport rates $\varphi \sim 40\,\theta^3$, giving $q \propto u_*{}^6$, but for large transport $q \propto u_*{}^2$.

Bagnold

In a lengthy and involved paper in 1956 Bagnold introduced the idea of bedload sediment transport based on the concept of the work done by the flow in moving the grains. This introduced an efficiency factor relating the work done on the grains to the energy available. Obviously the efficiency rises as the velocity of the grains approaches that of the water flow. He also argued that an upward directed stress is produced by the sheared layers of grains having to go up and around each other, and their interaction provides a frictional resistance to the flow when multiplied by the angle of friction. Yalin (1977) has produced a much more concise derivation of the result of Bagnold (1956) using basically the same arguments.

Let G be the weight of moving grains per unit area of the bed. If their mean velocity in the direction of flow is u_b, then the sediment transport rate q equals G times u_b. The frictional force opposing the motion is $G \tan \alpha$, where α is the dynamic friction angle. The work done on the grains by the flow is thus

$$W_b = u_b\, G \tan \alpha = q \tan \alpha$$

The work done equals the available energy times the bedload efficiency factor e_b. The available energy is the product of the stress times the velocity, but the energy available for moving the bedload is the excess shear stress multiplied by the velocity at the level of the grain motion u_D. Thus

$$E = (\tau_0 - \tau_c)u_D$$

and

$$W_b = e_b E$$

Combining these results gives

$$q_b = \frac{e_b}{\tan \alpha}(\tau_0 - \tau_c)u_D \tag{7.3}$$

In order to calculate u_D the rough turbulent velocity profile is used (Equation 3.24)

$$\frac{u}{u_*} = \frac{1}{\kappa}\, \ln\, \frac{z}{D} + 8.5$$

For $z = D$, $u = u_D = 8.5\,u_*$,

Consequently

$$q_b = \frac{e_b}{\tan \alpha} \frac{8.5}{} (\tau_0 - \tau_c)u_* \tag{7.4}$$

Because $u_* = \sqrt{\tau_0/\rho}$ and $\theta = \tau_0/(\rho_s - \rho)gD$, Equation 7.4 can be written as

$$\varphi = \frac{q_b \rho^{\frac{1}{2}}}{((\rho_s - \rho)gD)^{\frac{3}{2}}} \propto (\theta - \theta_c)\theta^{\frac{1}{2}} \tag{7.5}$$

Consequently $q_b \alpha u_*^3$ at high transport rates, providing e_b and $\tan \alpha$ are independent of θ or u_*. The value of $\tan \alpha$ was shown by Bagnold to range between the values of 0.32 and 0.75 depending on grain size and on the bed shear stress. However for coarse grains at relatively high transport rates and for fine grains at low transport rates $\tan \alpha$ is more or less independent of θ.

In a further paper, Bagnold (1966) extended this theory to include the suspended sediment. He considered that as the suspended solids are falling relative to the fluid at a velocity w_s, then the fluid must be lifting the solids at that velocity. The rate of work done on the suspended grains is thus

$$W_s = q_s \frac{w_s}{u},$$

whereas the work done by the fluid is $q_s w_s/e_s\bar{u}$, where $\bar{u}$ is the transport velocity of the suspension, assumed equal to that of the fluid. q_s is the suspended sediment transport rate and e_s the suspension efficiency. The work done per unit time is the power $W = \tau_0 \bar{u}$. The total stream power available is divided between the bedload and suspended load so that only $W(1 - e_b)$ is available for the suspension.

The total load

$$q = q_b + q_s = W \left(\frac{e_b}{\tan \alpha} + e_s \frac{\bar{u}}{w_s}(1 - e_b) \right) \tag{7.6}$$

Comparison with experiments gave values of 0.01 for $e_s(1 - e_b)$. Consequently Bagnold's total load formula is

$$q = W \left(\frac{e_b}{\tan \alpha} + 0.01 \frac{\bar{u}}{w_s} \right) \tag{7.7}$$

The values of e_b varied between 0.1–0.16 depending on grain size and on the flow velocity.

A drawback with this formulation is that no threshold of movement is included and movement is predicted at all shear stresses. Nevertheless for high transport rates $q \propto u_*^3$. Equation 7.6 can also be expressed as

$$q = KW = K\tau_0\bar{u} \tag{7.8}$$

Bailard and Inman (1979) have reconsidered Bagnold's approach for bedload transport and conclude that an additional factor should be included in the first term on the right hand side of Equation 7.6. This factor is the ratio of the velocity at the top of the bedload layer to the mean velocity within it. Also they considered that α should be equal to the internal friction angle φ. However, in practice, these differences become hidden in the value of K in Equation 7.8.

Ackers and White

The different sediment and flow variables are formed into three dimensionless groups.

Mobility Number F_{gr}

This is the ratio of the shear force on a unit area of the bed to the immersed weight of the grains. For fine sediments this number is equal to $\theta^{\frac{1}{2}}$, whereas for coarse sediments the velocity is used rather than the shear stress and the number also involves h/D. This takes into account the form drag effect of large grains in a shallow flow.

$$F_{gr} = \frac{u_*^{\,n}}{\left(gD \frac{(\rho_s - \rho)}{\rho} \right)^{\frac{1}{2}}} \left[\frac{\hat{U}}{2.46 \ln(10\,h/D)} \right]^{1-n} \tag{7.9}$$

where $\hat{U}$ is the depth averaged velocity. For coarse sediment n = 0, and for fine n = 1,

Dimensionless Grain Diameter D_{gr}

This can be derived from the drag coefficient and Reynolds number of a settling particle by eliminating the settling velocity.

$$D_{gr} = D \left[g(\frac{\rho_s - \rho}{\rho}) v^{-2} \right]^{\frac{1}{2}} \tag{7.10}$$

For $D_{gr} = 1$, D = 40 µm and for $D_{gr} = 60$, D = 2.5 mm.

Sediment Transport G_{gr}

The dimensionless expression was based on the stream power concept, for coarse grains using the product of net grain shear and the stream velocity as the power per unit area of the bed. For fine sediments the total stream power was used. The hypothesis was also made that the efficiency was dependent on the mobility number,

$$G_{gr} = \frac{qg\rho h}{\rho_s D} \left(\frac{u_*}{\hat{U}} \right)^n \tag{7.11}$$

The dimensionless sediment transport was presumed to be a function of F_{gr} and D_{gr}. Using a total of almost 1000 flume experiments in depths up to 0.4 m, the following relationship appeared

$$G_{gr} = C(F_{gr}/A - 1)^m \tag{7.12}$$

where C, A, m and n all vary with sediment size. For coarse sediments ($>$ 2.5 mm) A = 0.17, C = 0.025, m = 1.5 and n = 0. For transitional sizes (40 µm $<$ D $<$ 2.5 mm)

$$A = 0.23\,(D_{gr})^{-\frac{1}{2}} + 0.14$$
$$\log C = 2.86 \log D_{gr} - (\log D_{gr})^2 - 3.53$$
$$m = 9.66\,D_{gr}^{-1} + 1.34$$
$$n = 1.00 - 0.56\,D_{gr}$$

In Equation 7.12 the variable A represents the value of F_{gr} at which transport begins. This approach is valid for use over plane, rippled and duned beds, as well as with suspension and bedload, since the effects of these are implied in the calibration data. Comparison with field data suggested that D_{35} was probably a better measure than D_{50} of the hydraulic mean grain size.

Yalin

This approach is a combination of dimensional analysis and an extension of Einstein's analysis of saltation. The grain motion is entirely by saltation, which restricts the end result to bedload. An average saltation characteristic is assumed together with a threshold of motion. The hop length of the saltation increases with velocity, and the number of saltating grains is proportional to the excess shear stress. The dimensionless solids discharge is considered to be a function of four dimensionless variables: Du_*/ν, θ, ρ_s/ρ and h/D. From an analysis of the trajectory of the grain motion, the dimensionless solids discharge R defined as $R = q/(\rho_s - \rho)gDu_*$ is given by

$$R = 0.635 \left(\frac{\theta - \theta_c}{\theta_c}\right) \left[1 - \frac{1}{a\left(\frac{\theta - \theta_c}{\theta_c}\right)} \ln\left(1 + a\left(\frac{\theta - \theta_c}{\theta_c}\right)\right) \right] \qquad (7.13)$$

where

$$a = 2.45 \left(\frac{\rho}{\rho_s}\right)^{0.4} \theta_c^{\frac{1}{2}}$$

The constant 0.635 was obtained from experimental data. The presence of dunes was not considered and the formula is restricted to large h/D. Also since it was assumed that the bed roughness exceeded the thickness of a laminar sublayer, the result is applicable only to grains exceeding about 200 µm.

Consideration of the limiting conditions for Equation 7.13 gives an interesting comparison with the other theories. At low transport rates, if the group $(\theta - \theta_c)/\theta_c = s$, then

$$\mathrm{Lt\ as} \to 0 \quad \frac{1}{as} \ln(1 + as) = 1 + \frac{as}{2}$$

Thus Equation 7.13 becomes

$$q_s \propto u_*^5$$

184

At high transport rates

$$\text{Lt as} \rightarrow \infty \; \frac{1}{as} \ln(1 + as) = 0$$

Consequently Equation 7.13 becomes
$$q_s \propto u_*^{\;3}$$

Comparison between these formulae on a common data set does not give very close agreement despite the fact that each was separately proved against other data sets. However, there does seem to be a unifying concept that at high transport rates the relevant stream variable is the power $\tau\bar{u}$. As the trend of the formulae is similar it is better to calibrate the formula using *in situ* data in the area of interest. This philosophy has been adopted by Bagnold (1980) who is sceptical about understanding sediment transport in dynamical terms. Consequently he has suggested a purely empirical expression for bedload, being

$$q_b \propto (W - W_c)^{\frac{3}{2}} h^{-\frac{2}{3}} D^{-\frac{1}{2}} \tag{7.14}$$

and suggests *in situ* calibration to find the constant of proportionality before use. W_c is the fluid power at the threshold of motion. Inclusion of a depth dependency explains a great deal of the scatter of flume results and brings them together in a way that gives a good qualitative correlation. The depth dependency appears because, though the bedload is confined to a thin layer only a few grains thick, the stream power, i.e. the rate of kinetic energy supply to the flow as a whole, has to transport water as well as sediment. Consequently the proportion available to move sediment increases as water depth decreases. This effect is taken into account in the Ackers and White formula.

As the rate of sediment transport is sensitive to the shear stress, accuracy in the measurement of the shear stress is very important. When the bed is covered with ripples or dunes a large proportion of the total shear stress is form drag, and it is only the skin friction that moves sediment. The form drag is sensitive to the size and shape of the bedforms, as is shown by the variation of the ratio of skin friction to total drag (Figure 3.14). The maximum form drag occurs when the bedforms reach their maximum steepness. There are no satisfactory methods for predicting the proportion of drag occurring as form drag, though there are many empirical relationships.

Providing the bedforms are large enough, the velocity profile close to the bed can be measured and this gives a direct estimate of the local friction drag, or shear stress. However, the presence of ripples is then important since they are in general smaller than the elevation of the lowest current meter, and the form drag produced by them becomes unresolvable. Consequently the bed roughness length becomes a means of monitoring whether ripples are present or not. Since many studies do not have complete velocity profiles, but depend on the quadratic friction law and a current measurement at one height, then specification of the drag coefficient C_D, or alternatively z_0, will determine the absolute values of the calculated stream power. An error of 20 per cent in the value of C_D will mean an error of 20 per cent in the stream power irrespective of sampling errors in the velocity.

In summary, it appears that at high transport rates q is proportional to u_* cubed. At low rates near the threshold proportionality with a high power, perhaps u_*^5, is possible.

SEDIMENT TRANSPORT RATES IN THE SEA

Measurement of Sediment Transport

Bedload transport is very difficult to measure since the movement takes place within a few grain diameters of the sea bed. It is difficult to get samplers into the bedload layer without the likelihood of sampling some of the suspension on one hand, or even the bed itself on the other, and though bedload samplers are used in rivers they are of very variable efficiency. For use in the sea it is always advisable that the sampling should be carried out within the view of an underwater camera or television, so that the sampling conditions can be assessed.

A measure of the bedload sediment transport rate may be obtained from consideration of the volume of sediment involved in migration of ripples. The mass of sediment per unit width in a ripple of wavelength λ_r and height H is $\lambda_r C_o \rho_s H/2$, where C_o is the concentration of the bed sediment. The mean mass discharge per unit width per unit time at the crest is the result of a thin layer moving from the stoss to the lee slope, and thus

$$q_b = C_o \ \rho_s H U_r$$

where U_r is the ripple migration rate. At the trough, q_b would be zero. For the ripple to move a complete wavelength, the mean discharge per unit time past a point would be half the above rate, and is equivalent to moving the total mass of sediment in the ripple a distance λ_r. The value obtained is likely to be a minimum value since there is the inherent assumption that the grains are all always contained within the same ripple and that none jump from one ripple to the next. Kachel and Sternberg (1971) have used this approach in the sea, using stereocameras to measure the ripple migration rate, as well as the ripple height.

An alternative technique for measuring the progression of ripples has been described by Wilkinson et al. (1984). This comprised a rod about 1 m long directed into the flow from a sea bed rig about 10 cm above the bed. A light was shone onto the rod obliquely from one side, and a camera or television looked at the scene from above. The shadow of the rod appeared on the sea bed as a zig-zag line, and, knowing the geometry of the system, the size and shape of the ripples were measured from recordings and their progression rate measured. Unfortunately, only three or so ripples can be covered with this system which raises questions of representativeness of the results. McLean (1983) has used a traversing echo-sounder on a horizontal arm upstream of a sea bed rig to carry out a similar function.

For larger bedforms the movement can be measured by divers using sea bed reference stakes against which to measure the changing elevations, e.g. Jones *et al.* (1965), Langhorne (1982a). The drawback with this method is that profiles can normally only be obtained at slack water. To measure more frequent profiles over the maximum current requires a self-recording instrumentation system. Thorne *et al.* (1983/1984) describe a compact self-recording high frequency echo-sounder with a resolution of a few millimetres, and Langhorne (personal communication) has used several of these to measure the rate of movement of intertidal bedforms, with simultaneous measurements at a number of positions every minute.

With gravel particles it is feasible to measure transport rates by comparison of video recordings or time lapse photographs. Individual particles can be seen to move from the field of view and their sizes estimated. This is very tedious, however, and is too demanding to become a standard technique. It has been used, nevertheless, to calibrate an acoustic technique (Thorne *et al.*, 1983/1984). This technique depends on measuring the noise generated by the particles hitting against each other as they move. The frequency of the sand generated is an inverse function of the grain size, and for gravel particles is a few tens of kilohertz, ranging up to megahertz for sand. Quantitative measurements in the laboratory have shown that this technique can be used for determining the threshold of movement, as well as the sizes of the moving grains, and when calibrated against particle counting, the intensity of the sound generated by the movement is a good measure of the instantaneous transport rate. Because the sensor is passive and does not interfere with the movement it is measuring, this technique may have wide application for coarse grain sizes.

Simulation of the movement of gravel particles by an acoustic transponding 'pebble' of an appropriate size and density was proposed by Dorey *et al.* (1975). Though working prototypes were developed, surveying problems limited their detection by the side scan sonar system and they were not seriously used.

An alternative technique is to use radioactive tracers (Courtois and Sauzay, 1966; Courtois and Monaco, 1969). For simulation of sand, graded glass incorporating the isotope Scandium 46 can be used. This has a half-life of 84 days which enables tracing for considerable periods. The tracer is placed on the sea bed and its distribution mapped at intervals using scintillation counters, either towed along the sea bed or lowered at intervals. To turn this into a sediment transport rate requires knowledge of the depth of the mobile layer. The depth distribution of tracer can be measured in cores, and there are also techniques for calculating the response of the detector to buried tracer. The position of the centroid of the distribution can then be calculated for each survey. Of course the decay of the isotope has to be allowed for, and it has been found that a considerable time is required for the tracer to equilibriate with the sea bed. Experiments in Swansea Bay have suggested that in excess of seven days is required (Heathershaw, 1981). The use of fluorescent tracers for bedload measurement is reported by Lees (1979, 1981b). The techniques are very similar to those for radioactive tracers except that it requires direct sampling to detect the tracer, but fluorescent tracers are less hazardous.

The most commonly used bedload transport formula for the sea is that of Bagnold, in which the transport rate is proportional to the stream power (Equation 7.8). The stream power can be expressed in several ways; as the boundary shear stress times the mean velocity near the bed, as the cube of the friction velocity, or as the cube of a near bed velocity e.g. u_{100}^3. There is also considerable divergence of opinion on the inclusion of a threshold of movement. Consequently it is difficult to establish clear relationships, or values for the proportionality coefficient K.

Kachel and Sternberg (1971) were the first to examine the bedload transport rate in the sea. They determined the ripple migration rate and simultaneously measured the current profile within 1.5 m of the bed. Bagnold's formula was written as

$$q_b = K \cdot u_*^3 \tag{7.15}$$

When K was taken as a constant neither Bagnold's, Yalin's nor Einstein's formulae matched the experimental results particularly well. However, this was to be expected because of the form drag effect of ripples on the observed shear stress. A much better fit was obtained using a modified value of K. They found that the value of K varied with the ripple height which itself varied with the normalized excess shear stress. Reasonable comparison was also obtained with flume data using the modified coefficient.

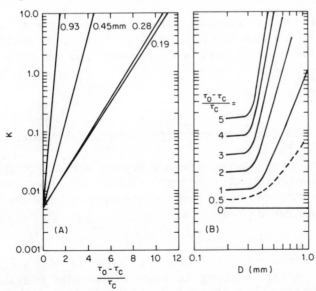

Figure 7.1 (A) Relationship between efficiency factor K and normalized excess shear stress for various sediment sizes in mm. (B) Relationship between K and the sediment diameter D for various excess shear stresses. From Sternberg, 1972, *Shelf Sediment Transport: Process and Pattern*

Sternberg (1972) used the same data to develop a predictive technique for bedload motion. Figure 7.1 shows the relationship that was presented between K, the grain diameter and the normalized excess shear stress. For grain sizes less than about 250 μm, K is given by

$$K = 0.005 \exp{(0.7 \, (\frac{\tau_0 - \tau_c}{\tau_c}))} \tag{7.16}$$

For D greater than about 300 μm

$$K = 0.005 \exp{((\frac{D}{140} - 1.5) \, (\frac{\tau_0 - \tau_c}{\tau_c}))} \tag{7.17}$$

The method comprised using the von Karman–Prandtl equation to calculate the excess shear stress. Then knowing the grain diameter, either Figure 7.1, or Equations 7.16 and 7.17 allow calculation of the value of K, and the bedload transport rate can be estimated from Equation 7.15.

A somewhat similar approach was made by Langhorne (1981). He reported measurements taken by divers on the changing crestal shape of a sand wave. The measurements were obtained at successive slack waters and the volume change during each half tide was considered to be proportional to the integral over the half tide period of the cube of the excess friction velocity. Linear regression gave the best fit equation as

$$q = 0.01 \, (u_* - u_{*c})^3 \tag{7.18}$$

Zero sediment transport occurred when the threshold friction velocity u_{*c} was 1.69 cm s^{-1}, (equivalent to a u_{100c} of 22 cm s^{-1}). When the results were considered in the form of Equation 7.15, the results were compatible with flume data and could be represented by

$$K \sim 0.18 \, \frac{\tau - \tau_c}{\tau_c} \tag{7.19}$$

which is a weaker dependence on normalized excess shear stress than that of Sternberg (1972).

The formulae of Einstein, Yalin and Bagnold were also considered in an estimate of sand transport on the New York Shelf by Gadd et al. (1978). Since the field measurements involved current measurements 1 m above the sea bed, the Bagnold equation was written as

$$q = K_1 \, (u_{100} - u_{100c})^3 \tag{7.20}$$

By consideration of flume data the value of K_1 was taken as $4.48 \times 10^{-5} \text{ gm cm}^{-4} \text{ s}^2$, being the mean value over the grain size range of interest (110–500 μm). K_1 was quoted as $7.22 \times 10^{-5} \text{ gm cm}^{-4} \text{ s}^2$ for 180 μm and $1.73 \times 10^{-5} \text{ gm cm}^{-4} \text{ s}^2$ for 450 μm sand.

Comparison was then made of predicted fluxes with transport estimates taken from radioactive tracer experiments. It was assumed that the drag coefficient C_{100} was 0.003 and the threshold velocity was 19 cm s^{-1}. Tracer results were in all cases within an order of magnitude of the predictions, but the predictions of the formulae varied between themselves by as much as an order of magnitude. Each of the formulae were then applied to a long series of current measurements at 18 locations. For each position the sand transport estimates showed up to an order of magnitude difference, but at 14 of the locations the directional variance of the three net transport vectors was less than 10 per cent.

It is possible that K may not increase indefinitely with excess shear stress, but attains a constant value at high currents. Siegenthaler (1981) analysed the thicknesses of individual tidal deposits in buried sand waves. Using these as a measure of the sediment transport rate together with present day currents, he suggested that Sternberg's formulation represented the data well, but reached a constant value of K at a limiting current. Values of 9.4 and 6.6 at limiting currents of 81 and 91 cm s^{-1} were obtained.

An alternative formulation of Bagnold's equation has been evaluated by Hardisty (1983), following a suggestion by Vincent et al. (1981). In this Bagnold's equation is written as

$$q = K_2 (u^2_{100} - u^2_{100c}) u_{100} \qquad (7.21)$$

Using the same flume data as that used by Gadd et al. (1978) the value of K_2 was determined

$$K_2 = (6.6 \; D^{1.23}) \times 10^{-5} \; \text{gm cm}^{-4} \text{s}^{-2} \qquad (7.22)$$

where D is in mm. In field measurements Hardisty and Hamilton (1984) have obtained a value of $K_2 = 0.39(\pm 0.32) \times 10^{-6} \; \text{gm cm}^{-4} \text{s}^{-2}$, a factor of about three smaller than the laboratory calibration.

A further formulation has been made by Bridge (1981b), with

$$q \propto (\tau - \tau_c)(u_* - u_{*c}) \qquad (7.23)$$

However he did not attempt to evaluate the coefficient of proportionality.

Heathershaw (1981) also compared a number of transport equations with transport rates calculated from the movement of two injections of radioactive tracer in Swansea Bay. The formulae of Einstein, Bagnold, Yalin, Engelund and Hansen (1967) and Ackers and White were used. Threshold velocities and roughness lengths were estimated from the known sediment distribution. Currents were measured nearby over a long period, and from these results the necessary values of u_*, τ, u_{100} and $\bar{u}$ were calculated. These were converted to tidally averaged sediment transport rate and direction, and compared with values obtained from the tracer experiments, by integrating over the necessary time intervals. The comparison is shown in Figure 7.2, and it can be seen that there is nearly a hundred-fold spread in the transport rates predicted by the five equations. Of all the predictions only Bagnold's fall consistently within limits, which are about a factor of 2 higher or lower than the observed transport rates, when the value of K_1 was taken as $7.22 \times 10^{-5} \; \text{gm cm}^{-4} \text{s}^{-2}$, appropriate to a grain size of 190 μm. This equation, together with current velocities measured from

190

a wide area, enabled a prediction to be made of the regional sediment transport circulation and magnitude within Swansea Bay. Heathershaw also carried out a sensitivity analysis in order to consider the effect of errors in z_0, D and h on the estimates of transport rate. This shows a very complex intercomparison with each formula being particularly sensitive to certain of these variables over different flow stages.

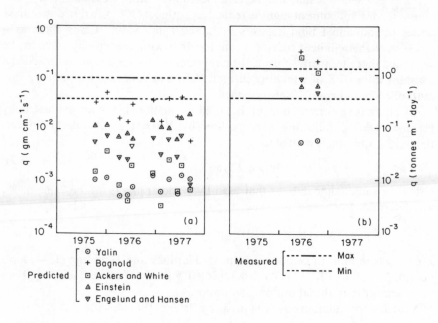

Figure 7.2 Comparison of measured and predicted net bedload transport rates. The maximum and minimum measured rates are from two radioactive tracer experiments, a and b. *From Heathershaw, 1981,* Marine Geol., **42.** *75–104. Reproduced by permission of Elsevier Science Publishers BV*

A similar comparison was carried out in the Sizewell–Dunwich area of the North Sea by Lees (1983), but using fluorescent dyed sand as the tracer. That work showed that the formula of Yalin gave the best comparison, with that of Bagnold overpredicting by nearly two orders of magnitude. The difference between these results and those of Heathershaw in Swansea Bay may be variations between the two areas in the proportions of the total load travelling as bedload. Direct measurements of the suspended load transport rate in the Sizewell–Dunwich area (Lees, 1981b) gave a tidally averaged rate of $5.66 \, \text{gm cm}^{-1} \text{s}^{-1}$, whereas the net rate from the tracer was very much less, $0.012 \, \text{gm cm}^{-1} \text{s}^{-1}$. The maximum instantaneous suspended sediment transport rate measured at spring tides was $30.82 \, \text{gm cm}^{-1} \text{s}^{-1}$. In Swansea Bay the net bedload transport rate of the order of $0.1 \, \text{gm cm}^{-1} \text{s}^{-1}$, was an order of magnitude greater than in the Sizewell–Dunwich area.

Transport Rate Under Waves

There have been many investigations of sediment movement under waves in the laboratory. They are almost invariably of regular sinusoidal waves of a single frequency, and consequently may not be particularly good representations of the natural situation., As we have seen the sediment movement occurs as bedload near the crests and as suspension in vortices shed from the crest and carried into the flow on the reverse stroke. The sediment transport varies very significantly during the oscillation. Sleath (1982) suggests that the maximum bedload transport rate is 8/3 times the average value. He also found that the maximum transport rate could be represented by

$$\frac{q_{max}\,\rho^{\frac{1}{2}}}{((\rho_s - \rho)gD^3)^{\frac{1}{2}}} = 5.2(\theta - \theta_c)^{\frac{3}{2}} \tag{7.24}$$

Generally the average transport rate over the wave cycle is a more useful quantity. Two of the most quoted laboratory experiments where this was measured are those of Kalkanis (1964) and Abou-Seida (1965). In both experiments a sand bed was oscillated within a flume, the beds containing recessed trays to collect the sediment moved in each half wave cycle. The results have been interpreted in a variety of ways.

Madsen and Grant (1976) have used the results of those two authors to demonstrate that the average sediment transport rate is proportional to θ^3. In particular for $\theta > 2\theta_c$ they show that the transport rate non-dimensionalized by the grain fall velocity and diameter can be written as

$$\frac{q}{w_s D} = 12.5\,\theta^3 \tag{7.25}$$

This is similar to the Einstein formulation at low transport rate and states that q is proportional to $u_m{}^6$.

Hallermeier (1982) has reviewed those two sets together with a large number of other laboratory measurements of oscillatory bedload transport and concludes that bedload movement appears to be fully developed when the peak flow velocity nears twice the threshold velocity. The results can be represented at this flow stage by the relationship

$$\frac{q}{\sigma D} = \left(\frac{M}{10}\right)^{\frac{3}{2}} \tag{7.26}$$

where q is the volume transport rate, σ is the oscillatory frequency ($= 2\pi/T$), and T is the wave period. M is given by Equation 5.5.

At lower transport rates it is necessary to modify Equation 7.26 to include the threshold of movement. This is difficult to do in a simple way since the threshold depends on the bed roughness, on the wave period, and on viscous effects near the bed.

An alternative interpretation of the same laboratory data has been made by Vincent et al. (1981). They found that the bedload concentration C_* at any

192

instant, i.e. the volume of sediment mobilized above a unit area of the bed could be represented by

$$C_* = 0.09 \ (\theta_t - \theta_c) \qquad (7.27)$$

where θ_t is the instantaneous Shields number using the wave friction factor. C_* has the dimensions of a length (cm^3 cm^{-2}), and can be thought of as a measure of the thickness of the bedload layer.

The instantaneous transport rate is then C_* times the instantaneous velocity, and integration over the wave cycle will give the average transport rate. Obviously this gives a cubic relationship similar to the unidirectional flow formulae.

Sediment Transport in Combined Waves and Currents

In shallow seas it is likely that the most significant occurrences of sediment transport will occur when the tidal movement is enhanced by wave motion.

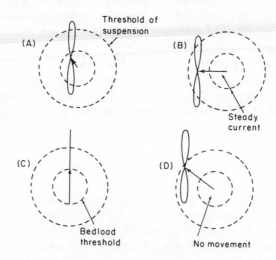

Figure 7.3 Diagrammatic representation of the interaction of waves with a steady current. The curves trace the tip of the combined velocity vector during the wave oscillation. (A) Sediment would be moved on the forward stroke much more than on the backstroke, whereas no movement would have been possible for the steady current. (B) Sediment movement equally on either stroke of the wave oscillation. (C) The codirectional case. (D) Suspended sediment movement on the forward stroke. In (A) and (D) Sediment movement would be to the right of the steady current direction. In (B) and (C) movement would be in the direction of the steady current

Consequently we have to consider how to combine the two processes. Wave motion by itself does not produce net sediment transport, except for the second order mass transport known as Stokes drift. However the waves are a very effective stirring mechanism and can create suspension at much lower equivalent velocities than a steady current. Once in suspension the sediment can be moved en masse by relatively small steady currents, currents which themselves may not exceed the threshold of movement (Figure 7.3A). Thus in one respect the wave orbital motion near the bed can be thought of as an additional turbulence, with the vortices acting in a similar manner to turbulent bursts in entraining the sediment. One of the major difficulties, of course, is that the wave direction can be at large angles to the current direction and, as a consequence, the sediment transport direction need not be that of the tidal current (Figure 7.3). Also since the wave boundary layer is only a centimetre or so in thickness, its presence may affect the boundary conditions, the roughness, seen by the steady current whose boundary layer is generally of a few metres in thickness. Consequently despite the enormous importance of the combined effects, there has been little progress in understanding them until comparatively recently. Laboratory studies, unfortunately, are of restricted help since flume studies can only consider the co- and contradirectional cases.

According to Harms (1969), combined flow ripples have crest length and pattern intermediate between current and wave ripples. They also have intermediate height characteristics. Asymmetry increases as the relative magnitude of the current increases and the crestal profile becomes more rounded. When the orbital diameter exceeds the ripple wavelength, more flattening of the ripple crests can be expected. His wave periods were low, however, the results being obtained from a flume study.

The asymmetry seems to have a strong effect on the sediment transport because it causes large differences in the strength of the vortex created in the lee of the ripple on each stroke. The vortices will then be shed at different velocities into the flow on the succeeding stroke, and penetrate to different heights above the bed. On the forward stroke the current and wave motions act together and create a very strong vortex on the lee slope (Figure 7.4), and this vortex is shed from the crest high into the flow on reversal of the current. On the backward stroke the combined velocity is less, but a small vortex may then be formed on the stoss slope if the wave orbital velocity is sufficiently strong and exceeds the tidal current by a large enough margin. The following forward stroke then lifts this vortex into the flow. As the original lee vortex is higher in the flow it can travel further downstream across the bed, but the sand suspended in it may have been deposited during the weak backstroke. The sediment movement produced by the wave–current interaction is complex, and in some circumstances could produce a movement of sediment in the up-current direction. This could result in fine sediment being carried offshore in suspension and coarser bedload material moving shorewards.

There seems to be a fairly significant distinction between the effect of waves that reverse the current flow and those that do not. In the former case, vortex generation can occur on both strokes of the wave oscillation, but with the latter, where the wave is a perturbation on the current, only one strong vortex is formed and this is not ejected into the flow in the same way since there is no reversal of the current. These effects have not been thoroughly described nor quantified yet.

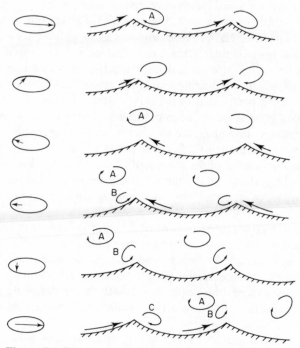

Figure 7.4 Diagrammatic sequence of vortex generation, separation and advection under combined waves and currents. The phase of wave is shown by rotation of the vector on the left

In order to establish a predictive sediment transport formula for combined currents and waves, it is necessary to calculate the fluctuating shear stresses at the bed and to integrate them over a wave cycle. Then the resulting mean shear stress can be used in one of the steady flow formulations of the sediment transport rate. This implies that no error is involved in not calculating the sediment transport at each instant, and integrating that over the wave cycle. Alternatively the maximum wave orbital velocity at the bed can be used to obtain the combined shear stress. This then makes the assumption that the transport will be controlled by the orbital velocity, but this is reasonable providing the wave motion is relatively much stronger than the steady currents. For suspended sediment transport it is additionally necessary to define the velocity profile of the combined flow near the bed.

One of the oldest approaches, and the one which is in most common use, particularly for engineering purposes, is that of Bijker (1967). He assumes that there is a viscous boundary layer present in the steady flow and he then proceeds to find the gradient of the combined velocity vector at this level. If u_b is the orbital velocity at the top of the viscous boundary layer, then the value of the orbital velocity at that height in the presence of the current is p times u_b, so that p is then a measure of the relative thickness of the wave boundary layer to the laminar sublayer. The resultant instantaneous velocity is obtained from the steady and orbital velocities by the cosine rule and the mean value over the wave cycle obtained by integration over the wave period. Prandtl's mixing length theory is then used to derive a shear stress from the velocity gradient, i.e.

$$\tau = \rho l^2 \left[\frac{du}{dz} \right]^2$$

where the mixing length $l = \kappa z$.

The component of the resulting mean bed shear in the direction of the current is

$$\frac{\tau_1}{\tau_{cu}} = 1 - N \left(\xi \frac{u_m}{\hat{U}} \right)^{1.5}$$

τ_{cu} is the shear stress under the current alone, $\hat{U}$ is the depth mean steady current, u_m is the maximum wave orbital velocity near the bed, and $N = 0.36 - 0.14 \cos 2\varphi$, where φ is the angle between the wave and current directions. The bed shear at right angles to the current

$$\frac{\tau_2}{\tau_{cu}} = M \left(\xi \frac{u_m}{\hat{U}} \right)^{1.25}$$

where $M = 0.205 \sin 2\varphi$. The resultant total bed shear stress

$$\frac{\tau_{wc}}{\tau_{cu}} = 1 + \tfrac{1}{2} \left(\xi \frac{u_m}{\hat{U}} \right)^2 \qquad (7.28)$$

where ξ is given by $\xi = p \ln \left(\frac{h}{z_0} - 1 \right)$, and τ_{wc} is the combined shear stress.

The direction in which the stress acts is given by $\tan^{-1} \tau_2/\tau_1$.

Laboratory tests were performed on stones and on sand with ripples, with relative angles φ of $0°$ and $15°$. The value of p was found to equal 0.45. Since these roughnesses are unlikely to actually have a laminar sublayer under steady flows of a reasonable magnitude, the factor p is a more imprecise empirical matching factor than its definition implies.

The right hand side of Equation 7.28 can be thought of as an enhancement factor ψ, and to calculate this factor needs a specification only of the roughness length z_0, providing the velocities are known. Figure 7.5 shows the variation of ψ with u_{100} and waveheight for waves of 8 s period in a water depth of 20 m. The value of z_0 was assumed to be 0.25 cm. The enhancement factor increases rapidly with decreasing current. If one assumes a u_{100} of about $0.35 \, \text{m s}^{-1}$ (equivalent to the threshold of a grain size of about 125 μm), then a 2 m high wave gives a tenfold increase in bed shear stress.

In order to calculate the total shear stress, the stress due to the steady current needs to be calculated, either from velocity profile measurements, or from a velocity measurement at some height above the bed, together with an assumed drag coefficient or roughness length.

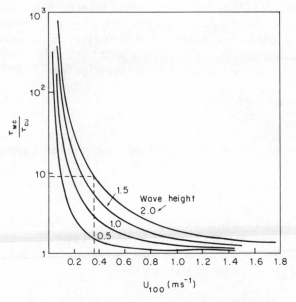

Figure 7.5 Enhancement of currents by waves according to Bijkers' formulation, Equation 7.28, for waves of 8 s period in water depth of 20 m. *After Heathershaw, 1981,* Marine Geol., *42, 75–105. Reproduced by permission of Elsevier Science Publishers BV*

Grant and Madsen (1979), in contrast to Bijker, consider that the wave and current motions near the sea bed cannot be treated separately and superimposed. There is a non-linear interaction which alters both from their pure form; the current in the region above the boundary layer experiences a shear stress which depends not only on the physical roughness but also on the wave boundary layer characteristics. Consequently they parameterize the influence of the wave on the current through an apparent increase in the roughness length. Additionally they assume that the environment is wave dominated; that the steady current velocities are relatively small. For this situation the shear stresses can be obtained from the near bed velocities by use of a wave–current friction factor. However, because of the non-linear interaction, the bed shear stress does not necessarily act in the same direction as the mean current. Also the steady current velocity above the bed is treated as an unknown, though the maximum near bed orbital velocity can be calculated from linear wave theory. Arbitrary wave/current directions are assumed.

Using a time-invariant eddy viscosity defined as $\mathbf{K_{wc}} = \kappa\, u_{*wc} z$, where u_{*wc} is the friction velocity under waves plus currents, they derive equations for the wave velocity within the wave boundary layer. For the steady current below the height of the wave boundary layer, the profile is given by the von Karman–Prandtl equation involving the physical roughness, but with the current reduced by a factor u_{*c}/u_{*wc}. Outside the wave boundary layer the velocity profile is given by the von Karman–Prandtl equation with a modified roughness. This roughness is found to be a complex relationship involving u_m, the amplitude of the near-bed motion A_b, the friction velocities of the wave, the current and their interaction, and the physical bed roughness. From experimental data they show the apparent bed roughness can be of the order 10 times the roughness in a steady current without waves.

The various formulae in the Grant and Madsen approach have to be solved by successive approximation on a computer to obtain values of the bed shear stress etc.

A somewhat similar approach to the problem is taken by Smith (1977). However he only considers the codirectional case, and assumes that currents are dominant. His approach basically assumes that the eddy viscosity based on the sum of the wave and current friction velocities is equivalent to that which would be produced by the shear of the combined motions, i.e.

$$K_{wc} = \kappa (u_{*cu} + u_{*m})z$$

where u_{*m} is the maximum oscillatory friction velocity, and u_{*cu} the friction velocity for the currents.

The presence of waves has the apparent effect of increasing von Karman's constant, and for a given velocity at a fixed height from the sea bed and a given roughness, a higher bed shear stress results. For a value of $u_{*m} = \frac{1}{4} u_{*cu}$ the combined boundary shear stress is increased by 56 per cent. Again the steady flow profile outside the wave boundary layer experiences an apparently rougher boundary.

The above approaches define the shear stress under combined waves and currents which can be used in standard bedload formulae. Those of Smith (1977) and Grant and Madsen (1979) also give the velocity profile which can be used, together with a profile of suspended sediment concentration to calculate the suspended load transport rate. Examples of mean velocity profiles obtained from a numerical model (Davies and Soulsby, in preparation) are shown in Figure 7.6, for three different wave orbital velocities. As can be seen the effect of waves is to reduce the mean value of the near-bed velocity. Laboratory investigations by Kemp and Simons (1982, 1983) support the increase in apparent bed roughness when waves are superimposed on currents. They also found that turbulence intensities and Reynolds stresses near the rough bed are increased by the presence of waves, particularly in the layer within two roughness heights of the bed. There the fluctuations are periodic and affected by vortices ejected from the roughness elements.

The idea that the wave velocities act as a stirring mechanism has been further exemplified by the results of measurements reported by Owen and Thorn (1978). They obtained time averaged suspended sediment samples by a pumped sampling technique, together with instantaneous velocities. At very low rms

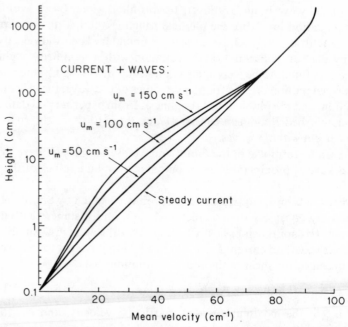

Figure 7.6 Vertical profiles of mean horizontal velocity for a steady current with a train of superimposed colinear waves of three maximum orbital velocities. From Davies and Soulsby, in preparation. The reduction in mean near bed velocity would produce an apparent increase in bed roughness length from measurements taken farther than 10 cm from the bed

velocity fluctuations, the sand discharge q in $gm\ m^{-1}\ s^{-1}$ was related to depth averaged velocity $\hat{U}$ in $m\ s^{-1}$ by

$$q = 1400\ \hat{U}^4$$

For higher wave velocities the results could be represented by

$$q = 1400 \left(\frac{U_{rms}}{0.05} \right)^2 \hat{U}^4 \tag{7.29}$$

where U_{rms} is the rms velocity fluctuations measured 0.05 m above the bed in $m\ s^{-1}$. The extra factor in Equation 7.29 can be thought of as a sand flux multiplier; the factor by which the transport rate exceeds that in the same current without waves. As can be seen from Figure 7.7 the factor is 25 for waves with rms velocities of $25\ cm\ s^{-1}$.

The wave velocities can also be thought of as additional turbulence, and as contributing to the turbulent kinetic energy. Grass (1981) has treated the combined effects of currents and waves in terms of turbulence energy. However the approach is likely to be restricted to the situation where currents are stronger than the wave velocities, and reversal of the current does not occur. The magnitude of the maximum turbulence kinetic energy is largely independent of flow roughness for steady flow and the ratio of the turbulent kinetic energy to the bed shear stress has a constant value

$$\frac{\frac{1}{2}\rho\overline{k}^{2}}{\tau} = \frac{1}{2}\left(\frac{\overline{k}^{2}}{u_{*}^{2}}\right) = \frac{1}{2} \quad (7.5)$$

where

$$\tfrac{1}{2}\rho\overline{k}^{2} = \tfrac{1}{2}\rho(\overline{u'^{2}} + \overline{v'^{2}} + \overline{w'^{2}})$$

Thus

$$\tfrac{1}{2}\rho\overline{k}^{2} = \tfrac{1}{2}\rho(7.5\,u_{*}^{2})$$

The effective turbulent kinetic energy of a steady flow producing the same average concentration of sediment as the combined flow is given by

$$\tfrac{1}{2}\rho\overline{k}_{s}^{2} = (\tfrac{1}{2}\rho\eta V^{2} + \tfrac{1}{2}\rho\overline{k}^{2})$$

where V is the root mean squared wave velocity fluctuation and η is a proportionality factor. Consequently the steady flow producing the same turbulence energy as the combined flow is given by

$$\frac{\tfrac{1}{2}\rho k_{s}^{2}}{\tfrac{1}{2}\rho k^{2}} = \frac{\tfrac{1}{2}\rho 7.5\,u_{*s}^{2}}{\tfrac{1}{2}\rho 7.5\,u_{*}^{2}} = \frac{\tfrac{1}{2}\rho(\eta V^{2} + \overline{k}^{2})}{\tfrac{1}{2}\rho 7.5\,u_{*}^{2}}$$

where u_{*s} is the bed friction velocity of the steady flow producing the same sediment transport as the combined flow.

Thus

$$u_{*s} = \left(\frac{\eta V^{2}}{7.5\,u_{*}^{2}} + 1\right)^{\frac{1}{2}} u_{*} = Nu_{*} \quad (7.30)$$

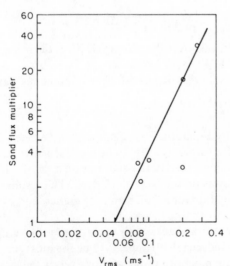

Figure 7.7 Sand flux multiplier M against the rms wave fluctuations in m s^{-1} measured 5 cm above the bed. *From Owen and Thorn, 1978,* Proc. 16th Coastal Eng. Conf., *1675–1687. Reproduced by permission of American Society of Civil Engineers*

Providing some estimate of η is available and the steady flow shear stress is known, then Equation 7.30 can be used in any formulae to calculate the sediment transport. There is an obvious comparison with the Equation 7.28 of Bijker.

In terms of the sand flux experiments of Owen and Thorn (1978), the sediment transport rate under the combined flows will be $q \propto N^{n-1} \hat{U}^n$, provided the depth mean velocity is directly related to u_*. Since n = 4, the multiplication factor in Equation 7.29 equals N^3. Comparison with the data of Owen and Thorn gave a value of η of 0.6. Though these results are unique to the particular circumstances of the measurements, they provide a useful empirical technique that may provide an order of magnitude answer for simple problems.

An alternative approach is possible when instantaneous velocity measurements are available, by appropriate modification of formulae developed for transport under waves alone. However the currents would need to be small compared with the wave oscillations.

Madsen and Grant (1976) have modified Equation 7.25 and propose that the instantaneous volume bedload transport rate is

$$\frac{q(t)}{w_s D} = 40 \, \theta^3(t) \frac{U(t)}{|U(t)|} \tag{7.31}$$

when θ exceeds the threshold value. The value of U(t) can be expressed as the vector sum of the near bed wave oscillatory current U_w and the steady current U_a. In Equation 7.31 the value of $\theta(t)$ is defined as

$$\theta(t) = \frac{0.5 \, f_w \rho}{(\rho_s - \rho)gD} \left| U_w^2(t) + U_a^2 \right|$$

Averaging Equation 7.31 over the wave cycle gives the mean sediment transport rate and direction.

Vincent et al. (1981, 1983) expressed the instantaneous transport rate as

$$q = C_*(t)(U_w(t) + U_a) \tag{7.32}$$

where C_* is given by Equation 7.27. Comparison between Equations 7.31 and 7.32 suggests that the former is likely to be more sensitive to the wave direction. Vincent et al. (1983) have used both equations on a set of velocity measurements obtained during storms on the New York shelf. The results differed by a factor of two in magnitude, Equation 7.32 being the higher, and by 10° in direction on one storm, but by rather more in a second.

There are inherent drawbacks with each of the above approaches, though that of Grant and Madsen (1979) appears to be the most complete so far. One of the biggest stumbling blocks to progress, however, is the lack of good field data on the combined wave/current problem. Laboratory experiments are of course generally constrained to the codirectional case. Also, because of the thinness of the wave boundary layer, it is difficult to obtain velocity profiles close to the bed for the estimation of bed shear stress, and, because of the phase lag through that layer, defining what is the effective sediment moving process is also difficult.

Complete profiles of the steady and oscillatory flow to within a centimetre of the bed are needed to establish the effect of the non-linear interaction on the profiles, and concurrent measurement of sediment transport rates would help to establish the sediment response.

CHAPTER 8

Cohesive Sediments

So far we have considered the movement of relatively coarse particles of silt, sand and gravel grade. These are generally fairly rounded, have a mass sufficient to ensure that grain inertia is important, and are transported as separate particles, only interacting in a hydraulic sense. With clay size particles, however, the situation is different. They are platelike with a diameter less than $2\,\mu m$, which is typically of the order 10 times the thickness. Their surfaces have ionic charges creating forces comparable to or exceeding the gravitation force, and these cause the particles to interact electrostatically. Consequently they do not act as separate individual particles but stick together. The degree of 'stickiness' —cohesion—rises with the proportion of clay minerals in the sediment and starts becoming significant when the sediment contains more than 5–10 per cent of clay by weight. Cohesive sediments have different threshold characteristics, packing structure and physical and hydraulic properties from cohesionless sands. Before considering the threshold conditions for deposition and erosion, we must consider the way in which the particles interact or flocculate, and the effect this has on the sediment characteristics.

FLOCCULATION

On the face of each platelet there is normally a negative charge due to the exposed oxygen atoms in the broken bonds of the crystal lattice. Additionally, negative charges will result from isomorphic substitution of positively-charged cations of a low valency within the lattice for other principal structural cations of a higher valency. At the particle edge, however, the charge is positive, because of the broken bonds of the silica tetrahedra. The overall particle charge is usually negative for clay minerals and its magnitude can be calculated by measuring the electrophoretic mobility of the particle, its speed of movement within a known electric field. From this the zeta potential of the particle can be calculated; the charge potential at a surface, the zeta plane, generally about 0.5 nm from the particle. The zeta potential varies from face to edge, and is also affected by the mineralogy of the particle, as well as by the pH and ionic concentration in the surrounding field.

In a saline fluid the free ions in the water interact with the charges on the particle, positive ions being attracted to the face and negative ones to the edges. A closely-held layer of cations such as sodium Na^+, potassium K^+ and magnesium Mg^{++} will be formed on the face by the action of the most negative residual ions and there will also be a more diffuse layer of hydrated cations neutralizing some of the remaining negative charge. These diffuse ions can be exchanged with the surrounding fluid depending on the concentration and valency of the ions in fluid. The total effect of the electrical double-layer is to reduce the magnitude of the negative charge on the particle.

If the charge on the face were the only factor, then the particles, being similarly charged, would continually repel one another, the electrostatic (Coulombic) force V_R being repulsive and decreasing exponentially with distance. However, there is also a molecular attractive force V_A, known as the London–van der Waals force. This varies inversely proportional to the square of the distance of separation and tends to largely counteract the repulsive force under certain conditions.

In river water, when the double-layer is not significant, the electrostatic repulsive forces are large and generally dominate, tending to prevent the particles flocculating. Nevertheless there is a possibility of the positively charged edges meeting the negatively charged faces and a very open 'house-of-cards' structure being formed. This is likely to be a very weak bonding which could easily be broken by turbulent shearing.

In saline waters the presence of the double-layer reduces the surface charge and the attractive force V_A dominates. Consequently the particles have a greater tendency to flocculate.

The sum of V_R and V_A gives the curve of the total energy of interaction with distance between two particles. This is shown in Figure 8.1 for kaolinite particles in saline conditions. The distance over which interaction is effective is of the order of 30 nm, but that, and the magnitude, depends on whether they are face–face or face–edge or other modes of interaction, and on the mineralogy, pH and salinity. When the particles are brought together they approach to the distance at which there is a maximum attraction, or minimum repulsion. It then requires energy to force the particles apart or push them closer together. At the primary minimum the attraction is greatest and the particle interaction would be at its most stable. For a face–edge situation there can be a small secondary minimum at some distance from the particles, as well as the primary minimum. Consequently particles can stably aggregate at that distance, but being a weaker bond it is fairly easily broken.

Increasing ionic concentration in the fluid and the presence of the extra cations causes the negative charge at the surface to decrease in magnitude. Thus the plane of a particular charge intensity is closer to the particle, leading to an apparent compression of the electrical double-layer. The decrease of charge

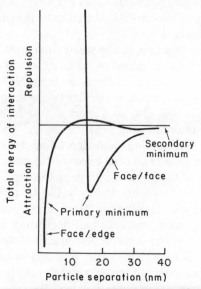

Figure 8.1 Total energy of interaction of kaolinite particles in sea water. *After James and Williams, 1982,* Rheol. Acta, *21, 176–183. Reproduced by permission of Rheologica Acta*

with increasing salinity is approximately exponential so that flocculation tends to very quickly reach an equilibrium situation at comparatively low salinities, providing the particle concentration is sufficiently high. Krone (1978a) considers that flocculation begins at salinities of 0.6‰ for kaolinite, 1.1‰ for illite and 2.4‰ for montmorillonite and is complete above 1–3‰. Similar results are obtained by Gibbs (1983) for illite and kaolinite, but flocculation of montmorillonite continued until much higher salinities. Whitehouse *et al.* (1960) also considers that flocculation of illite and kaolinite is complete above a salinity of about 4‰. Montmorillonite, however, flocculated gradually over the whole salinity range up to 35‰. For a natural mud from the Severn Estuary, Peirce and Williams (1966) found that flocculation was complete at a sodium chloride/mud ratio (gm/gm) greater than 0.0165.

As temperature increases, the thermal motions of the ions increase in magnitude and this leads to increased repulsion. Consequently flocculation is less effective as the temperature rises.

Organic material on the particles, such as mucal films caused by bacterial activity and organics adsorbed from suspension, have positive charges and significantly enhance flocculation. Organic binding makes the flocculates very much harder to break up. Whitehouse *et al.* (1960) found that the presence of complex carbohydrates in concentrations between 0.0005–1 g l^{-1} increased floc size, but more proteins decreased their size.

Flocculation is considered to be a reversible phenomenon and flocculated sediment placed in fresh water will be more prone to disaggregation processes. This occurs because the less firmly held cations in the double-layer will diffuse away because of the lower ionic concentration in the fluid and the locally high ionic concentration gradient. The degree of attraction between particles is thus reduced and they become more susceptible to being broken up, particularly by turbulent shearing.

For flocculation to occur in the first place the particles must be in very close proximity. Consequently the number concentration of particles in suspension is important. However, there are several mechanisms by which the collision potential of the particles is enhanced. These have been outlined by Krone (1978a), Spielman (1978), and McCave (1984).

1. Brownian motion is caused by bombardment of the particles by the thermally agitated water molecules. The motion tends to bring the particles occasionally into close proximity. Assuming the particles are the same size and are in the Stoke's regime, their collision rate is

$$K_B = \frac{4kTn}{3\mu} \tag{8.1}$$

where k is Boltzmann's constant, T is the temperature, n the number concentration of suspended particles and μ the water viscosity. This so-called perikinetic mechanism is only effective at very high particle concentrations and for particle size less than about 0.5 μm.

2. In the velocity gradient, particles will collide if they are spaced less than the sum of their radii apart in zones of the fluid being sheared one layer over the other. The collision frequency will be

$$K_v = \frac{4}{3} n R^3 \frac{du}{dz} \tag{8.2}$$

where R is the sum of the radii of the colliding particles and du/dz is the local velocity gradient. For large particles at small velocity shear or for smaller particles at larger shears, K_V will be much larger than K_B, but at very large shears the particles will be torn apart rather than continue flocculating. Consequently, in any particularly shear field there is a limiting floc diameter and particle number concentration above which particle size should be constant. Flocs formed under this orthokinetic flocculation mechanism should be relatively strong.

3. Inertial encounters. In a turbulent flow the grain inertia causes the larger particles to respond less quickly to local accelerations than small particles, and the particles can collide. Saffman and Turner (1956) have produced an analytical result for neutrally buoyant particles of radii r_1 and r_2 which are not greatly different. They obtain the collision frequency per unit volume as

$$K_1 = 1.294 R^3 \left(\frac{\Sigma}{v}\right)^{\frac{1}{2}} n_1 n_2 \tag{8.3}$$

where Σ is the turbulent energy dissipation rate and n_1, n_2 are the number concentrations of the two sizes of spheres. Equation 8.3 should only hold for particles smaller than the turbulent (Kolmogarov) microscale, which is about 100 μm.

The relative importance of velocity shear and inertia has been estimated by McCave (1984) and he suggests that internal encounters are only important when the grains differ in size by more than four times the microscale length. However comparison of Equations 8.2 and 8.3 suggest that inertial encounters may be important when there are few large grains in a suspension of much more numerous small grains.

4. Differential settling of particles can result in the larger, faster settling particles growing at the expense of the smaller, slower ones. The collision frequency is

$$K_s \propto \pi R^2 \delta_w n \qquad (8.4)$$

where δ_w is the relative settling velocity. This mechanism is effective at times of low velocity and with large particles, but leads to poorly-bonded flocs.

Small particles have a larger relative surface area and will flocculate most readily, though they prefer to interact with particles bigger than themselves. The number of particles per floc is proportional to the floc volume to the power $(-p)$, where p is between 2 and 2.33 (Smith, T. J., personal communication). A typical floc diameter is 60–150 μm, and each one would contain 10^5 to 10^6 particles, though Biddle and Miles (1972) show natural flocs up to 800 μm in size.

Kranck (1973) considers that flocculation progressively removes the finer particles from suspension and the settling velocity of the flocs forms the limit above which the grains occur as individual single ones. Since it is the finest particles that dominate the flocculation process, and the fine particles create the major part of the number concentrations, a small percentage by weight of montmorillonite is very important; because there are so many of these smallest grains.

Krone (1978b) from measurements of the viscosity of five sample suspensions at different concentrations derived a relationship between the volume concentration of the flocs and the mass concentration, from which he calculated the floc densities and their shear strengths. He considered that there was an order of aggregation in the flocs with the most compact zero-order floc having densities between $1.16–1.27$ gm cm^{-3} and shear strengths of 21–49 dynes cm^{-2}. Increasing order of aggregation was formed by flocs of the next lower order combining together in a less dense, weaker aggregate. The highest order of aggregate had densities of $1.06–1.07$ gm cm^{-3}, but shear strengths ranging between $0.20–1.9$ dynes cm^{-2}. It is unlikely in nature that such a simple picture would hold and flocs of many sizes and strengths would be present at any one time, some being broken down in patches of temporarily high shearing rate and others reforming elsewhere.

Since flocculation depends on the particle number concentration and on the velocity shear, it can be expected that the floc settling velocity will vary with these parameters. Also the strength of the flocs may determine other properties of the suspension and of deposits forming by settling from suspensions.

MUD CHARACTERISTICS

Suspension Rheology

A number of people have examined the effective viscosity in viscometers of mono-mineralic suspensions, mixtures in known proportions, or natural samples. These instruments have two concentric cylinders or a cone and a plate with the suspension placed between them. One part is rotated at a constant rate giving a steady shearing rate and the force on the other is measured, giving the shear stress. For clear water the resulting shear stress–shear rate curve is a straight line passing through the origin and with a slope equal to the molecular viscosity. This response defines a Newtonian fluid (Figure 8.2). Suspensions of quartz particles show the same response but with an effective viscosity increasing with concentration. A material with a Newtonian response throughout, but with a finite shear strength at zero shearing rate is a Bingham plastic.

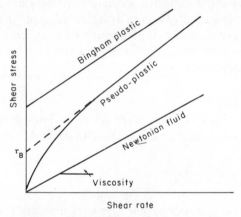

Figure 8.2 Shear stress versus shearing rate curves for materials with different rheological properties. The slope of the curves equals the effective viscosity

Suspensions of clay minerals, however, show a non-Newtonian response above a concentration of about $10 \, \mathrm{gm} \, l^{-1}$, and when the proportion of clay minerals exceeds about 20 per cent. At low shear rates the shear stress rises steeply, but as the shear rate increases the rate of increase in shear stress diminishes, eventually becoming linear (Figure 8.2). This type of material is a pseudo-plastic. The effective viscosity is high at low shearing rates and decreases with increasing shear rate to a constant plastic viscosity. Consequently these suspensions are often termed 'shear thinning'. The shear thinning effect is produced by the progressive breakdown of a network of flocs to a basic floc unit following, essentially, the idea of orders of aggregation. Most of the viscosity variation occurs at shearing rates less than about 200 s^{-1}, which is the important region as far as estuaries and coastal waters are concerned. It is extremely unlikely that the velocity shear would exceed 100 cm s^{-1} in 10 cm in the sea, a shearing rate of 10 s^{-1} except possibly in high energy turbulence. Migniot (1968) has shown that the effective viscosity at any particular shear rate varies with the fourth or fifth power of the concentration. Consequently there is a fundamental difference in the behaviour of suspensions of clay and those of sand, which are Newtonian at all concentrations.

The intercept of the extrapolated upper Newtonian region of the curves gives the residual stress τ_B, sometimes called the Bingham yield stress, which arises from the residual effect of the particle interaction, and which is related to the magnitude of the net attraction between particles. The slope of the curve at high shear rates gives the plastic viscosity arising from purely hydrodynamic effects. Both of these parameters are useful for characterizing flocculated suspensions. James and Williams (1982) have investigated mixtures of quartz and kaolinite at a volume concentration of 0.2 per cent. Their results show that pseudo-plastic effects become significant when the kaolinite exceeds 20 per cent of the mixture (Figure 8.3). The residual stress also increased with clay percentage. Additionally the results showed a decrease in τ_B with increasing pH in the range 6.0–8.0, which is normally experienced in the transition between fresh and salt waters in estuaries. Also the value of τ_B was proportional to the second power of the solids volume concentration,

$$\tau_B \propto C_V{}^2$$

This result can be obtained theoretically for dilute homogeneous suspensions. The solids volume concentration C_V is related to the mass concentration C by

$$C_V = \frac{1}{\rho_s} . C$$

Krone (1978a) showed, however, that $\tau_B \propto C^{\frac{5}{3}}$ represented the shear strength of a Philadelphia mud. Wan (1982) found $\tau_B \propto C^3$ for experiments with both kaolinite and bentonite (nearly pure montmorillonite), the constant of proportionality being 1.28×10^4 and 4.14×10^6 dynes cm^{-2} respectively, and Migniot (1968) found $\tau_B \propto C^4$ for a large number of natural clays. These three results may have given a higher power function because the experimental conditions were not dilute, or the muds may have contained several different clay minerals.

There have been many attempts to relate the residual stress value to the chemical properties of the particles, in particular the cation exchange capacity. General relationships are difficult to define, however, since there are differences in the way in which the cation exchange capacity has been determined. Its magnitude relates particularly to the montmorillonite percentage and varies with grain size—large particles having a different exchange capacity from a small particle of the same mineral. The pH and ion concentration of the ambient

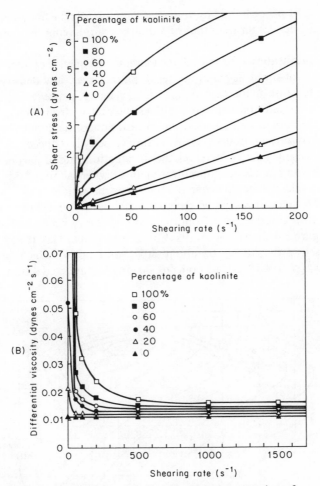

Figure 8.3 Variation of the rheological properties of a quartz/kaolinite mixture with proportion of the constituents. Concentration 2 per cent solids by volume. (A) Shear stress/shearing rate. pH 6.0. (B) Differential viscosity. pH 8.0. *From James and Williams, 1982,* Rheol. Acta, *21, 176–183. Reproduced by permission of Rheologica Acta*

fluid are also important. Consequently, though τ_B is related to the total interaction energy between particles, there is no way yet of predicting it for natural sediments in the sea.

Settling and Deposition

Though the individual particles may have densities of about 2.6 gm cm^{-3}, the flocs are very much less dense. Quoted values range between 1.06–1.8 gm cm^{-3}. A floc with a density of 1.15 gm cm^{-3} would have a porosity of 88 per cent fluid by weight, or 91 per cent fluid by volume. Consequently the floc size and surface area will be much larger than that of a quartz grain having the same fall velocity.

As the concentration is increased, the increased frequency of interparticle collision causes enhanced flocculation, resulting in larger, low density flocs. The net effect is to cause an increase in the settling velocity. However increasing concentration eventually means that the flocs interact hydrodynamically so that effectively the flocs in settling cause an upward flow of the liquid they displace. Thus hindered settling occurs and the settling velocity is reduced.

Since increasing salinity increases the flocculation rate, a particular settling velocity will occur at a lesser concentration in a higher salinity. Additionally the peak settling velocity should occur at a lower concentration.

Owen (1970a) has investigated the relationship in settling experiments in the laboratory with a natural clay from the Severn Estuary which had a high illite content. Figure 8.4 shows the results. For this particular clay the settling velocity increased with salinity to about 30‰, and above which settling was retarded. In excess of 10–20,000 ppm (10–20 gm l^{-1}) hindered settling occurred.

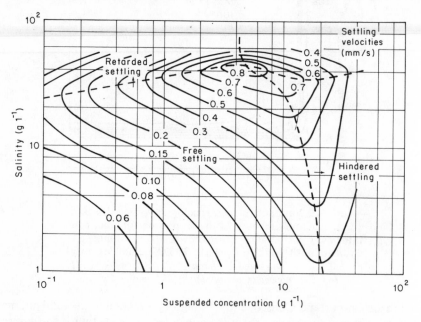

Figure 8.4 Relationship between settling velocity, in mm s^{-1}, salinity and suspended concentration. *From Owen, 1970a,* Hydraulics Research Stn Report INT 78. *Reproduced by permission of Hydraulics Research, Wallingford*

At any particular salinity the increase in settling velocity could be represented by the concentration to the power 1.1 to 2.2. Very similar results have been quoted by Odd (1982) for samples from the Severn Estuary (Figure 8.5). Tests

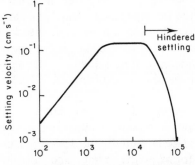

Figure 8.5 Variation of settling velocity with suspended concentration for mud from the Severn Estuary. *After Odd, 1982,* The Severn Estuary. *Reproduced by permission of Thomas Telford Ltd.*

on a sample from the Thames Estuary which was richer in montmorillonite, gave an exponent of about unity. Krone (1962) obtained a value of 4/3, and Migniot (1968) gave a range of results for several muds.

Above the peak settling velocity, in the region of hindered settling, Maude and Whitmore (1958) showed that the fall velocity was decreased according to

$$w_s = w_0 \, (1 - C)^m$$

where w_0 was the fall velocity of a single grain. The exponent m was a function of particle size and shape. For small particles m = 4.65 and for large m = 2.32. Peirce and Williams (1966) have shown for four different muds that the settling velocity agreed with the formula $w_s \propto (1 - C)^5$ and this relationship is fairly widely used. The constant of proportionality depended on the mineralogy of the muds. Consequently, the magnitudes of the settling velocities are likely to depend on salinity, concentration and mineralogy, and need to be determined separately for each sediment under consideration.

The above results were all obtained in still water and may not represent the settling in a naturally turbulent shearing flow because of the difference in the flocculation states. There are great difficulties in measuring the size and fall velocity in the sea, since sampling of the suspension will cause the velocity field to change and the flocs grow by settling, or are disrupted by pumping. Use of electronic particle sizing instruments can be questionable because of alteration of floc size due to the necessity of pumping the suspension through the sensor. Owen (1971) developed a tube for measuring settling velocity in as near a natural state as possible. This tube traps a horizontal volume of the flowing suspension and rotates it into a vertical position. Direct measurements of settling are taken immediately, before drastic change in the floc structure occurs. The velocities are about an order of magnitude higher than those obtained in the laboratory, presumably due to the different flocculation state and turbulence characteristics. Recently holographic techniques have been used in the ocean for measuring the size and the settling velocity of natural particles *in situ* (Carder *et al.*, 1982). For use of this technique in coastal and estuarine environments vertical mean water motions would need to be taken into account, as these are likely to be the same magnitude as settling velocities. The techniques of measuring, sampling and analysing suspended sediment have been reviewed by McCave (1979a).

When a column of suspended sediment settles in still water, there is a sequence of settling, deposition and consolidation. This process has been described by Been and Sills (1981). Initially the density is almost uniform, but subsequently an interface forms at the surface and falls linearly with time. The interface separates clear water from the suspension, which maintains an almost constant concentration (Figure 8.6). At the bottom a layer of relatively high

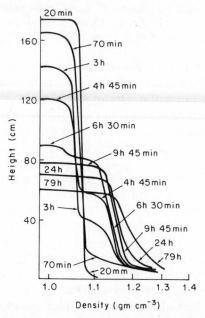

Figure 8.6 Density profiles in a settling suspension with initial density of 1.09 gm cm^{-3}. *From Been and Sills, 1981,* Geotechnique, *31, 519–535. Reproduced by permission of Thomas Telford Ltd.*

density forms, partly as a result of coarse particles settling quickly before being incorporated in flocs and partly as a result of the rapid consolidation of the layer when it is still thin. As the bottom layer thickens the consolidation rate declines because the pore water is expelled less readily, and the top of the layer has an intermediate density. Nevertheless there is a distinct step between the bottom layer and the suspension above it. It appears from the experiments of

Owen (1970b) and Been and Sills (1981) that when the initial suspension con-centration is in the hindered settling range, no distinct bed is formed, but there is a gradual gradient in density towards the base of the column and the density increases with time.

During the bed-forming period, the bed has been reported as rising linearly (Owen, 1970b) or with the reciprocal of time (Einstein and Krone, 1962). Even-tually, after about six hours, the rising bed meets the falling upper surface of the suspension. Thereafter, the surface continues to fall, but at a very much reduced rate caused by consolidation in the settled deposit (Figure 8.7). A plot of log time against log height of the interface shows two successive well-defined straight lines with a short transition region between. The initial forming period has been called the constant rate period, and the log–log phases the falling rate period.

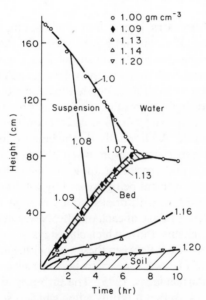

Figure 8.7 Settling of a suspension interface and growth of the bed with time. Initial density of 1.09 gm cm^{-3}. This figure is derived from the data on Figure 8.6. *From Been and Sills, 1981,* Geotechnique, *31, 519–535. Reprodu-ced by permission of Thomas Telford Ltd.*

The change of height of the suspension surface with time can be represented by

$$\frac{h}{h_\infty} - 1 = \frac{k}{t} \qquad (8.5)$$

where h is the height of the surface, h_∞ is the height after an infinite time, and k is a constant.

Einstein and Krone (1962) experimented with a clay settling from different initial concentrations in a 1 l cylinder. The volume of the suspension which is equivalent to the height of the interface, was plotted against $1/t$. From the pairs of straight lines from each test, the density of the settled deposit at the end of the settling phase and at the end of the consolidation phase were calculated as $1.11 \, \text{g cm}^{-3}$ ($167 \, \text{g} \, \text{l}^{-1}$) and $1.19 \, \text{g cm}^{-3}$ respectively. They also deduced that there was a reduction in inter-floc porosity of 0.44 during the consolidation phase.

Other examples of curves of settling of the upper surface of suspensions are given by Migniot (1968). Though obtained in still water the above results may represent the first stages of bed formation over a tidal slack water period, and in docks and harbours.

Experiments have been carried out in flumes to examine the deposition from a flowing suspension (e.g. Einstein and Krone, 1962; Partheniades, 1965). These generally entail achieving an equilibrium suspension, a suspension not producing deposition or erosion, then reducing the velocity, and monitoring the changing mean concentration in order to calculate the net deposition. There are difficulties about generalizing the results, however, since the suspensions are recirculated through pumps and it is possible that considerable flocculation and deflocculation takes place. Additionally the turbulence intensities may not be comparable between experiments even when the bed shear stress may be. Increasingly deposition and erosion studies are being undertaken in annular flumes, with the water movement being created by rotating paddles. Potentially there are problems here too, with brief instants of high turbulence and shear stress associated with the passage of each paddle. Nevertheless it has been found that there is a critical velocity above which clay stays in suspension and another below which clay is deposited. Between these two neither erosion nor deposition occurs. The deposition of flocs appears to be controlled by the near bed turbulence (Mehta and Partheniades, 1975). The turbulent shear controls the size and strength of the flocs, so that it is only a floc which is strong enough to withstand these forces which will settle onto the bed. Others will be disrupted and re-entrained into the body of the flow. Consequently the steady state suspension contains those flocs that are too small and weak to survive the near-bed shears.

Einstein and Krone (1962) show that at low flow velocities and with initial sediment concentrations less than 300 ppm ($0.3 \, \text{gm} \, \text{l}^{-1}$) the suspended sediment concentration falls exponentially with time, but with fall velocities equivalent to a Stokes diameter of $1.9 \, \mu\text{m}$. Consequently it appeared that, for these experiments, little flocculation was occurring. The concentration variation could be represented by the relationship

$$\frac{C}{C_0} = \exp\left[-\frac{ptw_s}{h}\right] \tag{8.6}$$

where p is a probability of a particle reaching the bed, w_s is the settling velocity, and h the water depth. From their data $p = (1-1.67\,\tau)$ which implies that no deposition occurrred when τ exceeded $0.6\,\text{dynes cm}^{-2}$. Consequently $p = (1 - \tau/\tau_d)$ where τ_d is a critical shear stress for deposition.

Differentiation of Equation 8.6 gives the deposition rate at the bed as

$$\frac{dm}{dt} = Cw_s \left(1 - \frac{\tau}{\tau_d}\right) \qquad (8.7)$$

where m is the mass deposited.

Equation 8.7 is a parameterization of the deposition rate which is widely used in mathematical models of mud transport, e.g. Ariathurai and Krone (1976), Cole and Miles (1983).

With an initial concentration of $20\,\text{gm}\,l^{-1}$ (20,000 ppm) Einstein and Krone found little deposition occurred initially as sediment cover formed on the floor of the flume and flocculation progressed. Then a rapid log-decrease in suspended concentration occurred until a concentration of $10\,\text{gm}\,l^{-1}$ below which the rate decreased. This lower rate was believed to occur when flocs settled independently to the bed, i.e. when the flocs were in the free settling regime. The limiting value seems to coincide reasonably well with the maximum settling velocity obtained by stationary settling experiments (e.g. Figure 8.5) even though a small amount of shear could aid settling by helping the flocs jostle past each other.

Within settling suspensions, in the laboratory at least, a degree of sorting of grain size has been noted (Been and Sills, 1981). Samples taken within the settled bed showed that nearly all the coarse and medium silt sizes were near the base of the deposit, whereas at the top the sample was 87 per cent clay. Consequently, despite the flocculation, a graded deposit could still be formed. This is often visible in muds thought to be formed from high concentration suspensions (Figure 2.13) (Kirby and Parker, 1981).

In nature there is often likely to be a concentration gradient in suspensions with concentration increasing towards the bed. During a decreasing flow the downward flux of settling mud increases the concentration and consequently the settling velocity. The downwards flux is the product of the concentration and the settling rate, and it reaches a peak at about $20,000\,\text{mg}\,l^{-1}$. At low concentrations there will be a gradually increasing flux of particles right to the bed. At higher concentrations, however, sediment can settle towards the near bed layers faster than it leaves them and a thick layer can build up with a concentration of about $20,000\,\text{mg}\,l^{-1}$. However, this would be unlikely unless the concentration of mud in the suspension throughout the whole water column initially exceeded about $2,000\,\text{mg}\,l^{-1}$.

Obviously with the higher concentrations difficulties are experienced because the suspensions are non-Newtonian and the settling becomes a complex function of the shear and the mineralogy. Even though the concentration near the bed increases with time, the near-bed zone is one of high velocity shear and this could lead to shear thinning and reduction of relative viscosity close to the bed. Consequently the high concentration near-bed layer can potentially flow along as a 'slug', maintaining its integrity for long periods and mixing little with the flow above, because of the high density contrast with the overlying water. Additionally because of the relatively high densities in these suspensions, they can move downslope under gravity as density currents.

Some aspects of the movement of cohesive suspensions have been examined by Gust (1976). He found that in clay suspensions the thickness of the viscous sublayer was increased by a factor of 2–5 and the friction velocity was reduced by as much as 40 per cent. For similar suspensions of quartz particles these effects were not seen, and the results were explained as being caused by agglomeration of clay particles. Also Gust and Walger (1976) have reported drag reduction at concentrations as low as $150 \, \text{mg} \, \text{l}^{-1}$. However, as discussed in Chapter 6, drag reduction appears to be a fundamental effect in suspensions, and further comparison between muddy and non-cohesive suspensions are required to resolve these discrepancies.

Consolidation

During bed formation flocs are continually being deposited on the bed surface, while the buried flocs are consolidating. The increasing rate of bed thickness is therefore the net rate of growth, the rate of deposition minus the rate of consolidation. Bed thickness does not begin to reduce until the rate of deposition is less than the rate of consolidation.

Within the settled deposit changes of density with depth are inevitable. Once the deposit is more than only a few flocs thick, the weight of overlying flocs causes those beneath to progressively collapse. The squashing causes a slow expulsion of pore water and a gain in shear strength of the mud. Partheniades (1965) has described this process as occurring in a number of stages (Figure 8.8). A freshly-deposited bed has a highly honeycombed structure with a large void ratio. At the bed surface flocs are grouped into floc aggregates which also join together to form aggregate networks. When the sediment is gradually consolidated, the weakest bonds will be broken first, or will deform, and the network will gradually collapse leaving the floc aggregates as the basic structural element. Next, the aggregate bonds become broken and the deposit will be formed of flocs with density equal to the original floc density but with voids between them. Further consolidation will cause the interfloc voids to disappear so that the entire sediment becomes equal in density to that of the flocs. Additional loading then requires fundamental re-arrangement of the particles, one between another, and breakdown of the integrity of the flocs.

Obviously this is a rather idealized picture, but it does describe some of the essential features of the consolidation process. The escaping pore water can cause 'piping' within the sediment so that the water, instead of escaping uniformly upwards, travels sideways towards zones of higher permeability. These develop into thin pipes which join together and at the surface appear as small fairly evenly spaced holes. Piping can lead to increased consolidation rates.

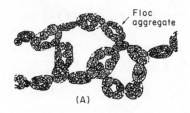

(A)

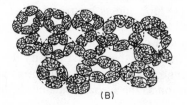

(B)

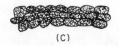

(C)

Figure 8.8 (A) Diagrammatic representation of network structure of a mud bed at its loosest state. (B) Aggregate structure of the bed, with network structure having collapsed. (C) Flocculate structure of the bed, with aggregate structure having collapsed. *From Partheniades, 1965, Proc. Amer. Soc. Civil Eng., 91, HY1, 105–139. Reproduced by permission of American Society of Civil Engineers*

Since we are dealing with a settled sediment it is sometimes convenient to consider density or moisture content rather than concentration. Conversion can be achieved using Equations 8.8 and 8.10.

The bulk density of the sediment ρ_m will be

$$\rho_m = \rho + \frac{C}{1000} \frac{\rho_s - \rho}{\rho_s} \qquad 8.8$$

where C is in $gm\,l^{-1}$. For quartz grains of density $2.65\,gm\,cm^{-3}$, Equation 8.8 becomes

$$\rho_m = 1 + 0.00062\,C \qquad (8.9)$$

The moisture content of a sediment is the weight of water in a sample divided by the dry weight of the sediment. This is given by

$$\text{Moisture Content \%} = \rho \left(\frac{1000}{C} - \frac{1}{\rho_s} \right) 100 \qquad (8.10)$$

For quartz grains this reduces to Moisture Content $= 10^5\,C^{-1}$—37.7.

Within a settled deposit formed as shown in Figure 8.7, there will be increasing density with depth. The surface of the deposit will maintain a constant density, but because consolidation can only proceed at a rate determined by the rate at which pore water can escape, the density at any depth below the surface gradually rises with time towards its final value, as shown in Figure 8.7. Initially during the bed formation phase the mean bed density changes very slowly indeed, but after a few hours rapid consolidation begins and the mean density increases with the logarithm of time. During the formation phase a wide range of densities is present in the bed even though the mean density is constant. At the surface densities are as low as a third the mean, and three times the mean occurs at the base. The surface density decreases quite rapidly with increasing initial suspension concentration. Owen (1970b) quotes figures of $55 \, g \, l^{-1}$ for $1000 \, mg \, l^{-1}$ to $20 \, g \, l^{-1}$ at $8000 \, mg \, l^{-1}$ and $16 \, g \, l^{-1}$ at $16000 \, mg \, l^{-1}$. This agrees with an increasing floc flux to the bed and a consequent reduction in flux of escaping pore water, as well as possibly implying that higher density flocs are deposited at low concentrations than at high. At the highest concentration the surface of the bed has the same concentration as the suspension, and the bed as such may cease to exist.

Migniot (1968) suggests there are three stages of consolidation, and Owen (1970b) found the lowest two stages separated at a density of about $120–150 \, g \, l^{-1}$. During the first stage lines of equal density propagate upwards through the bed at a rate inversely proportional to their density. At higher densities the variation is approximately inversely with the cube of the density. The development of density structure in a mud bed has been considered theoretically by Kynch (1952), but his approach considers the flocs as incompressible. A more complete theoretical approach is outlined by Been and Sills (1981). Figure 8.9A shows an example of density variation with depth achieved after consolidation is complete.

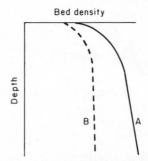

Bed density

Depth

Figure 8.9 Variation of density with depth in the bed for (A) Normal consolidation (B) Under consolidation

When the rate of sedimentation is slow enough for the expulsion of pore water and the consolidation process to keep pace with it, the mud becomes normally consolidated. With a higher sedimentation rate the water cannot completely escape and some of the load from the overlying sediment is borne by the pore water, rather than entirely on the sediment particle framework. These sediments are then underconsolidated (Figure 8.9B). Since the shear strength of the sediment is largely a function of the moisture content, as well as composition, underconsolidated muds are weak. Moisture content and shear strength are virtually constant with depth and the sediments are prone to become unstable on quite low angle slopes. On the Mississippi delta front the sedimentation rates are about 30 cm y^{-1}, the sediments are underconsolidated and would become unstable on $1\frac{1}{2}°$ slopes if their thickness exceeded 25 m (Moore, 1961). With normally consolidated sediments, however, the moisture content decreases exponentially with depth and there is a linear increase in shear strength with depth.

For normal consolidation certain relationships between overburden pressure, shear strength, void ratio and Atterberg limits are predictable (Skempton, 1970). The Atterberg limits are empirical, but effective, indices that compare the physical characteristics of clays, such as plasticity, that are dependent on the mineralogy. The liquid limit is the moisture content at which the mud can be considered liquid, and the plastic limit when it is plastic. For recently sedimented muds liquid and plastic limits may be around 100 per cent and 60 per cent respectively. The difference between the two is the plasticity index and this is related to the percentage of mud finer than 2 μm. With a newly sedimented mud the surface moisture content is much larger than the liquid limit. Skempton (1970) has found that the mean water content of the top 25 cm of a sea bed mud is 1.75 times the liquid limit. At depth the moisture content is normally between the plastic and liquid limits.

Erosion

Experiments in the laboratory and in the field have been aimed at relating a critical erosion velocity, or a critical erosion shear stress, to the properties of the mud, but there are difficulties in defining the threshold of movement. It is sometimes considered visually, but other workers have depended on measuring the increased concentrations in the water. Additionally many workers have not carried out sufficiently exhaustive measurements of the physical properties of the mud for comparison to be made. Nevertheless for any one particular mud the critical shear stress for erosion exceeds the critical shear stress for deposition.

In a flume containing a normally consolidated muddy bed sediment, an increasing velocity will eventually cause erosion of the topmost sediment. If the velocity is increased in small increments, the erosion will be manifest as an increase in the suspended solids concentration. Initially the concentration increases rapidly, but then gradually levels off to a constant value. A further increase in velocity will cause a further increase in concentration. This sequence

220

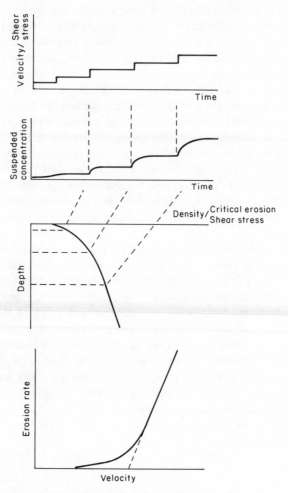

Figure 8.10 Schematic representation of erosion of a
mud bed with an incremental increase in velocity

is interpreted as the sediment being eroded down to a level at which the strength
of the sediment is sufficient to resist the shear (Figure 8.10). Since the strength
increases with depth then continual erosion will only occur when the shear stress
is considerably higher than the critical erosion shear stress of the mud. Consequently a plot of erosion rate against velocity, or shear stress, gives a rate that
increases approximately exponentially with increasing velocity (Figure 8.10).
The critical velocity obtained by extrapolating the high rate curve back to the
axis decreases with increasing water content, increases with increasing salinity
(Gularte *et al.*, 1980), and depends on mineralogy (Fukuda and Lick, 1980).
Mehta *et al.* (1982) also point to the differences between the erosion of
newly-deposited muds, and those that have undergone a degree of consolidation. The literature does not often distinguish between the two very clearly,

but that on consolidated beds is the more extensive, e.g. Einsele *et al.* (1974), Kamphuis and Hall (1983).

Partheniades (1965) considers that at low shear stresses individual flocs are removed from the bed, but at higher values lumps of the bed are torn away. He found that the erosion rates of San Francisco Bay mud depended on the average shear strength. Though the minimum eroding shear stress was approximately 1.0 dynes cm^{-2}, erosion increased very rapidly after a critical shear stress of between 4.8–13.4 dynes cm^{-2} had been exceeded.

Migniot (1968) examined the erosion characteristics of seven different clays and found that the critical erosion shear stress τ_e was related to the residual yield stress τ_B. For $\tau_B < 15$ dynes cm^{-2} he found the results could be expressed at $\tau_e = \tau_B^{\frac{1}{2}}$. Since for his experiments $\tau_B \propto C^4$ this implies that $\tau_e \propto \rho_m^2$. For $\tau_B > 15$ dynes cm^{-2} (~ 300 gm l^{-1}) $\tau_e = 0.25 \tau_B$, and the critical erosion friction velocity $u_{*e} \propto \rho_m^2$.

Experiments on Avonmouth mud have been reported by Owen (1975). He found that τ_e varied slightly depending on bed density and on salinity. More significantly, increasing bed density or increasing salinity both decreased the rate of erosion for a given shear stress. He wrote the erosion rate as

$$\frac{dm}{dt} = M_1 (\tau - \tau_e) \qquad (8.11)$$

where m is the mass of sediment eroded per unit area.

Owen obtained values of the constant M_1 varying from 1.07 to 2.04 gm dyne^{-1}s^{-1} for high and low densities respectively. For a given density more saline muds were more resistant to erosion and the value for M_1 decreased from 1.07 to 0.31 for high salinities. Linear extrapolation of curves of rate of erosion against shear stress gave critical erosion shear stresses of 9.3–11.4 dynes cm^{-2} at which erosion rates were zero.

Sheng and Lick (1979) have used Equation 8.11 in a model of suspended sediment transport in a lake. From laboratory measurements they obtained

$$\frac{dm}{dt} = 1.33 \times 10^{-6} (\tau - 0.5) \qquad \tau \leqslant 2 \text{ dynes cm}^{-2}$$

$$= 4.12 \times 10^{-6} (\tau - 1.515) \qquad \tau > 2 \text{ dynes cm}^{-2}$$

Other results have been presented by Thorn and Parsons (1980).

Equation 8.11 has also been written in a slightly different form using the normalized excess shear stress. This gives

$$\frac{dm}{dt} = M_2 \left(\frac{\tau}{\tau_e} - 1 \right) \qquad (8.12)$$

This formulation has been used in mathematical modelling by Odd and Owen (1972) and by Ariathurai and Krone (1976).

The coefficient M_2 in Equation 8.12 has the units of an erosion rate (mass per unit area per unit time) and varies from one mud to another, as well as with other factors such as temperature and the presence of organic matter. Ariathurai and Arulanandan (1978) have investigated the relationship between M_2 and cation exchange capacity, sodium adsorption ratio, pore fluid concentration and temperature. Values were generally in the range $M_2 = 0.005$ to 0.015, but varied particularly steeply with temperature, being greater at high temperatures.

As can be seen from Figure 8.10, an exponential relationship between the erosion rate and the bed shear stress may be more appropriate than two linear ones. This has been shown by the results of Gularte et al. (1980) and Fukuda and Lick (1980), and by the different formulations for the erosion rate which have been reviewed by Mehta et al. (1982); they present their own results in terms of an exponential dependence on the normalized excess shear stress. Consequently there seems to be little agreement on the best parameterization for the erosion rate, and this makes attempts to obtain unifying relationships to the physical and chemical characteristics of muds rather confusing.

Peirce et al. (1970) used a small annular flume on tidal mudflats to measure the erosion of natural muds and found that though the muds scoured differently there was little difference between any one mud in its natural state and the same mud remoulded in the laboratory, despite salt water and tap water being used respectively. General erosion of the bed occurred when shear stresses exceeded 16–160 dynes cm^{-2}. They did not measure the shearing properties or the density of the mud, however, and the results are somewhat at variance with those discussed above. Also Raudkivi and Hutchinson (1974) have examined mud erodibility in a flume and found no relationship with either zeta potential or with cation exchange capacity of the muds. This agrees with Ariathurai and Arulanandan (1978) who found little change of erodibility with cation exchange capacity above a value of about 10 milliequivalent per 100 gm.

Three muds with different and reasonably well-characterized properties have been tested under the same conditions by Thorn and Parsons (1980). The three muds had slightly different cation exchange capacities but significantly different clay mineral and organic contents. During the erosion tests, since the profiles of density within the bed were known, the shear stress just causing erosion at each level was compared to the density at that level. The results for the three muds could be represented by $\tau_e = 5.42 \times 10^{-5} \, C^{2.28}$ where C is the exposed concentration or dry surface density (g l^{-1}). The majority of the results were for $\tau_e < 10$ dynes cm^{-2} and, consequently, the relationship is similar to that of Migniot. Thus the results are not as dissimilar as one would have expected from the range of mud properties, and a correlation between τ_e and C may be a useful one for simply parameterizing the erosion characteristics of muds.

An alternative method of assessing erosion rates in the sea has been used by Lavelle et al. (1984). Currents and concentration time series were measured 5 m above the bed. From the currents a bed shear stress time series was calculated and, via a model of suspended sediment diffusion, comparison between observed and calculated sediment concentration made. From this, the erosion rate was calculated as $E = \alpha\tau^4$, where α had a value of 1.5×10^{-6} gm cm^{-2}s^{-1}. However these results were arrived at purely empirically, and comparison is needed with laboratory measurements and, ideally, also with absolute *in situ* erosion rate measurements.

In summary the thin surface layer of loosely held flocs appears generally to erode at a shear stress of about 1 dyne cm^{-2}. Once this has been suspended, the underlying deposit generally has a critical erosion shear stress depending on the square of the density, with a surface value of about 10 dynes cm^{-2}. However it is necessary in individual cases to carry out laboratory tests on representative samples to establish general relationships that can be used in prediction.

A number of authors have attempted to incorporate the results of experiments on the threshold of movement of cohesive sediments into a Shields type curve. Hjulstrom produced results in 1935 which were restated by Inman (1949) and are shown diagrammatically in Figure 8.11. The threshold rises with decreasing grain size below about 200 µm. Postma (1967) and various other authors have pointed out that the deviation from the curve for non-cohesive sediment varies with moisture content, with well-consolidated muds lying furthest from the Shields curve. However, though density correlates with moisture content, there is not a clear relationship between density and grain size. Consequently use of such a threshold curve for cohesive sediments is not to be recommended.

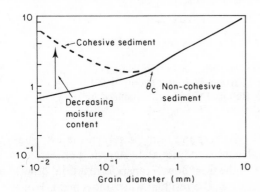

Figure 8.11 Threshold curve for muddy sediments. The dashed line is only approximate since the actual threshold for mud depends on mineralogy, concentration etc.

A very useful review of the state of knowledge of cohesive sediments pertinent to modelling was published by Owen (1976), and it seems that not a great deal of progress has been achieved since then.

TRANSPORT OF MUD IN TIDAL CURRENTS

In laboratory experiments the bed erodes to a level where the threshold stress equals the bed shear stress. At this stage there is an equilibrium suspended concentration. With a further increase in shear stress an increased equilibrium concentration is produced. Eventually continuous erosion occurs. When the shear

224

stress is reduced deposition occurs for a limited time, and the concentration falls to a new equilibrium value. There is, however, a considerable difference between the two curves (Figure 8.12). In a tidal current there will be shear stress fluctuations about the mean value, but at low shear stresses since the rate of deposition, shown by the gradient of the deposition curve, exceeds that of erosion, any material eroded by high shear stress events is deposited during the lower stress intervals. At intermediate shear stresses the rates are about equal and a continual exchange is possible around a mean bed level. At high shear stress erosion dominates. The difference between the shear stresses at which continuous deposition and erosion occur is known as scour lag.

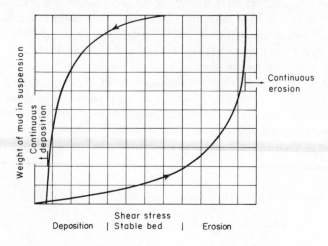

Figure 8.12 Curves representing erosion and deposition in a continuously changing current. *After Owen, 1975,* Hydraulics Research Stn Report INT 150. *Reproduced by permission of Hydraulics Research, Wallingford*

In a deep flow the suspension will not respond immediately in the way suggested above, and settling lag will cause additional hysteresis between the accelerating and decelerating stages of the tidal flow. Transport in a tidal flow is normally assumed to be in the direction of the residual water movement. However there are differences between the velocities at which sediment is eroded and deposited, and delays caused by settling lag. Consequently this assumption need not always be correct, and may in fact seldom hold. It is normally very difficult to prove one way or the other, because the residual suspended sediment transport is the small difference between large flood and ebb values, each with very large errors.

In many shelf areas there is a background turbidity, a concentration of suspended clay particles derived from eroding relict deposits, or from coasts and rivers. These are deposited as muds in often quite localized areas, generally of low maximum currents, and with convergent residual currents which provide a continuous supply of sediments. In one such area in the German Bight there is an incompatibility between the estimated deposition rate of about $2 \, mm \, y^{-1}$, and deposition calculated from observed turbidity values of between $2–10 \, mg \, l^{-1}$ (McCave, 1970). With the conventional view of some of the particles settling to the bottom at slack water and not being eroded at the next maximum current, the amount of sediment deposited in a half hour period each slack water, gave only a quarter of the estimated deposition rate. However McCave (1970) was able to reconcile the discrepancy by postulating a more continuous deposition whereby particles were exchanged by the bursting process through the buffer layer and into the viscous sublayer, where they were deposited virtually throughout the tidal cycle.

An alternative model is that the mud deposited at slack water has a chance to consolidate slightly before the current becomes strong enough to re-erode it. Because of the slightly enhanced resistance to shear at the base of the layer not all of the particles are eroded, and a small increment of deposition occurs. Terwindt and Breusers (1972) found that a 2 cm mud layer containing 37 per cent sand, consolidated for two hours could withstand a stress of 4 dynes cm^{-2}, a value higher than that for the erosion of sand. On the other hand, consolidated mud layers of 10–25 mm thickness are often observed even though maximum deposition of only 3 mm thickness of unconsolidation mud can occur during one slack water. Thus the deposits must be the result of several slack water periods without intervening erosion. Gust and Walger (1976) postulated that this would be assisted by the drag-reducing effect of the suspensions reducing the magnitude of the maximum shear stress and preventing re-erosion of the deposits. This is unlikely to occur in the low concentrations McCave discussed from the German Bight, however.

In estuaries there can be high concentrations of mud in suspension, involving the formation at times of layers with sufficiently high concentrations to become non-Newtonian. The large variations in turbulent energy during the tide causes similarly large variations in the suspended sediment concentration and this results in three basic forms of mud occurrence (Kirby and Parker, 1983):

1. Mobile suspensions. Where the suspension is moving freely under the tidal forces or downslope under the influence of gravity. The floc sizes are likely to be in balance with the shearing stresses.
2. Stationary suspensions. The suspensions are not moving horizontally but may be gradually settling. The floc sizes are likely to be increasing with time due to settling.
3. Settled mud. Mud forming part of the sea bed, resisting erosion for considerable periods and gradually consolidating. The overburden weight is predominantly supported by interparticle contacts.

During the tidal cycle, or the lunar spring–neap cycle, as the velocity of the current and consequently the shear stresses vary, the mud will preferentially appear in one or other of the above states. Based on extensive measurements in the Severn Estuary, Parker and Kirby (1977) have proposed the interlinkage shown in Figure 8.13. A crucial factor in the processes is the development of

226

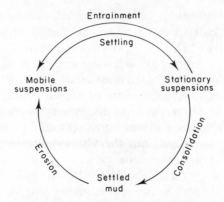

Figure 8.13 Schematic representation of
the states of occurrence of mud suspen-
sions, and the links between them. *From
Parker and Kirby 1977,* Proc. 2nd Int.
Symp. Dredging Technology, *B2-15–26.
Reproduced by permission of BHRA*

layering within the suspension. Over a single tidal cycle on spring tides a vertically homogeneous profile is present at maximum ebb or flood current with concentrations of 2–5000 mg l^{-1}. As the current diminishes, the load of sediment
cannot be supported and that near the surface commences settling. This forms a
sharp step, or 'lutocline', in the profile and this gradually gets closer to the sea
bed while the concentration beneath it gradually rises, rather like the hindered-
settling process in the laboratory. All the while the suspension would be still
moving along the estuary. Near to slack water the high concentration near-bed
layer, becomes stationary for a brief period. Concentrations of up to 60 gm l^{-1}
have been measured at this stage near the bed. As the current accelerates on the
opposite phase of the tide, erosion of the upper surface of the layer may occur
and the gradual re-entrainment will cause the lutocline to rise through the water
column towards the surface again. Alternatively, it is possible that the stationary suspension may be moved as a slug along the bottom for a while before
re-entrainment, depending on its rheological properties.

As the tidal range gradually diminishes towards neap tides the static suspensions formed at slack water survive progressively longer into the succeeding
tide, until they eventually resist erosion and remain throughout the complete
tidal cycle (Figure 8.14). The suspension then has several tidal cycles in which to
gradually consolidate. On the succeeding neap to spring cycle re-erosion of this
layer may not be complete, and a small increment of deposited sediment may
remain to become part of the seabed for a longer timescale. Though it has not
been proved, it is feasible that somewhat thicker units encompassing some centimetres of such a layer could be preserved in the settled mud of the bed. As the
tidal range diminishes the depositional and intermediate stages encompass a
greater proportion of the tidal cycle, giving a predominance of deposition over
erosion. Because of the consolidation that occurs, and settling lag in the water

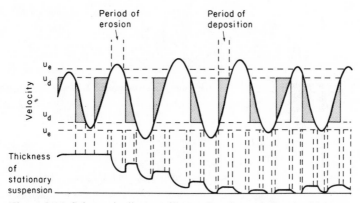

Figure 8.14 Schematic diagram illustrating the variation in thickness of stationary suspensions during part of a spring–neap cycle. u_e—threshold of erosion; u_d—threshold of deposition. Between the u_e and u_d no deposition or erosion takes place

column, there is likely to be a phase lag between the thickness of the stationary suspension and the tidal range.

At some stage during the development of the layering the interface can be detected by echo-sounders as indistinct 'ghost echoes'. The layers are then termed fluid mud, or fluff. Often several echoes can be distinguished within the fluid mud and comparison with the results of radioactive densimeters has shown that they correlate with density layering, with values ranging between 1.05 and 1.30 gm cm^{-3} (Figure 8.15) (Kirby and Parker, 1974). It is not certain what characteristics of the upper surface of the high concentration layer determines its detectability by echo-sounding. Both the vertical density and velocity changes are important in reflecting the sound signal, but, at high frequencies, so is the backscattering of energy by organic or other particles associated with the interface. Consequently layering may be present without necessarily being acoustically detectable, and detectability may vary with the frequency of the echo-sounder (Parker and Kirby, 1982a). Fluid muds are frequently observed in estuaries and ports and have been reported from the Gironde, the Loire, the Thames, the Severn, the Scheldt, from Baltimore Harbour and many others. Thicknesses of 2–4 m are common.

Settled mud may be gradually accumulating with time, in which case it is normally consolidated, or may be gradually eroding when it would show overconsolidation. The mud density is normally about 1.3 to 1.7 gm cm^{-3}. Settled muds show a complex internal structure of very fine laminations, often to sub-millimetre scales, but sometimes with thicker homogeneous units. Echo-sounding or seismic profiling also shows layering within the mud which tends to follow the contours of the underlying solid rock. However these layers depend on the acoustic properties and the degree of consolidation, which need not relate directly to the sediment layering. Because of their fairly high organic content settled muds often contain zones of gas bubbles which show up distinctly on seismic profiling records because of their acoustic blanketting effect (Schubel and

228

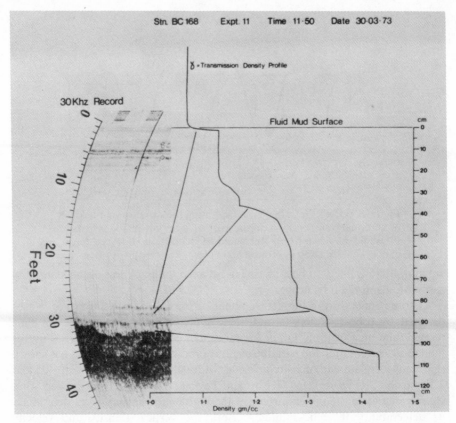

Figure 8.15 Gamma ray attenuation profile of density compared with a 30 KHz echo-sounder record of fluid mud. *From Kirby and Parker, 1974,* Dock and Harbour Auth, *54, 423–424. Reproduced by permission of Foxlow Publications Limited*

Schiemer, 1973). The bubbles can affect the consolidation characteristics of the muds, since they are compressible and will vary in size during the tidal cycle. This will cause stressing of the fabric of the sediment near the bubbles and a degree of pumping of pore water throughout the sediment above them.

There is an extraordinary dearth of information of the effect of waves on cohesive sediments. Migniot (1968) passed waves over mud beds of various viscosities and found that the orbital movement continued into the mud, but it was attenuated rapidly, being very small below a few centimetres depth except for the lowest viscosities. Displacement velocities of up to $0.5 \, \text{cm} \, \text{s}^{-1}$ were measured at the mud surface.

At some stage the shearing resistance of the clay will be exceeded and sediment will move into suspension. Thimakorn (1980) has shown that the near-bed concentration C_b normalized by the depth mean concentration $\hat{C}$ was related to the maximum wave induced shear stress by

$$\frac{C_b}{\hat{C}} = 1.227 \, \tau^{0.575}$$

where τ is given by Equation 3.57. However, the correlation was not very good, and further work obviously needs to be done.

Wells and Coleman (1981) describe field experiments in a muddy area off the Surinam coast. Because of the high turbidity and viscous effects the incoming waves were modified into solitary waves. The forward velocities under the crests, as well as the pressure forces, caused a steplike increase in turbidity with each passing wave (Figure 8.16), and the material was preferentially carried shorewards due to the wave motion. Unfortunately the measurements did not allow quantification of the effects.

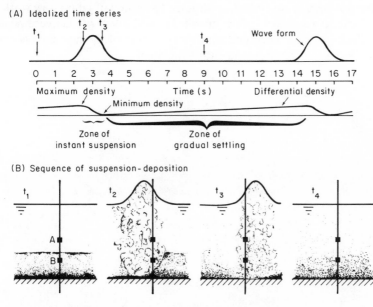

Figure 8.16 The suspension of mud by solitary waves. (A) Time series of density difference between two sensors A and B related to surface wave form. (B) Interpreted sequence of suspension–deposition. *From Wells and Coleman, 1981, J. Sediment. Petrol., 51, 1053–1068. Reproduced by permission of the Society of Economic Paleontologists and Mineralogists*

It should be fairly apparent from this chapter that the general state of knowledge of cohesive sediments is primitive compared with that of sand. The erosion of the mud is not simply a function of grain size, but depends on several chemical, biological and physical parameters, as well as varying with time. Flume studies have shown that prolonged sub-threshold flow over a mud bed tends to 'harden' the bed and make it less susceptible to movement. Additionally, whereas erosion of sand is a continuous process with each layer having the same threshold as that above, with mud the erosion resistance increases into the bed. Because of the non-Newtonian response of the bed the movement of cohesive sediment under waves could be fundamentally different from that of sand. One interesting problem that seems to have received little attention so far is the movement of mixed sediment. Muddy sands may contain only a small proportion of clay and yet act as a cohesive sediment at the threshold. However once the sediment is moving, the sand particles may travel as bedload. Mixed sediments are very common in coastal areas and estuaries.

Almost all of the research has been carried out in laboratories because of the difficulty of field work. Yet even so adequate quantification of the processes of flocculation, settling, consolidation, erosion and deposition has not been achieved, largely because the studies have been carried out without adequate characterization of the physical and chemical properties of the materials, and tests have not been in controlled conditions. Extension of laboratory work into the field will be necessary before knowledge will allow predictive modelling. Nevertheless, provided adequate laboratory testing of the materials is carried out, many situations can be empirically simulated.

CHAPTER 9

Estuarine Sedimentation

Within estuaries the river water is mixed with the sea water by the action of tidal motions, by waves on the sea surface and by the river discharge forcing its way to the sea. The tidal rise and fall governs the magnitude of the oscillatory currents, though high river discharge can have a considerable effect in modifying them in the upper estuary. The residual currents, however, tend to be dominated much more by the horizontal and vertical density differences between the river water and the sea, and these depend on the mixing processes. The difference in salinity between the river and sea water is about 35‰ which creates a density difference of about 2 per cent. Though small it is sufficient to create appreciable residual flows. Temperature differences normally being small are relatively less important in controlling the density distribution.

Salinity is readily measurable and makes a good tracer for the patterns of water movement and is an indicator of the intensity of the mixing processes. Because of the variations in tidal currents and in river discharge between estuaries of the same topographic form, they are likely to be classified into different types when salinity structure and the residual circulation of water is considered. The different estuarine types consequently have different sedimentation regimes. Classification of estuaries and their detailed physical characteristics are dealt with by Dyer (1973) and Officer (1976); only a brief outline is necessary here.

ESTUARINE TYPES

Salt Wedge Estuaries

If the river discharges into an almost tideless sea, the fresh water, being less dense than the sea water, will tend to flow outwards over the surface of the salty water that will rest on the bottom as an almost motionless salt wedge. The wedge occupies almost the whole water depth near the mouth, but will taper and thin towards the head of the estuary. Between the two water masses there will be

232

a narrow zone on the upper surface of the wedge with a very sharp salinity change, called a halocline. Because of the density gradient the halocline will tend to be rather stable and the two water masses will not mix together very readily. Because there is very little tidal flow, the bottom salty layer will be relatively stationary, but the river water will be flowing over the top of it at quite a high velocity. Consequently, close to the halocline there will be a velocity shear and, like the wind blowing over the sea surface, it will cause waves on the interface. These waves can break, ejecting small amounts of salt water into the more turbulent layer on the surface. This process is called entrainment and can occur anywhere on the halocline where the velocity shear is great enough. The entrainment is of the quiescent fluid into the more turbulent one. No fresh water is mixed into the bottom layer, the mixing is entirely upwards. Thus the bottom layer has a fairly constant salinity along its length but loses salt gradually into the surface layer. This loss must be made good by a slow inflow of new salt water from the sea. A further but more intermittent mixing mechanism can result from larger internal waves travelling on the halocline rather like surface swell waves from a distant storm. When these waves encounter a current moving against them at their own speed of travel they will slow down and stop. In so doing, however, they steepen and break and mix the salt and fresh water. This is a more effective mixing mechanism than entrainment, but is likely to be localized to certain sections of the estuary for relatively short periods.

Consequently the circulation of water and the salinity distribution is likely to be similar to that in Figure 9.1. The position of the salt wedge will vary with river flow. A typical example of a salt wedge estuary is the South West Pass of the Mississippi. When the flow is low the salt wedge extends more than 100 miles inland, but with high river discharge the salt wedge only reaches a mile or so above the river mouth.

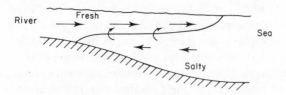

Figure 9.1 Diagrammatic representation of circulation in a salt wedge estuary. *Dyer 1980b, Essays on the Exe Estuary, 1–21, Special Volume, No. 2. Reproduced by permission of the Devonshire Association for the Advancement of Science, Literature and Art*

Partially-mixed Estuaries

When there are any appreciable tidal movements the whole water mass in the estuary moves backwards and forwards, the friction on the estuary bed causes velocity shear, and generates turbulence. The turbulence causes more effective mixing than entrainment, since not only does it mix salt water upwards into the fresher water, it also mixes the fresher water downwards. This dilutes the salt water near the bed and produces a salinity gradient towards the head of the estuary. Because of the two-way mixing the halocline will be much less abrupt than in the salt wedge estuaries, and though entrainment will still be active at the halocline, it will be relatively less important.

In the surface fresher layer there is still a residual discharge of water towards the sea, but now it carries with it the increased content of salt resulting from the enhanced mixing. This salt has to be replaced in the lower layer by a compensating inflow from the sea, which consequently is of much larger magnitude than that in the salt wedge estuary. The enhanced residual flow, caused by the density differences and the mixing, is called the vertical gravitational circulation.

If the cross-sectional area of the estuary is divided by the river discharge it becomes apparent that the mean velocity through the section would be small. For an estuary 2 km wide, of mean depth 5 m and with river discharge of $10 \text{ m}^3 \text{ s}^{-1}$, the velocity would only be of the order of 1 mm s^{-1}. Measurements show, however, that generally at the surface and in the saline layer near the sea bed the mean velocities are of the order $1-10 \text{ cm s}^{-1}$. The ten to one hundred fold difference is the result of the vertical gravitational circulation. These velocities are low compared with the tidal currents which can be of the order of 100 cm s^{-1} or more. Consequently it is difficult to measure them accurately.

The vertical profile of the residual velocity and tidally averaged salinity are shown in Figure 9.2. These 'classical' profiles have been measured in many estuaries. As can be seen there is a downstream flow of fresh water on the surface and an upstream flow of saltier water near the bottom. The residual water flows at the surface are manifest as a slightly stronger ebb current than flood current, with the reverse occurring near the bed. At about mid-depth there is a gentle halocline which coincides with a region of no mean movement, but across which

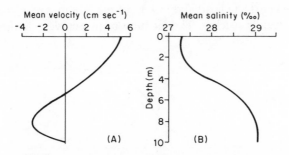

Figure 9.2 Tidal mean profiles in a partially mixed estuary: (A) Velocity; (B) Salinity. *Dyer 1980b*, Essays on the Exe Estuary, *1–21, Special Volume, No. 2. Reproduced by permission of the Devonshire Association for the Advancement of Science, Literature and Art*

234

there is considerable mean shear. Measurements at several places along the estuary show that the landward bottom flow diminishes, and the seaward surface flow increases towards the head of the estuary, and the level of no mean motion gets progressively deeper. At some point near the head of the estuary the level of no mean motion is at the sea bed (generally in the region of 2–5‰ salinity contour). Landwards of this null point the mean flow is seaward at all levels despite the fact that there is an oscillation of water backwards and forwards. This oscillation will even continue into a reach of river where there is no saline water even at high water, providing a weir does not limit the penetration of the tides.

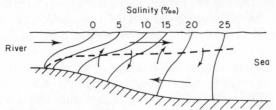

Figure 9.3 Diagrammatic representation of tidal mean salinity distribution and of circulation in a partially mixed estuary. – – – Level of no net horizontal motion. *Dyer 1980b*, Essays on the Exe Estuary, *1–21, Special Volume, No. 2. Reproduced by permission of the Devonshire Association for the Advancement of Science, Literature and Art*

A typical longitudinal salinity distribution is shown in Figure 9.3. The isohalines are much steeper than in the salt wedge estuary, the implied increase in potential energy being derived from the kinetic energy of the tidal flow via the turbulence. The stratification normally tends to increase towards the head of the estuary as the narrowing makes the river flow relatively more important than the tidal flow. Taking the depth mean of the tidally averaged salinity, the distribution along the estuary typically looks like Figure 9.4, with low horizontal salinity gradients near the head and mouth of the estuary and a zone of higher gradient in the middle reaches. The location of this curve will alter with river discharge.

Figure 9.4 Longitudinal distribution of the tidally averaged depth mean salinity in a partially mixed estuary. *Dyer 1980b*, Essays on the Exe Estuary, *1–21, Special Volume, No. 2. Reproduced by permission of the Devonshire Association for the Advancement of Science, Literature and Art*

The intensity of the turbulent mixing also alters during the spring–neap cycle, because in many estuaries there can be a doubling of the tidal range on a fortnightly cycle. The spring tide currents enhance the exchanges of salt and fresh water. The extra amount of salt water mixed from the bottom layer and discharged at the surface is replaced by the compensating landward inflow, the result being a decreased stratification, an enhanced mean circulation and an apparent shift in the mean salinity towards the sea. At times of river flood the partially-mixed estuary will take on many of the properties of a salt wedge estuary providing the relative importance of river flow to tidal flow is altered sufficiently far.

Partially-mixed estuaries are common on the coasts of north west Europe and north east America.

Well-mixed Estuaries

As the strength of the tidal currents increases relative to the river flow, the mixing becomes more and more intense until it is sufficiently strong to effectively mix the water column completely. These well-mixed estuaries are likely to be shallow, with a high tidal range, and probably with intertidal mud flats and banks. Because of the shallowness the vertical circulation present in partially-mixed estuaries cannot develop to any great extent, but there can be lateral variations in salinity. The residual flow of the fresher water tends to be preferentially down one side of the estuary and the saltier water enters the other side. Consequently a horizontal residual circulation is developed (Figure 9.5). Sometimes the ebb dominated and flood dominated parts of the circulation can be separated into topographically distinct ebb-and-flood channels, with a bank between.

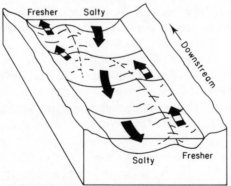

Figure 9.5 Diagrammatic representation of mean salt transport in a wide vertically homogeneous (well-mixed) estuary. *From Dyer (1977). Reproduced from Estuaries, Geophysics, and the Environment by permission of the National Academy of Sciences, Washington, DC*

Much of the mixing in these estuaries is carried out by water being trapped in embayments and in the shallower areas, and slowly bleeding back into the main body of the flow at a later stage in the tidal cycle.

In long shallow estuaries there is an additional factor affecting the water flow. The tidal wave within the estuary will take some time to travel inwards. Consequently at high water, water is still flowing into the estuary to raise the levels further upstream. This results in high water slack current being delayed after high water and low water slack being similarly delayed after low water. Thus at high water more water passes landward per unit velocity than is discharged at low water for the same unit velocity, because of the change in the estuarine cross-sectional area. To ensure that the accumulated water is removed, the peak velocity of the ebb is increased and that of the flood tide decreased. Further into the estuary however, the tidal curve itself is altered by the narrowing of the estuary and by the fact that the tide travels faster in deeper water. Thus the high water travelling into the estuary gradually overtakes the water in front and the tidal curve becomes slightly asymmetrical with a shorter higher velocity flood and a longer low velocity ebb. In extreme situations the asymmetry can develop into a steep fronted bore.

Examples of well-mixed estuaries are the Severn and the Gironde.

Fjords

In fjords, because of their tremendous depth, some of them are over 600 m deep, the effect of tidal flow in the mixing must be negligible. The mixing between the fresh and salt water is consequently carried out by the entrainment process. The salinity of the almost motionless bottom layer will not vary significantly from mouth to head, and the surface fresh water layer will be only a few tens of metres deep. Thus fjords can be considered as salt wedge estuaries with an effectively infinitely deep lower layer. Fjords often have a shallow bar of rock, or a 'sill' near their mouths. Tidal flow over this can cause turbulent mixing within a limited area, and the mixed water penetrates into the basin at intermediate depths. Often the inflow of salt water into the deep basin can be restricted and replenishment of the deeper water can occur only intermittently, sometimes only yearly. The normal circulation pattern tends to a rapid surface outflow and a slower inflow into the top of the saline water.

Estuaries can also be classified by their tidal ranges as: microtidal, with range < 2 m; mesotidal, with range between $2-4$ m; and macrotidal with range > 4 m. These categories often correspond roughly to the salt wedge, partially-mixed and well-mixed estuarine types respectively.

SEDIMENTATION IN A SALT WEDGE ESTUARY

The essential components of a salt wedge estuary are a large river flow and a small tidal range. Allied to the large river discharge there normally is a large discharge of sediment which, when deposited at the river mouth, forms a delta. The topography of the delta depends on the degree to which the tides, coastal currents and waves are capable of redistributing the deposited sediment. When these agents are negligible, elongate 'birds foot' deltas such as the Mississippi are formed. In areas such as the Nile, before the building of the Aswan High Dam, though the tidal range is similar to that off the Mississippi, coastal currents and wave action smoothed the outline and produced the more usual arcuate delta. Now the sediment source in the Nile has been cut off by the dam, the delta-front is undergoing considerable erosion and modification. Additionally, especially in subtropical areas, the river discharge can be seasonally extremely variable. Thus deposits brought down during floods can be redistributed during the rest of the year when discharge may be almost negligible. The mouths of these rivers tend to be very variable in position with extensive lagoon systems often extending behind the barrier beaches. Long spits develop across the mouth during the dry season and they are breached during the floods. During floods there is often no saline water within the estuary, the river debouching directly into the sea.

The idealized deltaic sedimentary deposit has been derived from steady flow studies in lakes and in the laboratory, and consists of a prograding delta-front upon which successive layers of sediment are deposited (Figure 9.6). The layers are thickest on the steep slope of the delta-front where they form the fore-set.

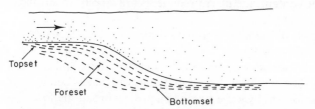

Figure 9.6 Schematic diagram of the development of a small-scale prograding delta front

Each layer thins into shallower and into deeper water, in the top-set and bottom-set beds respectively. Though this structure is seen frequently in small-scale features in the geological column, it is unlikely to form an accurate model of large modern deltas.

By far the best described delta is that of the South Pass of the Mississippi, both in sedimentological (e.g. Russel and Russel, 1939) and dynamical terms (e.g. Wright and Coleman, 1974; Wright, 1978). Many of its general features can be used to illustrate deltaic processes.

The Mississippi mouth is subject to a tidal range of 33 cm and it has a river discharge varying between 840–2800 m^3s^{-1}. The sediment discharge averages about 500×10^6 tons y^{-1}, of which between 10–20 per cent is bedload. The suspended load is typically 40 per cent silt, 50 per cent clay and 5–10 per cent very fine sand. The bedload is mainly very fine sand. Near the mouth there is a continual subsidence of about 2.5 cm y^{-1}, and as the whole delta also subsides at much the same rate, shallow flanking depressions are formed. These do not receive much sediment directly from the river, but their presence causes intermittent shifting of the position of the delta and subdeltas. The delta contains several distributary channels and mouths (passes).

The most fundamental landform is the levee, the high banks confining the river channel. The height of the levee is determined by floods and reaches several metres above the surrounding flood plain. At flood stage, overtopping is fairly general along the crest and sediment deposition occurs on the flanks. Levees are composed of fine sand to clayey silt, coarsest on the crest, where silt or very fine sand occurs. The mean grain size decreases rapidly down the back slope where clayey silt occurs. This is very poorly sorted because of the abrupt decrease in velocity after overtopping. The levees are broken by occasional channels, known as crevasses. These relieve much of the pressure of flood water and distribute sediment over the interchannel areas, but generally fill with sediment on the decreasing discharge stage. Crevasses only become permanent when they are formed at a small angle to the current in the main channel.

The crevasse deposits are very fine sand and coarse silt, generally fairly well sorted, which spread out as fans or splays onto the surrounding marshes. They probably contain a large proportion of the bedload of the river. Within the marshes deposition is slow in the quieter water, mainly comprising poorly sorted fine silts and clays, layered with coarser crevasse-derived sediments. Due to the high organic content, the marsh sediments are largely anaerobic and dark blue to black.

The deposits of the lakes and bays depend on size, exposure to wave action, and distributary distances. Where wave action is active the deposits are extremely well sorted very fine sands. Offshore bars and beaches contain the majority of the coarsest fine sand material, with shells being locally abundant. The bars are formed by wave action and protect broad shallow lagoons, some of which collect very fine jelly-like oozes of clay and organic colloids.

Submarine levees extend for over a mile seaward of the ends of the Passes. These form on the edge of the jet-like river effluent. With continued deposition the levees inhibit expansion of the jet, the flow becomes largest at the edges, and a middle ground shoal develops leading to bifurcation of the channel. The dynamics of the jets have been examined by Wright and Coleman (1974). They conclude that the inertia of the jet is balanced by friction, which causes the less dense water to thin and spread laterally. The initial rapid expansion leads to a reduction in velocity, deposition of sediment and the formation of a bar a few kilometres seawards of the mouth, and to submarine levees.

The position of the salt wedge varies between the various passes depending on their proximity to deep water. For the South West Pass the salt wedge is only flushed clear of the channel during floods and at high discharge during the ebb tide.

During floods the river flow pushes the toe of the salt wedge seawards of the bar crest (Figure 9.7). At that time there is very strong downstream transport within the channel when current velocities exceed 1 m s^{-1} a metre from the bed.

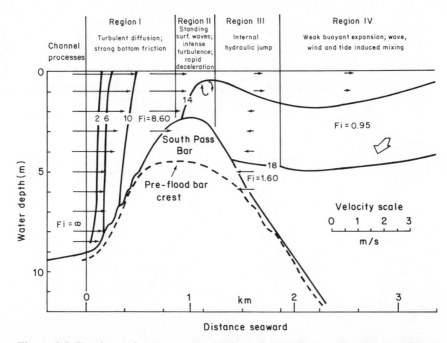

Figure 9.7 Density and current cross-section of the turbulent flood-stage effluent from South Pass. *From Wright and Coleman, 1974,* J. Geol., *82, 751–778. Reproduced by permission of the University of Chicago Press*

This downstream bedload transport causes rapid accretion of poorly sorted sand and silt at the bar crest, since, seaward of the bar, there is a weak landward bottom flow within the salt wedge. The very intense shear at the toe of the salt wedge, together with the small flow depth leads to internal Froude numbers in excess of unity, and an internal hydraulic jump, which causes an abrupt slowing and thickening of the surface flow. Consequently the strong turbulent mixing at the interface near the bar crest must temporarily suspend some of the bedload sediment, and this, coupled with flocculation processes, causes a maximum suspended sediment concentration near the interface. The turbid surface layer is only a metre or so in thickness and the underlying clearer more saline water can often be seen as patches at the surface in the wakes of ships. Seaward of the bar the reduction in velocity as the flow expands laterally causes increasingly fine grains to settle into the lower layer and the weak currents distribute them over a broad fan. Consequently there is a gradual decrease in grain size seaward of the bar.

During low and normal river flow the tip of the salt wedge is well within the mouth of the South Pass, but the lateral expansion and lateral thinning over the bar causes the internal Froude number to increase rapidly to exceed unity with consequent intense vertical mixing (Figure 9.8). Seaward of the bar the surface flow decelerates to subcritical values, and again there is a weak landward flow within the salt wedge.

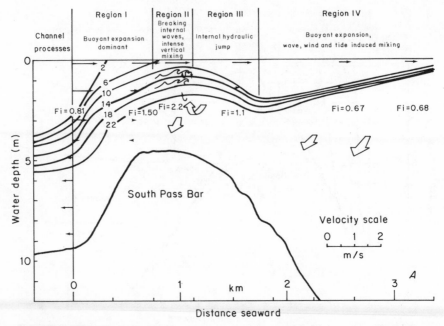

Figure 9.8 Density and current cross-section of the stratified low-stage effluent from South Pass. *From Wright and Coleman, 1974,* J. Geol., *82, 751–778.* Contours are of salinity ‰. Fi is the internal Froude number. *Reproduced by permission of the University of Chicago Press*

There is an appreciable tidal movement of the salt wedge, despite the small tidal range. During ebbing tide the salt wedge is displaced seawards, the thickness and the velocity of the upper layer increasing. The flooding tide does the opposite. The surface flow is still seaward, but less than on the ebb, and some of the impounded river flow exits via the crevasses. The result being that the densimetric Froude number F_i is maintained near to unity at the mouth at all times.

A combination of the normal and flood events leads to the idealized model (Figure 9.9). The bar crest will oscillate backwards and forwards depending on the river discharge, but will gradually move seawards. The sedimentation rate is greatest on the upper slope, and this causes underconsolidation leading to slumping.

The East Pass of the Mississippi is described by Wright and Sonu (1975). The usual picture of a delta containing top-set, fore-set and bottom-set beds finds little support in the Mississippi since subsidence is a major factor. Thus the delta structure has been likened to a pile of superimposed leaves, the veins of which are the levee system. Since subsidence is most rapid near the ends of the distributaries, the delta is a thick lens, thinning towards both land and sea.

THE TIDAL RESPONSE IN ESTUARIES

When there is a significant tidal rise and fall within an estuary, the tidal wave is affected by several factors; convergence of the estuary sides concentrates the

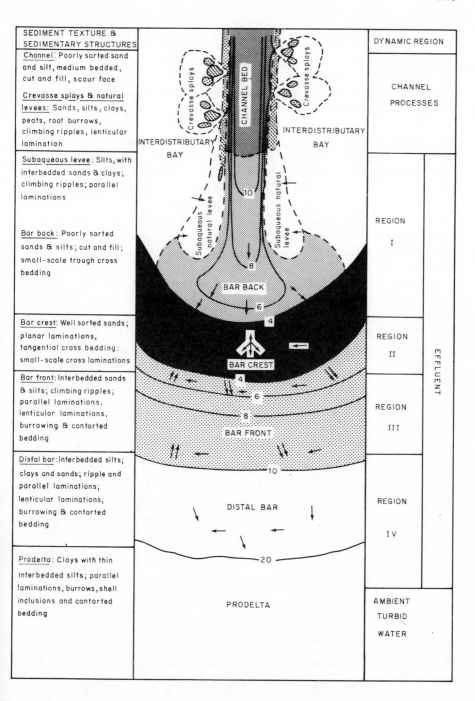

Figure 9.9 Idealized morphodynamic model of highly stratified mouth system. *From Wright and Coleman, 1974,* J. Geol. *82, 751–778. Reproduced by permission of the University of Chicago Press*

242

tidal wave energy into a smaller crest length, the energy is reflected by the sides and head of the estuary, and the shallow water causes frictional dissipation of energy. The former two mechanisms cause an increase of tidal range towards the estuary head, and the last a decrease. The balance of the three is complex, depending on the estuarine depth and topography, and consequently is difficult to predict without fairly detailed numerical models.

In a prismatic rectangular estuary a pure standing wave can occur when the estuary length is equal to an odd number of quarter wavelengths of the tidal wave, and when there is a node with minimum range at the mouth. For the simplest situation with a node at the mouth and an antinode at the head, the length would be $(T/4)\sqrt{gh}$, where T is the tidal period and h the depth. For a depth of 10 m the length would be about 112 km. The peak tidal velocities would be maximum at the mouth and minimum at the head, with a $\pi/2$ phase lead over the elevation, i.e. the maximum flood current would occur midway through the rising tide. High water and low water would occur simultaneously throughout the estuary.

Because of the tidal conditions in the neighbouring seas, estuaries do not normally have a node at their mouths, but form co-oscillating components of the offshore tidal system. They show most of the characteristics of a standing wave response, but with some modification caused by friction. Friction leads to a slight progressive element, with high water and low water occurring later towards the head of the estuary (Figure 9.10). These delays are commonly of the

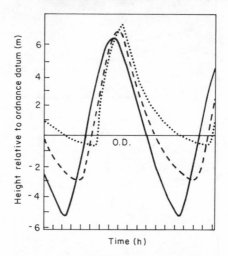

Figure 9.10 Progression of a tidal wave in a macrotidal estuary (Severn Estuary), showing delay in times of high and low water, and development of tidal asymmetry. *After Glen, 1979,* In, Estuarine Hydrography and Sedimentation. *Reproduced by permission of Cambridge University Press*

order of a minute or two per kilometre, and involve a delay of the slack water after the maximum and minimum elevation. On the flood tide, for instance, since the water level is still rising landward of a particular position at high water, a flow of water is necessary through the cross-section, leading to slack water being delayed after high water.

Friction also causes the crest of the tidal wave to travel faster than the trough. Consequently, especially in macrotidal estuaries, the flooding tide rises quickly and falls slowly, giving a long ebb and saw-tooth tidal curve (Figure 9.10). The flood currents are thus of higher velocity than the ebb currents. The asymmetry becomes more marked towards the head of the estuary and can eventually give rise to a bore, when the rise during the early flood becomes dramatic.

The covering and uncovering of intertidal flats during the tide can produce some interesting effects on the velocity distribution. As the tide rises to cover the flats a large volume of water has to flow through a small cross-sectional area with, obviously, a high velocity. Once the flats are covered the velocity will be reduced because of the increased cross-sectional area. The reverse will happen during the ebb tide and the result is a very asymmetrical velocity curve, even though the tidal elevation change is quite regular (Figure 9.11). If the tidal flats

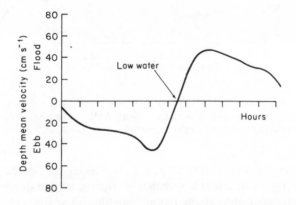

Figure 9.11 Illustration of the asymmetry produced in the depth mean velocities during a tide by the covering and uncovering of intertidal areas near low water mark. *Dyer 1980b*, Essays on the Exe Estuary, *1–21, Special Volume, No. 2. Reproduced by permission of the Devonshire Association for the Advancement of Science, Literature and Art*

are extensive near the low water mark then the maximum flood and ebb currents will be very close to low water, whereas, if the intertidal flats are close to high water, the maximum currents will be on either side of high water.

In mesotidal estuaries friction is of limited importance and there is little asymmetry of the tidal curve, but in macrotidal estuaries friction is particularly important at the head. There is obviously a wide spectrum of responses.

244

In most estuaries there is an increase in tidal amplitude towards the upper estuary because of convergence, but near the head of the estuary, where friction becomes important, the tidal range diminishes. Additionally, tidal range becomes reduced where the elevation of the estuary bed rises above the general low tide level. In this zone in macrotidal estuaries, the low tide level at spring tides can be above that at neap tides. This is caused by friction preventing discharge on the ebb of all the water brought in on the spring flood tide.

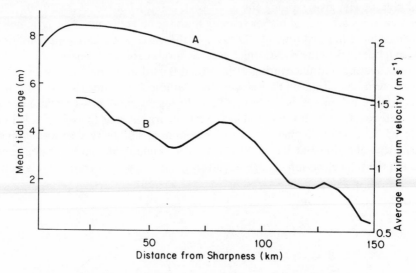

Figure 9.12 Variation of tidal range (Curve A) and tidal velocity (Curve B) along the Severn Estuary. *After Shaw, 1980,* An Environmental Appraisal of the Severn Barrage. *Reproduced by permission of T. L. Shaw*

Figure 9.12A shows the tidal range variation along the Severn Estuary. The effect of convergence dominates over that of friction, so that diminution of tidal range does not occur until about 15 km from Sharpness when the sea bed rises to low tide level. The tidal velocities will vary according to the changes in width and depth along the estuary. At any particular section the velocity will be given by the rate of change of estuary water volume landward of the section, divided by the channel cross-sectional area. Thus a shallow or narrow area will have relatively faster currents. If the estuary converges rapidly, but maintains a relatively large volume between high and low tide marks (tidal prism), then the velocities will increase towards the head of the estuary. This occurs in the Severn Estuary (Figure 9.12B) and in many other macrotidal estuaries. In most mesotidal estuaries, however, the tidal prism decreases relatively rapidly and the maximum tidal velocities decrease towards the head.

Many estuaries show an exponentially varying width, depth and cross-sectional area with distance from their heads. This must be the result of the complex interaction between the morphology, the tidal range, the tidal prism and friction. If this can produce a condition of equal work per unit area of the estuary bed, then preferential erosion or deposition in any particular section is restricted, and an equilibrium can become established.

In Cambridge Gulf, Western Australia, the tidal range rises towards the mouth of the Ord River where it reaches over 5 m on spring tides. Within the river itself the range decreases steadily over the 60 km to the head of the estuary. Wright *et al.* (1973) have shown that the width and depth can be represented by the exponential relationships:

$$b_x = b_0 e^{-4.28(x/L)}$$

and

$$h_x = h_0 e^{-2.76(x/L)}$$

where the suffix zero denotes the value at the mouth ($x = 0$) and L is the estuary length.

TURBIDITY MAXIMUM

One of the most distinctive features of sediment transport in meso- or macrotidal estuaries is the presence of a turbidity maximum. The energetic tidal flow is capable of maintaining quite high concentrations of suspended sediment in the upper estuary, higher than occur either in the river or in the sea. This feature is called a turbidity maximum and there are two dominant mechanisms which contribute to its maintenance. For mesotidal estuaries the residual circulation of water is the most commonly cited cause. For macrotidal estuaries the tidal asymmetry produces a net landward movement of sediment. Both processes involve a sequence of deposition, erosion and transport of sediment during the tidal cycle, with significant variations in timing and intensity along the estuary. Because the balance between residual circulation and tidal asymmetry changes along the estuary, and since the degree of stratification changes during the spring–neap cycle, both processes may be active in the same estuary at different positions and at different times. In macrotidal estuaries residual circulation should be important at neaps with high river flow, but may not be significant at spring tides.

The presence of the turbidity maximum has also been linked to flocculation–deflocculation processes, as discussed on page 205, but due to our present inability to measure the size of flocs *in situ*, this has not been proved. However it is likely to be a process of minor importance.

The peak concentration of suspended sediment in the turbidity maximum varies between wide limits. However, despite the differences due to sediment availability, low tidal range estuaries have maxima with concentrations of the order 100–200 ppm ($mg l^{-1}$), whereas high tidal range estuaries have much higher concentrations, of the order 1000–10,000 ppm ($1–10 gm l^{-1}$).

246

Residual Circulation

Suspended sediment is brought into the head of the estuary by the residual downstream transport in the river. In the upper estuary the mixing transfers suspended sediment from the bottom more saline layer into the surface layer, where there is a seaward residual flow carrying downstream transport. In the middle estuary, the sediment settles into the lower layer, in areas of less vigorous mixing, to join sediment entering from the sea on the landward residual flow. It then travels in the salt intrusion back to the head of the estuary where it becomes trapped at the convergence, or node, of the bottom residual flow. The process is illustrated in Figure 9.13, together with an example of the turbidity maximum from the Rappahannock Estuary.

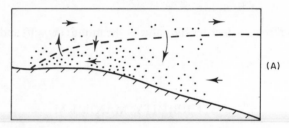

(A)

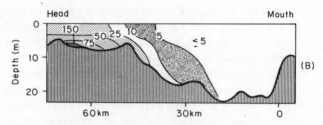

(B)

Figure 9.13 Turbidity maximum caused by residual circulation. (A) Residual circulation in a partially mixed estuary with
vertical exchanges across the level of no net motion (dashed
line). (B) Example of a turbidity maximum—Rappahannock
Estuary, distribution of average total suspended sediment concentration in mg l^{-1}. *After Nichols and Poor, 1967,* Proc. Amer.
Soc. Civil Eng., **93**, *WW4, 83–95. Reproduced by permission of
American Society of Civil Engineers*

During the tidal cycle, of course, the concentration of suspended sediment in the turbidity maximum varies because of erosion and deposition so that it is not simply the residual water circulation that causes the sediment circulation. It is thus of interest to consider these tidal variations. Figure 9.14A shows the velocity and suspended sediment concentrations for a station in the upper part of Chesapeake Bay during the spring high river discharge period (Schubel, 1969). This station was on the landward end of the turbidity maximum, and the residual water flow was downstream at all levels despite reversal of the flow for part of the tidal cycle. The maximum suspended sediment concentration occurs

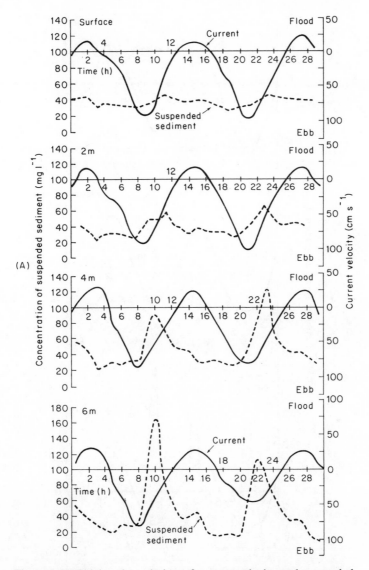

Figure 9.14 Tidal cycle variation of current velocity and suspended sediment concentration at two stations in Chesapeake Bay. *From Schubel, 1969, Chesapeake Bay Institute,* Tech. Report 60, Ref. 69–13. *Reproduced with permission.* (A) At the landward end of the turbidity maximum

near the bottom and because of hysteresis effects, it occurs from one to two hours after maximum current. The delay is greater with increasing height above the bed. Consequently during the ebb tide there is a net downstream flow of suspended sediment. On the flood tide the currents are too weak to resuspend the bottom sediment.

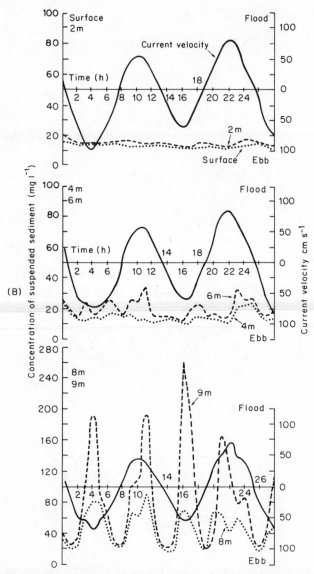

Figure 9.14 (B) On the seaward side of the turbidity maximum.
See p. 247 for caption

The conditions on the seaward side of the turbidity maximum are exemplified by measurements taken further downstream at low river discharge (Figure
9.14B). The current velocities show considerable diurnal variation, but the
residual water flow is seaward near the surface and landward near the bed. Peak
concentrations are now associated with both flood and ebb currents, though
still with a lag. There is consequently a net seaward flux of suspended sediment
near the surface, and a net landward flux near the bed.

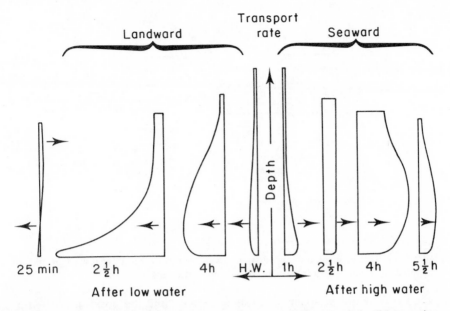

Figure 9.15 Typical vertical distributions of transport rate of suspended solids at various stages of the tide, 19 miles below London Bridge, River Thames. *From Inglis and Allen, 1957,* Proc. Inst. Civil Eng., *7, 827–868. Reproduced by permission of Thomas Telford Ltd.*

The variation of sediment flux during a tidal cycle on the seaward side of the turbidity maximum is shown in Figure 9.15. On the flood tide the large near-bed velocity erodes more material giving a near-bed suspended sediment flux. On the ebb tide the slightly lower bed velocity erodes less sediment, but the higher surface current leads to the maximum flux being higher in the water column. Additionally, the presence of increased salinity stratification on the ebb tide reduces the mixing of the suspended sediment into the surface layer.

At the extreme landward end of the turbidity maximum the largest suspended sediment concentration should occur at about high water, because of the landward advection of the sediment eroded from the bed further seawards. At the seaward end of the maximum the converse should be true, with the largest concentration occurring at low water. In the area of the peak concentrations of the turbidity maximum entrainment of sediment should be active on both ebb and flood tides with little preferential movement.

Consequently, though the turbidity maximum is present throughout the tidal cycle, its magnitude and position varies during the cycle (Figure 9.16). At high water the maximum is at its most landward position in the estuary, and the concentrations are low because the material has largely settled out. During the ebb tide the suspended concentrations in the water mass increases as it is advected down the estuary. Towards low water the concentrations again diminish.

As river flow varies, so the turbidity maximum will adjust its position within the estuary. Gibbs (1977) shows that data from representative estuaries suggest

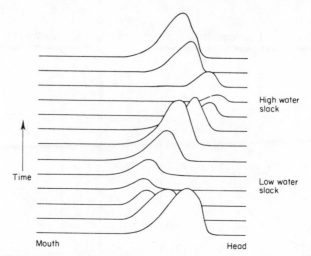

Figure 9.16 Evolution of the turbidity maximum during a tidal cycle with profiles of suspended sediment concentration at about hourly intervals

a good correlation between the flushing velocity and the position of the turbidity maximum. The flushing velocity is the river discharge divided by the cross-sectional area. In the Thames increased tidal range and changes in river flow can push the maximum seawards by up to 12 miles (Inglis and Allen, 1957).

Maximum deposition occurs near the null point. Though this occurs near the head of the estuary in the centre of the channel, in cross-section there is often a seawards sediment flux in the shallows on the estuary sides, with the landward movement concentrated in the deep channel (Figure 9.17). This is a consequence of a two-layer residual flow within a triangular cross-section. Ideally this would lead to a ∧-shaped zone of maximum deposition, with high rates occurring on the edges of the deep channel where the direction of mean sediment transport changes, and with the apex at the mean position of the null point.

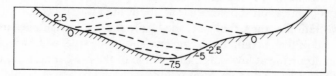

Figure 9.17 Transport rate of total suspended sediment in a cross-section of the Rappahannock Estuary. Rates in 10^3 gm m^{-2} hr^{-1}. Negative values, upstream transport. *After Nichols and Poor, 1967, Proc. Amer. Soc. Civil Eng., 93, WW4, 83–96. Reproduced by permission of American Society of Civil Engineers*

The maintenance of the turbidity maximum has been investigated by Festa and Hansen (1978) using a tidal averaged two-dimensional numerical model. This model represents the residual flow field, with the tidal mixing included as constant exchange coefficients. A turbidity maximum was created in the model by the residual flow, and its magnitude depended on the particle settling velocity. A simple box-model of the turbidity maximum has been described by Officer (1980) based on similar principles.

Examples of estuaries where residual circulation is important in sustaining the turbidity maximum are the tributary estuaries of the Chesapeake Bay system (Schubel, 1972; Nichols, 1972; Nichols and Poor, 1962); San Francisco Bay (Conimos and Peterson, 1977); Thames (Inglis and Allen, 1957); the Mersey (Price and Kendrick, 1963); the Tay (Buller *et al.*, 1975), and several estuaries of the Atlantic seaboard of America (Meade, 1969).

Tidal Asymmetry

In macrotidal estuaries considerable deformation of the tidal wave occurs as it propagates upstream in the estuary and the flood currents become very much stronger than the ebb currents. This asymmetry is considerably accentuated when bed shear stress is considered. Consequently there is a preferential movement of sediment landwards towards the head of the estuary, until the point where the ebb current due to the river flow becomes dominant in transporting sediment. This process is illustrated in Figure 9.18. In addition the slack water

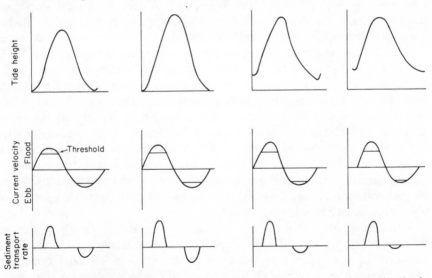

Figure 9.18 Schematic diagram of the transport of sediment towards the head of a macrotidal estuary. *After Allen* et al., *1980,* Sedimentary Geol., *26, 69–90. Reproduced by permission of Elsevier Science Publishers BV*

period at high water is longer than that at low water, enabling a greater proportion of the material to settle to the bed. Thus the combination of the greater velocity on the flood tide, and the settling at high water is responsible for the turbidity maximum being more pronounced on the flood than on the ebb. The landward flux over the flood tide is then balanced by a seaward flux during the ebb tide, which, though it has a lower concentration and a lower velocity, persists far longer. Allen *et al.* (1980) have shown that the total amount of suspended sediment in the Aulne estuary varies from a minimum of 2000 tons to a maximum of almost 12,000 tons during a tide. In the Gironde estuary the tidal excursion of the turbidity maximum is between 10–20 km (Allen *et al.*, 1977), but it is likely to migrate by 30–40 km between the periods of low and high river discharge.

A feature of macrotidal estuaries is the large difference between spring and neap tides. Because of the considerable variation of energy there are corresponding changes in the turbidity maximum. This effect has been examined by Allen *et al.* (1980) and Gelfenbaum (1983). During the neap–spring cycle the tidal prism varies significantly, changing the ratio of river flow to tidal volume. Consequently though the estuary may be well mixed at spring tide, at neap tides it can be partially mixed, or even well stratified. At spring tide the turbidity maximum will have its highest concentration, the currents being able to erode and sustain more sediment in suspension, and it will be further up the estuary. This is due to the fact that there is a higher mean sea level in the upper estuary at springs than at neap tides, arising because the increased range at spring tides involves a large extra volume of water at high tide, but only a slight volume difference at low tide, relative to the neaps. During decreasing tidal amplitude towards neaps, the peak currents decrease, and less and less material is capable of being re-eroded and suspended. Additionally the durations of slack water increase, enhancing deposition. In the Severn estuary (Kirby and Parker, 1983) and the Gironde (Allen *et al.*, 1977) this leads to the formation of fluid mud at neap tides. At slack water at spring tides, settling of the suspension leads to the formation of a high concentration layer, near the bed (see page 226). But this only persists for a short while into the following high current. As the range decreases towards neaps, these layers persist longer, until eventually they survive from one slack water until the next. It is unclear whether an increasing increment at the base of the suspension survives on subsequent tides, or whether nothing survives until a critical point at which the whole layer remains. When the concentration is high enough, the suspensions may be pseudo-plastic, in which case the latter situation may hold.

During neap tides the fluid mud has time to consolidate slightly so that on increasing range tides not all of the deposited sediment is re-eroded. A small net increment of deposition then occurs over the neap–spring tidal cycle. No measurements have been made which confirm this hypothesis, and in some estuaries it is plausible that the complete layer could survive as a deposit sometimes. However the turbidity maximum is generally associated with net deposition, which dredging records have shown to be of the order of a few centimetres per year in many estuaries. In addition to spring–neap movement of the turbidity maximum, and seasonal changes due to varying river flows, alteration of the estuarine topography can have an effect. In the Seine estuary there has been progressive marsh reclamation and landfill, coupled with dredging of the channel, over the last 130 years. This has caused a progressive migration of the turbidity maximum about 50 km down the estuary (Avoine *et al.*, 1981). Consequently the area of maximum deposition now occurs in harbour areas at the mouth.

In well-mixed estuaries, though there is little vertical salinity stratification, there can be lateral differences resulting in considerable horizontal water circulation. These lateral effects can also be apparent in the turbidity maximum. In the Severn estuary the most turbid water is on the southern, English, side of the estuary, with clearer water on the Welsh side. Separating the two water masses is a distinct 'front', a zone of steep change in turbidity, which is present under all conditions, and which appears to be independent of the salinity distribution (Kirby and Parker, 1982). The front is maintained by a southward cross-channel flux of suspended sediment near the bed, and a northward flux near the surface. This lateral circulation is opposite to that of the water, and arises because of the hysteresis in the sediment response to the water flow, the combined effects of suspension and settling.

In the Gironde Estuary, Allen *et al.* (1977) have demonstrated considerable lateral movement of suspended sediment within the estuary from analysis of radioactive tracer tests.

Examples of macrotidal estuaries, where tidal asymmetry is an important factor in maintaining the turbidity maximum, are the Severn, the Gironde and the Aulne.

Particle Size Distributions

The mechanisms maintaining the turbidity maximum are continually active and will tend to sort the particles very effectively depending on their erosion and settling characteristics. In many estuaries there is a natural background of $10-20$ mg l^{-1} of suspended sediment, of an equivalent diameter of 3–4 microns, present at all stages of the tide. These particles may be finely-divided organic matter, living planktonic creatures and fine clay particles, with fall velocities of the order of 10^{-3} cm s^{-1}. These fall velocities are of the same magnitude as the mean velocity of vertical secondary currents and the motion associated with vertical diffusion. As the particles would fall less than a metre a day in still water they must be almost perpetually in suspension, and thereby form the washload fraction of the sediment. This fraction will therefore be susceptible to being swept out of the estuary on the mean river discharge. The sand size particles will move as bedload and will not be suspended to form part of the turbidity maximum. Consequently the size range of particles in the turbidity maximum is narrow because of the sorting process (Schubel, 1969). However, during the tidal cycle there may be changes in the particle size distribution of the suspended sediment. The coarsest particles are nearest the bottom, but Postma (1961) and Sheldon (1968) report little variation in the particle size distribution during the tidal cycle. This implies that increasing currents do not preferentially pick up more of one size than another. Schubel (1971), however, reports a size distribution displaced to larger sizes with increasing velocity. The size distribution near the bottom was positively skewed at all times, but skewness decreased near times of maximum current as more large particles were resuspended.

Flocculation is an important process in controlling the settling velocities of particles in estuaries. Kranck (1981) has used the smooth, progressively changing nature of the size spectra during the tide and along the estuary to suggest that the alternative process of biological agglomeration is not important in the Miramichi Estuary. Samples from the turbidity maximum were more highly flocculated than in freshwater or further seaward, but concentration was thought to be more important than salinity in controlling the flocculation. Without flocculation she concludes that the particles, as they are so fine, would probably be swept out of the estuary.

The concentration of suspended sediment in the turbidity maximum varies with river discharge and with tidal range. From a long time series of turbidity measurements in the Humber Estuary, Jackson, W. H. (1964) has also shown that temperature is important. A 14°C change in temperature was related to almost trebling of the silt concentration. The viscosity of the water only changes by about 50 per cent over this range, so that the turbidity increase cannot solely be due to the effect of viscosity on settling velocity. The effect of decreasing temperature on increasing flocculation potential may be an important additional factor, as may biological factors. The decreased temperature in the autumn and winter is likely to reduce biological and microbiological activity, thus making the mud easier to erode, both subtidally and intertidally.

Frostick and McCave (1979) have shown for an estuary in Suffolk, England, that there was an accretion of ~ 5 cm between April and September on the intertidal mudflats during algal growth, and erosion of that much during the autumn and winter. Since the amounts involved could not be accounted for by transfers with the sea, it must have passed via the turbidity maximum to the channel bed. A similar temperature dependence of concentration has also been found in the Mersey Estuary (Halliwell and O'Connor, 1966).

In mesotidal estuaries the decreasing tidal velocity towards the head of the estuary limits the penetration of sandy bedload material from the sea. Consequently most estuaries show a decrease in grain size of the bed sediment towards

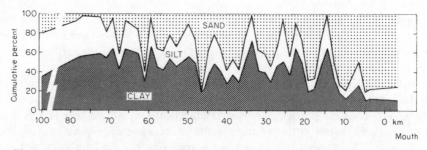

Figure 9.19 Variation in bed sediment texture along the James River. *From Nichols, 1972, in* Environmental Framework of Coastal Plain Estuaries. *Reproduced by permission of the Geological Society of America*

the estuary head (Postma, 1961; Nichols and Poor, 1967), and the position of the turbidity maximum is marked by a 'mud reach'. Figure 9.19 shows an example of the average sediment distribution along the James River (Nichols, 1972). Price and Kendrick (1963) report large variations in the proportions of sand present in suspension across the Mersey Narrows, which may also indicate lateral variations in the bed sediment characteristics. A surprising feature of many estuaries is the relatively narrow zones occupied by mixed sediment, sand giving way very quickly to mud, reflecting the possible dichotomy between the processes of bedload and suspended load transport.

In macrotidal estuaries the fact that the maximum tidal currents are directed landward and increase in strength towards the head makes it possible for sand to be carried almost to the tidal limit. In the Severn Estuary, for example, in the area of the turbidity maximum there are active sand waves which are moved at maximum current, but which are draped with mud over slack water. However, macrotidal estuaries generally exhibit horizontal water and sediment circulations which complicate idealized models.

SOURCES OF SEDIMENT AND SEDIMENT BUDGETS

There are several potential sources which can contribute sediment to the turbidity maximum. They are: rivers, streams and outfalls, the sea, erosion of estuary coastline and tidal flats, erosion of the seabed, biological production and the atmosphere. The relative importance of these sources will vary between estuaries, and will depend to a great extent on the seasonal cycle of river discharge. Meade (1969) has argued strongly that the majority of sediment in estuaries of N.E. Amercia is derived from the sea, and this conclusion seems to be valid for many temperate estuaries. For instance there has been a fall of about 10 per cent during the last 100 years in the volume of the Mersey estuary despite the dredging of over 400 million cubic yards of material. As the volumes of incoming solid sewage matter and of material suspended in the river flow were not a major source, the material must have moved in from the sea (Price and Kendrick, 1963).

The riverborne and marine sources of sediment can often be distinguished from examination of the clay mineralogy or heavy mineral content of the sediments. Nichols (1972) has shown that illite, montmorillorite, kyanite and sillimanite are indicative of riverborne sediment in the James River, and kaolinite, chlorite and staurolite are of marine derivation. Similarly in Garolim Bay in Korea, illite and smectite indicate a marine source and kaolinite, chlorite and amphiboles a land source (Song et al., 1983).

Biggs (1970) has examined the sources and distribution of suspended sediment in the inner Chesapeake Bay. As can be seen from Figure 9.20 the river contributes over 80 per cent of the suspended material in the upper bay. Lower down, the organic contribution is larger, but shore erosion becomes the main sediment source. Of the suspended sediment, 4 per cent of the upper bay load escaped to the middle bay and 23 per cent of the middle bay load escaped to the

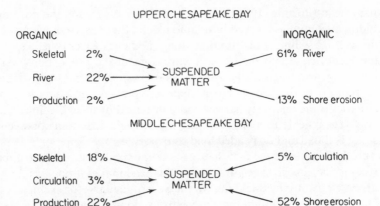

Figure 9.20 Contributions to total suspended sediment load by organic and inorganic sources in the Upper and Middle Chesapeake Bay. *From Biggs, 1970*, Marine Geol., *9, 187–201. Reproduced by permission of Elsevier Publishing BV*

lower bay. The net amounts sedimented out in the upper and middle bays, if evenly distributed, would have amounted to 3.7 and 1.1 mm y^{-1} respectively.

Schubel and Carter (1977) have obtained a sediment budget for Chesapeake Bay as a whole by defining the mean fluxes through the sides of a simple single-segment box model. The total yearly averaged input of sediment was 1.89×10^6 tons of which 57 per cent came from the Susquehanna River, 32 per cent from shore erosion, and the rest from the ocean. They estimated that 92 per cent of this material was deposited giving a bay-wide sedimentation rate of about 0.8 mm y^{-1}. The residue was contributed to the residual circulation.

In the Severn Estuary sandy sediment has moved into the estuary in the past, but the source in the Bristol Channel now appears to be exhausted (Parker and Kirby, 1982). About 10^7 tonnes of suspended mud is present at spring tides and this is probably two orders of magnitude greater than the annual river input. However there are two large areas of intertidal and subtidal mud, one of which contains at least 2.7×10^8 tonnes. Within this area there are three zones; one shows evidence of fairly steady deposition, one shows evidence of erosion, and between the zones is a stable seabed (Kirby and Parker, 1981). Of the 30 km^2 of subtidal mud about 10 km^2 is thought to be accretionary and 7.5 km^2 erosional. Presumably the mud cycles through the turbidity maximum between the zones of erosion and deposition.

This illustrates a simplified concept of the turbidity maximum as a feature through which sediment is continually cycled from one part of the estuary to another with small amounts of material being added from the rivers and the sea, these inputs being balanced by deposition. Individual particles may spend a considerable time moving within the turbidity maximum between initial insertion and final deposition. In between, the particles may undergo many periods of temporary deposition and re-erosion.

The relative importance of the different sediment sources in different areas can be quantified by drawing up a sediment budget. This is difficult because of the large regular seasonal and longer period intermittent variations in inputs, and becasue of the lack of precise information on sediment deposition or erosion rates. Nevertheless estimates of potential deposition rates have been made by assuming they would be equal to actual dredging rates. There are many estuaries that appear to be effectively full of sediment and are in the balance, so that little further reduction is possible and the net sedimentation is zero, despite much sediment being in motion. These include the Solway Firth and the Exe in England. In the Clyde, little material is derived from the sea (Fleming, 1970). The rivers discharged 203,758 tons of suspended load and 9,189 tons of bedload annually. Sewage discharge and spillage from harbours and docks accounted for a further 29,829 tons. These sources totalled 242,776 tons a year. Dredging removed 317,769 tons and as 75,000 tons of this was capital dredging, the sediment entering from the rivers and depositing in 12 miles of the upper reaches was balanced by dredging. Krone (1979) has established the budget of San Francisco Bay shown in Figure 9.21, but Conomos and Peterson (1977) point out that within the bay there are marked differences in deposition, with the southern bay floor undergoing erosion. Consequently there may be considerable exchanges within the estuary which are not shown up in the budget.

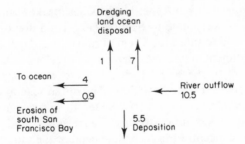

Figure 9.21 Sediment budget for San Francisco Bay. Values in millions of cubic yards. *From Krone, 1979, in Conomos, T. J. (Ed.)* San Francisco Bay: The Urbanized Estuary, *85–96. Reproduced by permission of Pacific Division AAAS*

Allen *et al.* (1977) have considered the budget of the Gironde estuary using results from a radioactive tracer experiment. The suspended sediment influx from the rivers was 2.0×10^6 tonnes per year. About 8 per cent of this is sedimented or escapes to the sea under high river flow conditions. Presumably very much more is sedimented as the flow diminishes.

An alternative way of defining the budget is by measuring the fluxes rather than the total quantities. The difficulty of doing this within an estuary was illustrated by McCave (1974). If the area of mudflats within an estuary is 100 km^2 and they are building up at 1 mm y^{-1} to match sea level rise, then the net input of sediment is $10^5 \text{ m}^3 \text{ y}^{-1}$ or 1.4×10^5 tonnes y^{-1}. This is equivalent to an average of about 200 tonnes tide^{-1}. If the tidal range is 1 m then the tidal prism would be 10^8 m^3. The concentration difference between the flood and ebb to create the necessary sedimentation is then $2\,\mathrm{mg\,l^{-1}}$. Much larger differences than this are measured, but they are differences between two large numbers, both with very large errors, one a flood tide average flux and the other the ebb tide average flux. Additionally there are considerable difficulties in obtaining an adequate cross-sectional average. Consequently well-quantified sediment balances are difficult to obtain and this makes the predictive models of sedimentation in estuaries difficult to validate.

Storm Effects

Most investigations in estuaries consider the normal mean river discharge situation. During storms and floods this may be rather unrealistic for, as far as sediment is concerned, a single brief flood may introduce more sediment than many years of normal flow. However, these extremes could by-pass the estuary and discharge directly onto the shelf. There have been few studies which adequately cover these events. Hayes (1978) reviews the impact of tropical storms on estuarine and lagoonal sedimentation on the eastern seaboard of America.

In June 1972 Tropical Storm Agnes hit the Chesapeake region. The response of the Susquehanna was reported by Schubel (1974) and that of Rappahannock by Nichols (1977). As a result of the 30 cm of rain in two days, the Susquehanna discharged more sediment in one week than in the previous half century, exceeding 50×10^6 tonnes. In the Rappahannock there was a sediment discharge exceeding 50,000 tonnes d^{-1} at peak flow. The initial response of the Rappahannock was a rapid change from partly mixed to salt-wedge regime with the saline limit pushed 20 km downstream (Figure 9.22), and with maximum suspended sediment concentration above the head of the salt wedge. In the salt wedge the strong seaward current throughout the upper layer carried sediment seawards. The return current on the sea bed was weak. After this 'shock' stage there was a 'rebound' period during which decreasing inflow allowed the salt wedge to creep landward and the landward residual flow to strengthen and a turbidity maximum to become re-established. The flood passing down Chesapeake Bay then complicated the recovery period. Nichols concluded that 90 per cent of the sediment was trapped within the estuary, because the saline intrusion was not pushed clear of the estuary and the residual circulation retained the sediment inside the estuary. As the saline intrusion moved back up the estuary, the sediment also moved back, with the result that final deposition at the head was 7.5 mm and at the mouth 0.5 mm.

This result indicates how deposition of riverborne sediment in a tidal estuary can lead to a sedimentary balance. The deposition at the head of the estuary, if carried out intermittently over a long period, would eventually lead to a shortening of the estuary. The major floods would then be able to push the saline water clear of the estuary, and the sediment would be discharged into the sea, with less chance of being drawn back into the estuary as discharge

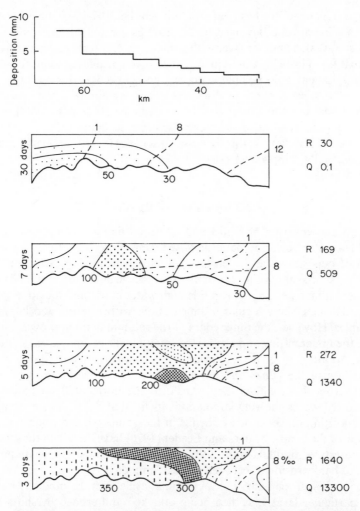

Figure 9.22 Response of the Rappahannock Estuary to a flood event. *After Nichols, 1977,* J. Sediment. Petrol., *47, 1171–1186. Reproduced by permission of Amercian Society of Civil Engineers.* Days after the storm. Contours of suspended sediment concentration are in mg l^{-1}. Salinity contours (dashed, ‰). R is the river flow, m^3 s^{-1}, and Q is the sediment discharge, tonnes day $^{-1}$.

diminished. The deposition would thus slowly reduce until eventually any sediment deposited under normal conditions would be eroded and swept out of the estuary during floods, resulting in a quasi-equilibrium.

In Chesapeake Bay itself several hundred acres of new intertidal flats were formed during Agnes. This temporarily reversed the coast erosion prevalent in the Bay, but the material presumably then is re-eroded under normal conditions, cycles through the turbidity maximum and is deposited sub-tidally. Post-storm coring showed that an average of 17 cm of sediment was deposited in the upper reaches of the bay (Zabawa and Schubel, 1974) and it has been concluded that Chesapeake Bay aged by 50 years as a result of this single storm. Coring also revealed another layer with a thickness of 30 cm that was correlated with a great flood in 1936. Consequently these two episodes account for at least half of the sediment deposited in the upper Chesapeake Bay this century.

In some estuaries extremes of discharge occurring annually can give results similar to floods. In the Fraser River, for instance (Milliman, 1980), 80 per cent of the annual suspended sediment discharge occurs in the 2–3 months of spring and early summer. At this time river discharge is sufficiently high that the estuary is virtually completely fresh.

INTERTIDAL AREAS

The drying and covering of banks and intertidal flats in the estuaries creates complicated water flows and patterns of sediment dispersal. Often signs of erosion and of deposition can be observed in closely adjoining areas, and these form both a source and a sink for material in the turbidity maximum. The surface of the salt marshes above neap tide high water mark, and the upper part of the intertidal flats, show regular sedimentation, with layering and laminae in core samples. However, the outer edges of the salt marsh often show erosion by 'cliffing', the undercutting and erosion of small blocks of compacted salt marsh sediment. Additionally gullies and meandering channels cross the flats and show active erosion of the banks and migration of the meanders. As the channels meander across the mudflats they transform the horizontally stratified sediments into sequences showing laminations inclined at 7–15°. These are produced by deposition on the inside of bends in the gullies while erosion occurs on the outside of the bends. Bridges and Leeder (1976) have described these processes from intertidal mudflats in the Solway Firth, Scotland, and they quote average rates of bank erosion of between 3 and 15 mm day^{-1}.

Consequently one can envisage a continual cycle with the mudflats building up to a particular level, and then being attacked and eroded by shifts in the channels and by gullying. The eroded sediment is exchanged via the turbidity maximum to other areas of temporary deposition. These movements of the main channels of the estuary cause fluctuations of as much as 5 per cent in estuary volume (Inglis and Kestner, 1958).

An important feature of salt marshes are the plants, which can actively trap sediment and enhance deposition. Generally there is a zonation, as one goes from neap tide to spring tide high water marks, of flora which are tolerant to decreasing frequency of immersion. In sheltered areas a height variation of only a centimetre or so is needed to clearly distinguish between two zones. In England a common zonation is:

1. *Salicornia* (Annual Glasswort), together with *Suaeda maritima* (Annual Seablite) or *Aster tripolium* (Sea Aster). These need to be above high tide mark every day over neap tides.

2. *Halimione portulacoides* (Woody Sea Purslane), together with Sea Lavender and Sea Mugwort. Only covered by the highest spring tides.
3. *Agropyron pungens* (Sea Couch Grass) very rarely flooded.

During the summer these plants are effective traps for settling suspended sediment at high water, and during the winter sediment accumulates readily in broken stems of hollow reeds which are then incorporated into the sediment.

In Southern England the introduction of *Spartina townsendii*, which has rapidly colonized previously bare soft mud, led to extensive increases in the areas of salt marshes because of the entrapment of suspended mud. In the tropics, of course, the same function is performed by the stems and rootlets of the mangrove trees. Algae have also been associated with deposition of up to 5 cm during the summer (Frostick and McCave, 1979).

Erosion on the mudflats can be effected by ice, rain and waves (Anderson, 1983). Small amplitude waves (less than 5 cm) are capable of suspending fine grained sediment in shallow water and can increase the suspended sediment concentration by a factor of three over calm conditions. Anderson (1972) showed that the results were linearly dependent on wave height on the flooding tide, but there was no relationship on the ebb.

Lower down the intertidal flats there is often a more sandy area near low water mark. This may be the result of more active and prolonged wave activity, or, on the other hand, the stronger tidal currents winnowing the finer material away. This is a common feature in the Wash, in eastern England (Evans, 1965), where there is also a biological zonation.

Rain causes erosion particularly during low water not only because of the impact but because the run-off is controlled by the gradient and can achieve quite high velocities.

Because the tide progresses faster in the deep channels than in the shallows, the water surface slope is almost normal to the shore. Consequently, though the maximum velocities are directed parallel to the shore in deep water, they are onshore–offshore in the shallows. Flood tidal velocities are controlled by the rate of tidal sea level rise. Ebb velocities, however, are only controlled by the sea level fall until the sea level reaches a point lower than the water surface in the channel, then a gravity component enhances the flow. Velocity measurements in the creeks have shown the maximum flood velocity occurs after the upper marsh is covered (Figure 9.23), but the maximum ebb current occurs when water is present only in the creeks. There are, of course, differences in the cross-sectional areas through which the flow takes place that allow the discharges to balance. However, because of shallowing the peak velocities diminish towards high water mark (e.g. Evans and Collins, 1975). The movement of water within a tidal creek has been modelled by Boon (1975) and Pethick (1980). These models show that it is a reasonable assumption that the form of the tidal curve is hardly modified within the creek system, but imposed largely by the tidal dynamics of the estuary as a whole, and that the peak velocities diminish landwards.

The results of Healey *et al.* (1981) suggest that storm tide, or surge, could

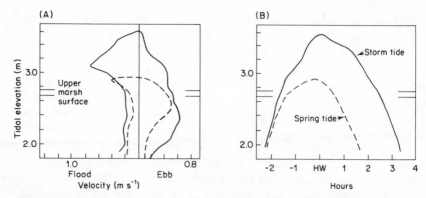

Figure 9.23 (A) Curves of velocity against tidal elevation for an intertidal creek for storm and spring tides. (B) Curves of tidal elevation against time for storm and spring tides. *From Healey* et al., *1981,* Estuarine Coastal and Shelf Science, *13, 535–545. Reproduced by permission of Academic Press Inc. (London) Ltd.*

cause major sedimentary effects, particularly because of the enhanced flood currents on the upper flats. This has also been suggested by Stumpf (1983) for Delaware Bay. Storms are considered to provide the major inputs of sediment to the upper marsh, where biological trapping is the main depositional process. Consequently the flood current carries suspended sediment landward out of the creek and onto the flats. At high water the sediment has only a small water depth to settle through before reaching the bed. The ebb flow is concentrated in the winding creeks and channels, where there is a larger mean depth at low than at high water. Additionally low water slack tide is comparatively short, giving the particles little time to settle out. Thus there is a progressive movement of fine sediment onto the upper mudflats, with the coarser sediment being deposited lower down.

A qualitative model describing the landward transport of sediment in shallow tidal flat areas, has been developed by van Straaten and Keunen (1958) and Postma (1961). If the maximum current velocity decreases towards the shore, particles drifting with the current will undergo a changing velocity with distance along the channel during the tide (Figure 9.24). The exact form of the distance–velocity curve will depend on the asymmetry of the tidal curve, caused by the faster progression of the tidal wave in deeper water. The total envelope of the distance–velocity curves will be the shorewards decreasing maximum current. The concept of settling lag and scour lag are now introduced. Settling lag is the time taken for a particle of sediment to reach the bottom when the decreasing current cannot maintain it in suspension. Scour lag is the delay due to the difference in the current necessary to erode a particle from the bed, and that occurring at final deposition. Assuming scour lag is zero, and following a particle during the tide (Figure 9.24); at 1 the particle is lifted off the bed. It then travels with the water until 2, when it starts to settle. Because of settling lag it reaches the bed at 3. On the following ebb tide it will not be entrained until later in the tidal cycle when the threshold velocity is reached at that position, and will

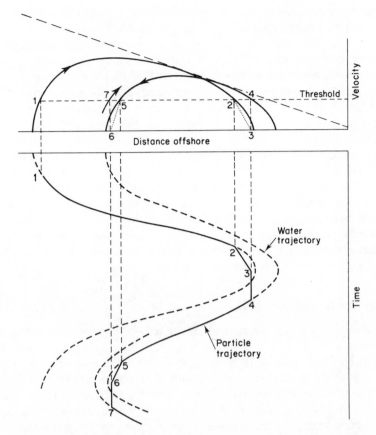

Figure 9.24 Qualitative model illustrating the tidal transport of suspended matter shorewards by an asymmetrical tidal current. After Postma, 1961. See text for explanation

follow a different trajectory to be deposited on the bed again at 6. Thus the particles will migrate shorewards and accumulate in the area where the peak tidal current equals their threshold velocity. This will be at the inner ends of the tidal flats. Deposition there will tend to increase the slope of the mudflats and this in turn will reduce the asymmetry of the current and eventually the deposition.

This model has been quantitatively examined by Groen (1967). Providing the ebb and flood currents were asymmetrically disposed about slack water, even with equal maximum currents, suspended sediment could be made to move preferentially in one direction. In one simple example, the displacement of suspended matter on the flood exceeded the displacement on the ebb by 38 per cent. This model could be applied to the generation of the turbidity maximum by tidal currents, described earlier.

An identical model will hold for scour lag, except that section 2 to 3 then corresponds to movement of the particles during the time when the velocity is between the thresholds of erosion and deposition. Normally settling lag would dominate for suspended particles and scour lag for bedload.

There is considerable controversy about the role that tidal creeks and mud-flats play in the budgets of organic carbon, nitrogen and other substances (Nixon, 1980). Some measurements suggest that they are sources and others that they are sinks, but some of the discrepancies may result from extreme difficulty of establishing budgets, as discussed earlier. Many aspects of work on salt marshes in America have been reviewed by Frey and Basan (1978).

Sandy intertidal areas have had particular attention from the point of view of the movement of sand waves and other bedforms. One difficulty inherent in these studies is that the bedforms can only be examined after the ebb tide. Nevertheless reversal is indicated by flood oriented features still being visible. Neap–spring variations in size as well as variation in migration rate have been reported. Many of the studies have been complemented by coring and sampling of the bedforms in attempts to link the internal structures with the hydraulic set-ting. Since these studies are complementary to those on subtidal sand waves, they will be discussed more extensively in Chapter 10.

Estuary Mouths and Tidal Inlets

The mouths of many estuaries are restricted by the presence of spits. These arise because local coast erosion has supplied a large amount of material, and this is brought to the mouth by processes within the breaker zone. The spit builds out to the stage where the constriction increases velocities to the extent that they erode as much material from the tip of the spit as is supplied. The same hydrodynamic balance is attained at tidal inlets present in areas, such as south east America, which join the extensive coastal lagoon systems to the sea through the fringing barrier beaches. This is illustrated by the linear relationship that has been demonstrated between the cross-sectional area of the inlet and the tidal prism.

The tidal currents flow through the entrance with flow smoothly converging like a potential flow on the upstream side, and a jet on the downstream side. This means that on the upstream side water is drawn evenly from all sectors, accelerating towards the mouth. On the downstream side, however, the flow separates during the maximum flow and there is reverse flow close to the shoreline (Figure 9.25). In reversing tidal flow an almost mirror-image flow situation occurs on either side of the mouth on ebb and flood, and tidal averaging leads to strong residual flow away from the inlet in the centre, but towards the inlet at the sides (Figure 9.25). A similar circulation of sediment is to be expected. An obvious consequence of this residual circulation is the development of flood and ebb channels. In the flood channels the dominant flood current carries sediments inland. Towards the head of the channel the peak velocities and the residual flow lessens as the water fans out and the channel shallows. Water flowing up the flood channel on the flood tide tends to move across towards the ebb channel near high water. The ebb channel shallows seawards, and ebb and flood

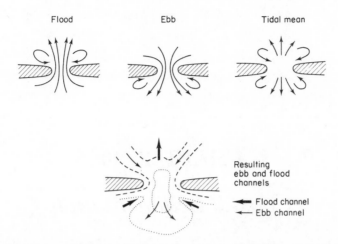

Figure 9.25 Diagrammatic representation of the generation of
ebb and flood channels in a tidal inlet

channels interfinger, with flood channels truncated by ebb channels. On the
banks between the channels there is obviously a circulation of sediment. Consequently, though a large amount of sediment may be mobile at any position, the
cross-sectionally averaged movement into, or out of the estuary is small.
Because of the continual circulation the sediment becomes well sorted, the finer
grains being winnowed away in suspension, and the coarser sands and gravels
being restricted to the deep channels where the peak velocities are higher.

Since the tidal currents diminish in strength rapidly away from the inlet,
particles swept from the spits are quickly deposited and form either a flood or
an ebb tidal delta. On the flood delta inside the mouth, ideally the channel bifur-
cates, the flood current jet impinges directly onto the junction driving the sediment up the flood ramp and onto the flood shield, a nearly flat area just below
mean high tide level. The ebb channels are formed on either side, recirculating
the sediment back towards the mouth. The ebb tidal delta will be the direct
analogue on the outside of the mouth, but its form will be modified greatly by
wave processes.

The morphology of tidal deltas is described by Oertel (1972) and briefly summarized by Hayes (1975). Aspects of inlet hydrodynamics and their stability
have been described by Bruun (1978), Mehta (1978) and Mehta and Christensen
(1983).

Within well mixed estuaries flood and ebb channels often exist, separated by
elongated banks. In this case the flood and ebb flow are directed to opposite
sides of the estuary by bends. The flood flow is affected by the topography
downstream, and the ebb by that upstream. Consequently sediment is driven
down the ebb channel and up the flood channel, in the process circulating
around the bank. As we shall see later similar circulation of sediment occurs
around coastal banks.

CHAPTER 10

Coastal Sedimentation

There have been many descriptions of the sediment distribution and circulation in various coastal and continental shelf areas. The routes by which sediment is transported have been postulated from the grain size distribution, from flow residuals, and by assessment of the distribution of bedforms. As we have already seen, measurements of the rates of sediment transport in these areas and quantitative prediction of the overall circulation depend to a great extent on understanding the dynamics of the bedforms, their effect on the flow and in turn the effect of the flow on them, as well as the overall availability of sediment.

There is a sequence of bedforms that occur in sandy sediments in areas of progressively changing flow velocity. This has been outlined by work on the European continental shelf, summarized by Stride (1982), and the sequence is shown in Figure 10.1. Near the source, in areas of strong tidal and residual currents, rock or gravel is exposed as a result of sand moving away, and sand is still occasionally lost from the gravel interstices during intermittent movement of the coarse grains. Under certain conditions longitudinal furrows occur and these seem to form a route by which sand is transported in the direction of the maximum tidal shear stress. In slightly weaker currents, further from the source, longitudinal sand ribbons occur. These gradually broaden down the velocity gradient and transverse sand waves form on the ribbons. The ribbons eventually merge, and on a seabed entirely covered with sand, larger sand waves are present. Initially the sand waves are strongly asymmetric, but they grow in amplitude and become more symmetrical. At the distal, outer edge of the sand sheet they reduce again in size, where the sand is only occasionally moved by currents barely reaching the threshold, and they give way to relatively flat depositional sand patches.

The various bedforms will be discussed in order of their occurrence on the transport path.

266

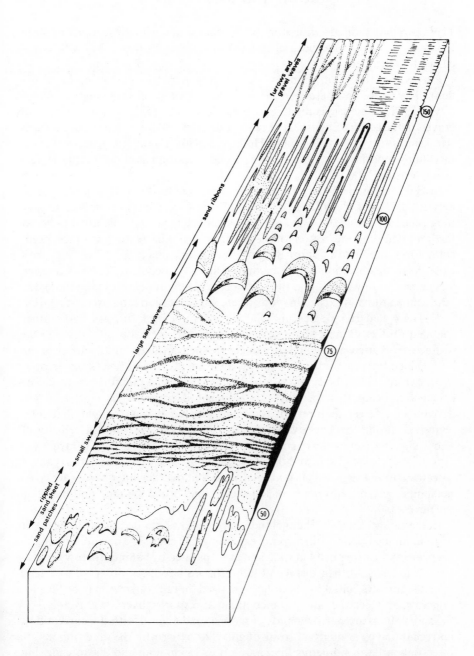

Figure 10.1 Block diagram of the main lower flow regime bedforms made by tidal currents on the continental shelf, with the corresponding mean spring peak near-surface tidal currents in cm s^{-1}. *From Stride, 1982,* Offshore Tidal Sands. *Reproduced by permission of Chapman and Hall*

268

FURROWS AND SAND RIBBONS

These two longitudinal features are both oriented in the flow direction and since the same mechanism has been postulated to account for them, they will be discussed together here.

Furrows, as their name implies, are long flow-parallel indentations in the sea floor. They were first described in the estuarine muds of Southampton Water (Dyer, 1970c), and have subsequently been found in weakly resistant rocks and gravels in the central English Channel (Stride *et al.*, 1972), and in the deep ocean (Hollister *et al.*, 1974; Flood and Hollister, 1980). Flood (1983) has reviewed their known occurrence and proposed a classification based on morphological characteristics.

In Southampton Water the furrows were up to 4 km long with depths between 0.5–1 m and widths of up to 5 m. Furrow spacing was variable, but they were generally about 10–25 m apart and extended from water depths 1 m below low water mark to the bottom of the channel. A notable feature was their branching form; the furrows bifurcating in 'tuning-fork junctions'. These junctions close down the direction of the maximum flow. Consequently furrows become less numerous downflow, and they also deepen. According to Flood (1981), the best-defined furrows existed in muds where sedimentation rates exceeded 3.6 cm y^{-1}. These had steep sides and flat floors. Ill-defined furrows with rather rounded profiles existed where sedimentation rates were 1.8–3.6 cm y^{-1}. One distinctive feature was the presence of cockle shells in layers on the floors of the best defined furrows. In the interfurrow areas minifurrows often occurred originating from cockle shells, but it was thought that these minifurrows would be ephemeral, being swamped by the sediment deposition before they could develop into larger features. In the main furrows it was thought that mud deposited during the slack water covered the cockle layer until a velocity of 20 cm s^{-1}, measured at 15 cm above the bed. When the cockle layer was exposed by erosion of the mud the consequent increase in bed roughness caused a decrease in the near bed current relative to that over the interfurrow areas. At maximum current, movement of the shells could erode the mud and maintain the furrows.

In the central English Channel individual furrows can be traced for as far as 9 km in an area with peak currents exceeding 150 cm s^{-1}. They are about 1 m deep, are of the order of 14 m wide and are separated by flat interfurrow areas at least 25 m wide. At their distal end sand ribbons have been observed to extend from the furrows, and it has been hypothesized that the furrows may be the passageways by which the sand is preferentially transported over the gravel.

Sand ribbons are narrow bands of mobile sand aligned with the current flow and separated by a rough substrate of gravel or bare rock. The sand ribbons are often only a few centimetres thick, are up to 15 km long and 200 m wide, but broaden down the velocity gradient. Their surface can be rippled or covered with megaripples and sand waves. On the British continental shelf they are found in areas of mean spring currents of about 100 cm s^{-1}. Kenyon (1970) gives a very comprehensive description of their features and occurrence. He relates the variation in ribbon spacing to velocity and to sand supply, rather than to water depth, and found closer spaced ribbons in higher velocity regions. Werner and Newton (1975), however, show that the spacing of wide ribbons depends on water depth.

The most generally accepted cause for regularly spaced sedimentary features aligned with the flow is the presence of secondary circulations which result from the water flowing in a spiral fashion. These circulations take the form of longitudinal helical vortices that are present in contra-rotating pairs, and which create slight cross stream components in the flow, leading to flow convergences and divergences at the sea bed and sea surface (Figure 10.2). The maximum lateral component of the velocity in the secondary circulation is only of the order of 2 per cent of the mean flow, and the helical vortices produce only a small angular deviation of the flow from the longitudinal direction. Consequently it is very difficult to measure their presence. Laboratory experiments have shown that the vortices have a circular or slightly elliptical cross-section, and ideally the ratio of water depth to the spacing between convergences should be 1 : 2 to 1 : 4. If stratification were present, however, the circulation would be confined within the lower layer. At the sea bed a flow convergence would cause an alignment of the mobile material which is swept away from the zone of erosion as a result of the flow divergence.

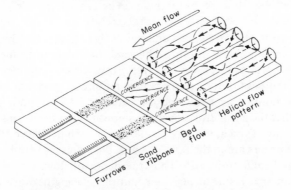

Figure 10.2 Relationship between helical secondary circulations, zones of convergence and divergence at the bed, and the appearance of furrows or sand ribbons. *From Flood, 1981,* Sedimentology, *28, 511–529. Reproduced by permission of International Association of Sedimentologists*

Helical vortices are thought to be common features of boundary layers. In laboratory experiments over flat uniform beds the vortices are random, intermittently appearing and disappearing at intervals over the bed surface. A small scale version may be the low speed streaks present in the turbulent boundary layer. However mean streak spacing is $\lambda = 100 \, v/u_*$. Though this may account for the centimetric scale parting lineations seen in sedimentary deposits, and the linear features seen by Allen (1969) in flume experiments with mud, it is too small for the larger scale features.

Two other mechanisms exist which are most likely to produce larger scale helical vortices in the sea: flow around obstacles and spatial distribution of bed roughness. When the flow encounters an obstacle, such as a small boulder, separation occurs and a wake will be created downstream in which there will be a pair of vortices producing divergence and scour on the bed (Figure 10.3). Once generated the vortices are likely to propagate downstream. Obstacle marks have been observed in the Baltic where strong unidirectional currents are intermittently present (Werner and Newton, 1975), and similar features are often visible in the lee of wrecks (Caston, G. F., 1979).

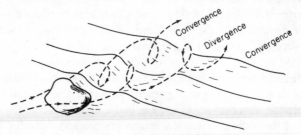

Figure 10.3 Schematic diagram illustrating the creation
of obstacle marks by separation of flow round a pebble

When there are spatial roughness differences there is a higher shear stress over the rough bed than over the smooth. This leads to a higher turbulence intensity over the rough bottom which would drive a transverse component of the flow towards the smoother floor. Continuity then requires a downward flow over the rough bed and upward over the smooth. McLean (1981) found, in laboratory experiments, that the shear stress over the smooth bed is as much as 50 per cent less than the mean. Typical velocity profiles are shown in Figure 10.4. Since the sand transport occurs along the sand ribbons, their presence tends to reduce the overall potential sand transport. Thus the depth of flow, or boundary layer thickness sets the mean turbulence scale and the spacing between the sand ribbons sets the scale of the lateral turbulence gradients. These two factors interact to control the spacing. McLean developed a model for the situation based on a perturbation expansion, which showed that the strength of the secondary circulation reached a maximum when the spacing was four times the depth. When the underlying substrate was hydraulically smooth he found that no ribbons developed. This is comparable to the results of other workers who have observed ribbons developing only when the initial sediment is poorly sorted, enabling a rough stable substrate to form with finer grains moving over it. There is also the possibility that as ripples and larger bedforms develop on the ribbons, they become hydraulically rougher and more comparable to the gravel between the ribbons. The strength of the longitudinal vortices would then diminish, and transverse vortices become dominant.

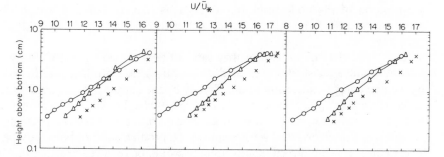

Figure 10.4 Velocity profiles obtained in three flume experiments over a bed composed of longitudinal rough and smooth strips. ○ Profiles taken over the rough bed; △ Profiles taken over the smooth bed; × Profiles taken over the border between rough and smooth strips. *From McLean, 1981,* Marine Geol., *42, 49–74. Reproduced by permission of Elsevier Science Publishers BV*

In Southampton Water the cockle shells form a rough, more mobile surface· than the cohesive mud, and Flood (1981, 1983) considers that longitudinal helical vortices caused their alignment in a similar way to the sand in sand ribbons. Scour around the moving shells prevents deposition of mud in the furrows from suspension, so that once an equilibrium profile is attained the furrow floor has the same deposition rate as the interfurrow areas. Flood (1983) has extended this model to account for furrows in other muddy environments.

Helical vortices may account for the growth and maintenance of longitudinal features under steady flow. However, for oscillating tidal flow a mechanism is necessary to establish the distribution of roughness which the vortices will continue to lock onto despite the variations in velocity and water depth. Under an oscillatory tidal flow furrows and ribbons may have originally been created as obstacle marks. A pattern of flow disturbance would then form on both flood and ebb tides, and interaction between the wakes would tend to emphasise those circulations which are naturally resonant with the flow conditions. The original triggers could then become eroded when the pattern was established and self-sustaining.

Dyer (1982) has suggested an alternative mechanism that could account for the initial stabilization of a roughness distribution over the furrow area in Southampton Water by the action of internal standing waves. Internal waves on the salinity interface (halocline) are produced by a lateral surface seiching of the water and these are reflected from the estuary sides, causing a regular standing wave pattern with nodes and antinodes extending up and down the estuary. Beneath the standing waves frictional effects lead to residual circulation cells with flow upward beneath the antinodes and downwards beneath the nodes. At the bed the to-and-fro motion under the internal waves is superimposed on the residual circulation and, in combination with a longitudinal current, leads to the streamline pattern shown in Figure 10.5. The oscillatory motion is a maximum under the nodes and is zero under the antinodes. Since the circulation is fixed by the position of the nodes and antinodes, a pattern of sediment distribution could be formed, from which a fixed pattern of helical vortices could arise naturally.

Karl (1980) has hypothesized that sand ribbons found on the Californian shelf, normal to the shelf, are created by helical vortices produced in the water by the prevailing winds, rather than directly by the currents. These vortices are called Langmuir circulations and are coupled to rougher windrows on the sea surface. They have very similar depth to spacing ratios, and relative circulation velocities as those vortices discussed above. To produce sand ribbons, however, a very constant wind direction and strength would be required. Further discussion of these Langmuir circulation effects is made by Pantin *et al.* (1981). Alternatively the ribbons could be produced by shelf edge-waves; standing waves created with crest line normal to the shelf by obliquely incident long oceanic waves.

Furrows and sand ribbons remain the least understood of the sedimentary bedforms, but it is apparent that a very poorly-sorted mobile sediment is necessary for their formation, in a flow field capable of moving one fraction of the sediment more frequently than the other. Secondary flow must then align the moving grains into narrow zones, before any sort of equilibrium is established.

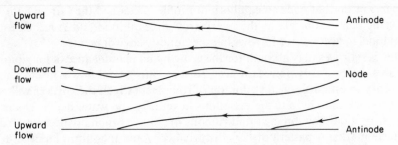

Figure 10.5 Streamlines at the sea bed produced by a lateral standing wave superimposed on a longitudinal current. *After Dyer, 1982,* Sedimentology, *29, 885–889. Reproduced by permission of International Association of Sedimentologists.* Compare with Figure 10.2

SAND WAVES

The definition of sand waves and comparison with bedforms in unidirectional flow has been considered in Chapter 4.

Sand waves have a variety of cross-sectional and plan forms. van Veen (1935) distinguished four forms of sand wave (Figure 10.6): trochoidal, asymmetric trochoidal, cat-back, and progressive. These were related to the tidal currents, the symmetrical forms occurring in areas of equal ebb and flood currents and the asymmetric forms in areas dominated by one tidal stage. Variation in form of sand waves between flood and ebb stages has been reported by several authors. Cat-back forms may also be indicative of crest bifurcation.

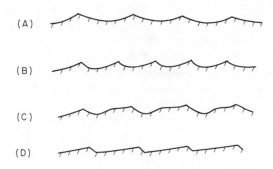

Figure 10.6 Forms of marine sand waves. (A) Trochoidal; (B) Asymmetric trochoidal; (C) Catback; (D) Progressive. Vertical scale exaggerated. After van Veen, 1935

In the asymmetric form the total slope (crest to trough) angle of the steeper lee slope seldom exceeds 10°, and that of the stoss slope is generally half this value (Langhorne, 1973). In mobile areas the lee slope can reach avalanche angles. In spite of having lower lee slopes, the steepness of the sand waves is similar to that of dunes in unidirectional flow, typically being about 1/30. This means that the stoss slopes must be steeper for sand waves. Symmetrical sand waves often show greater steepness than asymmetric ones.

The variation in plan form has been detailed by Kenyon and Stride (1968). They categorize long-crested, sinuous, and barchan-shaped sand waves, and suggest a sequence of form with sediment supply and with the shape of the tidal current ellipse. For sinuous sand waves it is suggested that there is a plentiful supply of sediment and an elongated tidal current ellipse. For straight crested sand waves there is a more restricted sediment supply. Barchan-shaped features are formed where the current ellipse is almost circular and the sediment supply is restricted.

Kenyon and Stride (1970) consider that sand waves occur in areas where maximum surface tidal streams reach between 0.6 and 1.3 m s^{-1}, which makes sand waves common features on continental shelves, near coasts and in estuaries, and they form extensive fields in some areas. The field off the coast of Holland has an area of about 15,000 km^2 (McCave, 1971b). Sand waves are associated with banks (Kirby and Kelland, 1972; Langhorne, 1973) and the continental shelf edge (Cartwright, 1959); they occur in hollows (Harvey, 1966), as intermediate stages on a bed load transport path (Stride, 1963; Belderson and Stride, 1966), and intertidally. They can reach large sizes, with heights of 18 m and wavelengths of 900 m (Stride, 1963).

Sand waves in intertidal areas have been widely studied in order to relate the characteristics of the bedforms, and the flows creating them, to the internal sedimentary structures. Those in the Minas Basin, Nova Scotia have been particularly intensively studied. There the sand waves occur on a series of sand bars (Klein, 1970) and are controlled by both flood and ebb tidal currents. On the steep faces of the bars the sand waves are controlled by both flood currents, whereas on the gentler slopes the ebb current is dominant. During the late stage of the ebb tide the currents tend to flow along the sand wave troughs, draining downslope. The sand waves exist only where the mean grain size exceeds 274 µm, where average maximum currents exceed 57 cm s^{-1} and water depth is greater than 3.5 m. The sand waves average 0.81 m in height and 37.9 m in wavelength, and heights and wavelengths are well correlated (Dalrymple, 1984) through

$$H = 0.0635 \lambda^{0.733}$$

The lee faces of these intertidal sand waves commonly have inclinations of only 10–20°, and they carry megaripples on their backs. Tracer studies have shown that the sand waves have a net sediment discharge averaging 189 kg m^{-1}. However short term migration rates undergo a cyclic variation related to the movement of the overlying megaripples. When the megaripple crest is coincident with the sand wave crest the instantaneous migration rate is high. As the megaripple trough approaches the sand wave crest, the migration rate decreases, or even reverses (Dalrymple, 1984). Because of this effect the sand wave migration rates cannot normally be correlated with tidal range.

Elsewhere, however, sand wave migration rates have been correlated with tidal range. Allen and Friend (1976a), for instance, found what they call 'major dunes' (of height about 0.15 m and wavelength 5–15 m) moved only when the tidal range exceeded 2.60 m. This was equivalent to a maximum ebb velocity of 0.6 m s^{-1} at 0.60 m above the bed. The bedforms ordinarily travelled only a few tenths of a metre per tide, but on the maximum high tide of 2.97 m a movement of 1.2–1.3 m was measured between low waters. For sand waves in a Dutch estuary, Boersma and Terwindt (1981) used a lower threshold for migration of 0.2 m s^{-1}, and found a maximum sand transport rate of ~ 0.8 m^3 m^{-1} tide^{-1}.

Because of the changing flow velocities during the tide it is difficult from the above observations to know precisely what conditions form the bed features. However the largest bedforms are likely to reach equilibrium when the sediment transport is highest, and the lag of the bedform shape behind the flow is less with increasing flow. Allen and Friend (1976b) have termed this lag the 'relaxation time'.

The shape and movement of subtidal sand waves have been studied by many people, mainly by repeated echo-sounding surveys, but the studies generally suffer from problems with accuracy of survey positioning and intermittent coverage.

It is normally assumed that the degree of asymmetry of the sand waves is an indicator of the relative intensity of sediment transport, and some surveys have shown differences between flood and ebb of the crestal shape, e.g. Hawkins and Sebbage (1972). Both Langeraar (1966) and Terwindt (1971) found the movement of sand waves in the southern North Sea over several years to be less than the errors of navigation. Kirby and Oele (1975), however, measured movements of up to 40 m in three years on the Sandettie Bank. Langhorne (1973) reports intermittent movements near the Thames Estuary of up to 25 m y^{-1}, but these were more a flexing of the crest than a steady widespread advance. Salsman *et al.* (1966) working in a tidal bay in Florida measured rates of advance averaging 1.35 cm day^{-1} and Ludwick (1972) obtained movements of 35–150 m y^{-1} at the mouth of Chesapeake Bay. Ludwick also found that the migration of symmetrical sand waves was insignificant. On the other hand some sand waves are ancient structures formed in an environment which no longer prevails (Kirby and Kelland, 1972). Consequently a wide range of migration rates is possible, depending on the local flow conditions.

To overcome the surveying problems some studies have used divers. By diver measurements Jones *et al.* (1965) obtained rates of advance of sand wave crests of 50–100 cm day^{-1} on a sand bank south of the Isle of Man. Langhorne (1982a) has also used divers to measure the crestal shape relative to sea bed reference stakes, though the study was limited to only one sand wave. Surveys were completed at slack waters to cover the differences between flood and ebb tides, as well as the spring to neap variations and the effects of storms. He found that at neap tides no movement occurred. Once the threshold was exceeded sediment was transported up each flank of the sand wave by successive flood and ebb tides and deposition occurred immediately downstream of the crest (Figure 10.7). The newly-deposited sediment had a rounded crest and an avalanche slope with a distinct break of slope at its base. The upstream slope achieved an almost planar form on each half cycle. The crest varied in position by up to 2 m on either side of its mean location at spring tides, but there was a gradual progression of the mean crest position in the direction of the maximum ebb stream totalling 3.5 m. With reduction of the flow after spring tide the sediment transport diminished and crestal oscillation decreased. The sediment movement became increasingly restricted towards the crest, with a consequent increase in crestal height and a gradual retreat in the mean crest position. At the second neap tide the crest was 0.3 m higher than the spring height.

During storms, however, the crest of the sand wave was drastically modified. The wave induced currents caused enhanced erosion at the crest which became considerably reduced in height and much more rounded, but the effects were dependent on the magnitude, duration and direction of the winds. The observed changes are shown in Figure 10.8. The regular movements deduced from this study agree qualitatively with predictions from the theoretical work of Richards and Taylor (1981) described below.

Using measured velocities and the calculated volumes of sediment moved across the crest each half tide, Langhorne (1981) derived a threshold u_{100} of 22 cm s^{-1} and found that the sediment transport rate $q = 0.01 \, (u_* - u_{*c})^3$. In this case the equation represents the total sediment transport, since fluorescent tracer studies showed that the movement did not extend from one wave to the next. More generally, however, sediment may be carried off the crests of sand waves into suspension. Also the presence of megaripples superimposed on the stoss slope of sand waves may indicate sediment travelling as a carpet over the

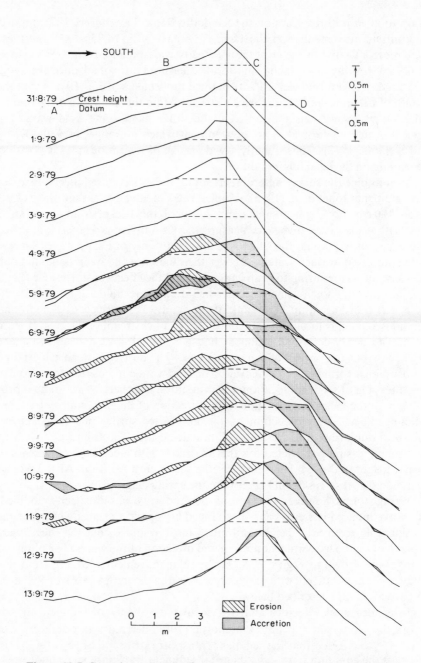

Figure 10.7 Crestal movement of a marine sand wave for a neap–spring–neap cycle. Profiles obtained at successive slack waters from 5 September, but only daily until 4 September. *From Langhorne, 1982a,* Sedimentology, **29,** *571–594. Reproduced by permission of International Association of Sedimentologists*

form of the sand wave. This could produce higher transport rates than would be calculated from the movement of the sand wave itself and in some situations any change in form or position of the sand wave may be the result of changes in the megaripple distribution.

Attempts have been made to understand the processes controlling the shape of the sand waves, and their growth and movement, by application of numerical modelling. Taylor and Dyer (1977) and Richards and Taylor (1981) have used bedforms comparable in shape with natural symmetrical and asymmetrical forms, and considered flow from both directions over the wave form. Because of the small bed slopes the flows were non-separating. The paper of Richards and Taylor (1981) has given the more comprehensive results.

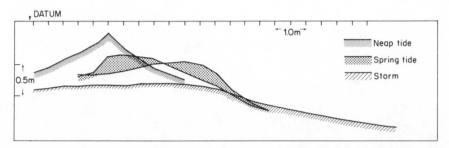

Figure 10.8 Variation in elevation of the crest of a sand wave over neap and spring tides and after storms. The two curves for spring tides show the extent of movement between successive slack waters. *From Langhorne, 1982b,* Int. Hydrog. Rev., **LIX,** *79–94. Reproduced by permission of International Hydrographic Bureau*

The bed shear stresses were calculated from the modelled flow using appropriate bed roughness lengths, and the bedload transport rate was calculated using Bagnold's approach, modified for the effects of bed slope. Equation 7.6 thus becomes

$$q = \frac{e_b \tau^{\frac{3}{2}}}{(\tan \varphi + \tan \beta) \cos \beta} \tag{10.1}$$

The growth rate of the bed is related to the local bed load transport using the sediment continuity equation

$$\frac{dq}{dx} = -(1 - n) \frac{d\eta_b}{dt} \tag{10.2}$$

where n is the porosity of the bed and η_b the bed elevation. Two values of tan φ were considered, tan $\varphi = 0.6$ and 0.3, simulating relatively low and high shears respectively.

For a sine wave, a trochoidal wave, and an asymmetric wave in unidirectional flow with the same amplitude to bed roughness length ratio, the form drag was 27 per cent, 35 per cent and 37 per cent of the total drag respectively. The max-

278

imum bed shear stress was 23°, 15° and 6° upstream of the crest, and the phase difference decreased for relatively smoother beds. For the sinusoidal wave the inclusion of bed slope in the transport formula decreased the phase difference so that maximum transport occurred 12° upstream of the crest (Figure 10.9A) and this decrease was larger with smaller values of tan φ, i.e. for relatively higher shear stresses. The maximum erosion occurred approximately 34° upstream of the crest and the maximum deposition 22° downstream (Figure 10.9B). Flow in the reverse sense produced the same effects on the opposite side of the sand wave, and for an oscillating flow the net effect would be to cause deposition within 38° of the crest, with the maximum at the crest, and maximum erosion at 57° on either side (Figure 10.9C). The crest would tend to sharpen and the trough become flattened, leading to a more trochoidal form.

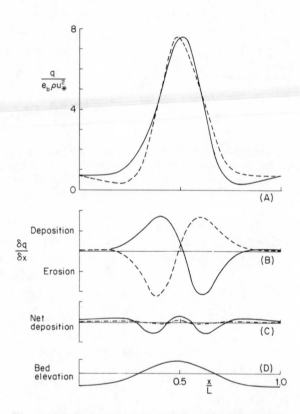

Figure 10.9 Bed load transport over a symmetric wave. Solid line flow right to left, dashed line flow left to right. Tan φ = 0.3. Relative roughness $a/z_0 = 200$. (A) Bedload transport rate; (B) Deposition/erosion rates; (C) Net deposition/erosion over a tidal cycle; (D) Bed elevation. *From Richards and Taylor, 1981,* Geophys. J. R. astr. Soc., **65,** *103–128. Reproduced by permission of Blackwell Scientific Publications*

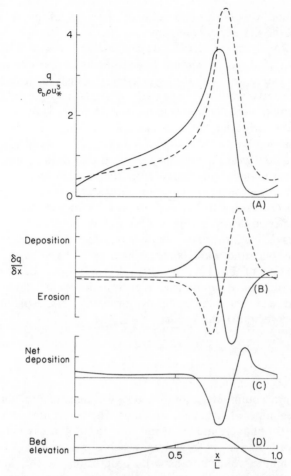

Figure 10.10 Bedload transport over an asymmetric wave. Solid line flow right to left, dashed line flow left to right. Tan $\varphi = 0.6$. Relative roughness $a/z_0 = 200$. (A) Bedload transport rate; (B) Deposition/erosion rates; (C) Net deposition/erosion over a tidal cycle; (D) Bed elevation. *From Richards and Taylor, 1981,* Geophys. J. R. astr. Soc., **65**, *103–128. Reproduced by permission of Blackwell Scientific Publications*

For the trochoidal wave a net erosion curve very similar to the sinusoidal wave was obtained for the same flow characteristics and tan $\varphi = 0.6$. With tan $\varphi = 0.3$, however, the maximum bed load moved to 2° downstream of the crest so that for mean conditions over the tidal cycle a reduction in sand wave height would occur. For a slightly different bed roughness Richards and Taylor (1981) were able to achieve a condition when the net erosion was zero over the entire wave. These results lead to the conclusion that there may be a range of stable symmetrical sand wave forms depending on the value of tan φ; the higher the value of tan φ, the sharper the wave crest.

The results for the asymmetric wave are shown in Figure 10.10. The wave steepness was 1/25 with a maximum slope 0.16. As can be seen it was close to separation when flow was with the asymmetry, from left to right. In a unidirectional flow the wave would become sharper crested and the lee slope steeper. With the reverse flow the crest would move back towards the left, and the net conditions over the tidal cycle would lead to the wave becoming more trochoidal. The only way a progressive asymmetric form could be simulated was by making the flow with the asymmetry exceed that in the opposite direction, i.e. having an asymmetric tidal curve. Even then further factors, such as introduction of a threshold shear stress, or an efficiency factor e_b variable with shear stress, may be necessary to achieve progression of the sand wave without steepening and eventual separation becoming inevitable.

One conclusion from the above study is that for many circumstances, oscillatory flow over a sand wave tends to increase the amplitude, whereas for unidirectional flow of the same intensity a decrease in amplitude would occur. Consequently sand waves would tend to grow in the sea at velocities lower than those required to produce equivalent bedforms in rivers, explaining the lower Froude number regime of the former.

Megaripples

Megaripples are commonly about 5–10 m wavelength and about 30 cm in height, with lee slopes at avalanche steepness. Their crest lines may intersect the sand wave crests obliquely and reach angles up to 60°. Since the sand waves produce a significant amount of form drag, which is proportional to the square of the velocity, and the sediment transport is proportional to the cube of the velocity excess over threshold, the megaripple orientation may reflect the true sediment transport direction. In the troughs of sand waves associated with sand banks, megaripples sometimes occur with crests normal to the sand wave crests, suggesting sediment transport at right angles to the sand wave orientation, as will be discussed further on page 282.

On the backs of the megaripples are normal ripples whose wavelength is considered to be proportional to the sediment grain size. Richards (1980) has suggested that the megaripples may be controlled by the roughness produced by the ripples. He finds that the ripple mode forms with a wavelength between $50z_0$ to $1000z_0$. A rippled bed of wavelength 20 cm would create a $z_0 \sim 0.5$ cm, and a megaripple of 0.25 to 5 m wavelength is possible. Megaripples would then be the second stage in a hierarchy of ripples built upon the roughness of the first. Certainly megaripples do not show close correspondence with the depth of flow, and Langhorne (1977) has even shown the reverse, with megaripples longer on the crest of sand waves than in the trough. This may be the result of grain size variation over the sand wave creating ripples of different sizes, but it would require coarser sediment at the crest than in the trough. Langhorne (1977) has also reported very significant changes in megaripple wavelength produced by gale conditions. Short wavelength megaripples were found immediately after gales which, as time passed, were gradually modified and increased in wavelength to their tidal equilibrium values. Near the crest 50 per cent of the equilibrium wavelength was attained in approximately 60 hours and 90 per cent in 340 hours. However the trough took about twice as long to recover the same amounts. These results are a good measure of the relaxation time of megaripples. The recovery is diagrammatically shown in Figure 10.11. Values for the mean megaripple wavelengths 35 hours after the storm maximum were 1.72 m, 1.67 m and 1.55 m for crest, mid-slope and trough respectively. Wave measurements showed a significant wave height of 1.6–2 m and period of 5 s during the storm. These conditions would lead to near-bed particle orbital diameters of 0.3–0.5 m in a water depth of 20 m. Consequently considerable recovery had already occurred from purely wave-induced megaripples, if they were originally formed.

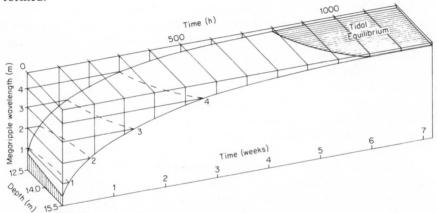

Figure 10.11 Recovery of megaripples to tidal equilibrium after a storm. *From Langhorne, 1977,* Int. Hydrog. Rev., *LIV, 17–30. Reproduced by permission of International Hydrographic Bureau*

Megaripples are observed intertidally and Dalrymple *et al.* (1978) distinguish two distinct types. Both types have lee faces at the angle of repose, and their average height is 0.18 m and wavelength 2.8 m. Type 1 megaripples are simple in appearance, with straight to smoothly sinuous lee faces. Crest lengths are long and height is very regular. Type 2 megaripples are more complex and sinuous. The troughs contain well-developed scour pits, and height varies considerably along the crest line. The Type 1 megaripples occur in sands coarser than 180 μm and at lower velocities than the Type 2 megaripples, and they are not as steep. Up the stoss side of a sand wave, Type 1 megaripples would give way to Type 2, with both height and wavelength increasing towards the crest. The megaripples apparently reverse their orientation, at least at their crests, with reversal of current.

The intertidal megaripples are also oriented at an oblique angle to the sand waves, averaging 23° (Dalrymple, 1984). Tracer experiments showed that megaripples were oriented at a smaller angle to the direction of net sediment transport than the sand waves, and responded very much more to the local currents. The sand waves progressed unevenly along their length, as a result of their large lateral extent which permitted them to span marked hydraulic gradients.

There are very strong similarities between subtidal and intertidal megaripples, and their relationships to the underlying sand waves. There is little information on subtidal megaripple rates of movement, since echo-sounding surveys have depth and positional inaccuracies greater than megaripple dimensions, but intertidal megaripples travel 10–50 times faster than sand waves (Boothroyd and Hubbard, 1975). Thus an understanding of megaripple dynamics is crucial when considering sediment transport through sand wave areas.

SAND BANKS

Sand banks form significant features on many continental shelves and coastal regions, and they often store the major proportion of the available sand. There are two main categories of sand bank: longitudinal sand ridges, or linear sand banks; and headland associated, or banner banks. Though the latter type is closely controlled by the flow pattern in the vicinity of the headland, there are also close similarities between these and the linear sand banks found in more open water. Generally the banks are formed from medium or coarse sand, but when the tidal currents are sufficiently strong and there is a source of gravel, gravel banks of a very similar form can be found.

It is evident that not all sand banks are presently active. Sand waves occur on active banks and the banks have an asymmetrical cross-sectional form with the steeper face having a slope up to 6°. The moribund banks have lower slopes, are more symmetrical in cross-section, and do not possess sand waves. In many cases moribund banks have crests that do not approach the water surface.

There is normally an asymmetry in the current strengths across the banks with maximum currents being in the ebb direction on one side and the flood direction on the other. Correspondingly the sand wave asymmetry is in different directions on either side of the bank, while at each end there is normally a zone of symmetrical sand waves. The megaripples are active indications of the sediment movement and they show a gradually changing orientation towards the bank crest, indicating a substantial component of sediment transport across the bank axis (Houbolt, 1968; McCave and Langhorne, 1982). Caston, V. N. D. (1972) showed that sand wave orientation also varies, with angles of 40° between their crests and the bank axis on the gentle slope side, and 30° on the steep. At the ends of the banks, in the zone of symmetrical sand waves, the megaripples trend at very high angles, 55° to the sand wave crests according to McCave and Langhorne (1982), indicating transport effectively along the troughs. The obvious conclusion is that sediment is circulating round and over the bank and this process helps to maintain the bank (Smith, 1969). As banks generally stand on a floor of less mobile gravel or rock, there must be components of the water flow which help to keep the sediment from dispersing more widely.

Internal structure in the banks observed on seismic profiles indicates that the banks gradually progress in the direction of the steep slope. Sand brought across the crest of the bank tends to build the steep slope gradually outwards. Since the crests of the banks are often at or a few metres below low tide mark, wave processes must be important in enhancing the sediment transport across the crest and limiting its vertical growth. The grain size is normally finest and best sorted at the crest.

McCave and Langhorne (1982) have constructed an illustrative box model to describe the main features of sediment transport around a sand bank. This model is shown diagrammatically in Figure 10.12. It assumes that the bank is effectively a closed system, that the highest fluxes are about $250 \, \text{m}^2 \, \text{y}^{-1}$, and that the cross bank flux implies a lateral migration rate for the bank of 2.5 m y^{-1}. In order to have equal transport along the two sides, the steeper northern side requires much higher fluxes per metre than the southern side. Using the model, it was estimated that it takes 550 years for sand to travel round the bank. However Caston, G. F. (1981) found no evidence for sand circulation though sand waves faced in opposite directions on either side of the banks, and she suggested that sand enters the banks through a low apron at the broad head of the bank and leaves via the narrower tail.

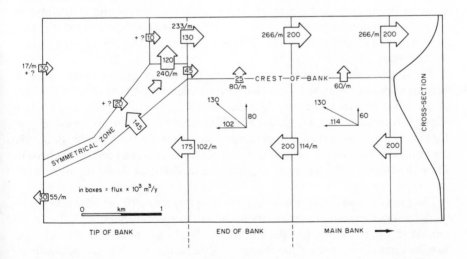

Figure 10.12 Box model of sand circulation around the end of Haisborough Sand. Fluxes in boxed arrows are annual values in thousands of m^3 passing across the side of the box which they cut. Fluxes written alongside arrows are values of m^3 m–width^{-1} y^{-1}. The two vector diagrams are labelled with fluxes as m^3 m^{-1}. For further explanation see text. *From McCave and Langhorne, 1982*, Sedimentology, **29,** *95–110. Reproduced by permission of International Association of Sedimentologists*

Caston, V. N. D. (1972) has suggested that there is an evolutionary sequence in the development of linear sand banks. First slight kinks develop in them which become amplified into S-shaped meanders and which eventually separate into two or three parallel banks. Also Ludwick (1975b) has described how sand banks have originated in the entrance to Chesapeake Bay by growth and separation from the coast during transgression of the sea. Similarly, banks have been formed on the open coast of eastern America by inundation of spits formed where the longshore transport of sand was perturbed (Swift, 1976).

Linear Sand Banks

Nearly all shallow tidal seas, where currents exceed about 0.5 m s^{-1} and where sand is present, have ridges. They can be up to 80 km long, and typically 1–3 km wide and tens of metres high. Their spacing tends to be proportional to their widths. The characteristics and distribution of ridges was first examined by Off (1963) and he suggested that they could be formed by the presence of longitudinal helical vortices. This mechanism was also invoked by Houbolt (1968) who further suggested that asymmetrical banks could be caused by vortices with unequal strengths. An alternative explanation was put forward by Stride (1974), who pointed out that the majority of linear banks have their steep slope corresponding with the sense of rotation of the tidal current vectors. He reasoned that, since the rate of sand transport is a steeply increasing function of shear stress, and there is a lag in the transport of suspended sediment, the maximum sediment transport occurs after the maximum current and at an angle to it, in the direction of rotation of the tidal ellipse. It is not clear, however, that suspension transport is generally sufficiently important for this to be a significant process.

The validity of the helical vortex explanation was put in doubt when it became apparent, from more detailed measurements, that the linear banks were at an angle to the maximum tidal currents (e.g. Caston, 1972) rather than aligned with them. McCave (1979b) made current measurements specifically to try and measure any helical components of flow near the West Hinder Bank. He could find no evidence for helical flow but obtained Eckman veering throughout the water column. The whole tidal ellipse was at 10° to the direction of the bank axis.

The axes of sand banks are commonly offset by up to 20° with respect to the peak tidal current, and the majority in the northern hemisphere are offset anticlockwise, i.e. to the right of the current. Only a minority are offset clockwise, and it is not clear whether the dominant direction is reversed in the southern hemisphere. The ends of the banks turn more obliquely across the currents, however.

When a tidal current crosses a ridge, the flow will accelerate because of the decreasing water depth, but the increased friction tends to reduce the current. Which effect dominates depends on the roughness and steepness of the slope, but normally the current over the top of the bank is greater than that in deep water on either side.

The only current measurements taken simultaneously at several positions across a linear bank appear to be those reported by Venn and D'Olier (1983). Three positions were occupied on a transverse line up the western slope of the long straight South Falls Bank. Between 40 m and 20 m water depth the on-bank component at a height of 10 m above the bed nearly doubled, but between 40 m and 10 m water depth the velocity 2 m above the bed increased by a factor of 3, rather than 4 which would be expected from continuity considerations. The longitudinal component, however, was approximately constant at the same heights above the bottom at the measurement stations. Combining these results indicates that the resultant current turned towards the bank as the crest was approached.

Let us consider a parcel of water approaching an anticlockwise ridge (a ridge at a slight angle to the right) in the northern hemisphere (Figure 10.13A). Because of the increasing velocity towards the crest, the Coriolis force at the leading edge of the parcel is greater than the mean, and at the trailing edge less. This gives a clockwise torque on the water. Likewise the frictional force on the right hand edge is greater than the mean friction, and that on the left hand side less, also giving a clockwise torque. On the lee slope the opposite rotations occur. Thus the stream-line will be deflected to the right as it passes over the bank, being almost normal to the axis at the crest, but, because of fluid inertia it is displaced slightly downstream. When the current reverses, the same pattern of rotations is produced together with a downstream displacement. Averaging over the tidal cycle gives a clockwise residual circulation of water and sediment. Because of the overall decrease in the current strength due to the presence of the bank, the sand transport onto the bank will be greater than that off it, leading to an asymmetrical cross-section, and this is amplified if one phase of the current is stronger than the other.

With a clockwise ridge (Figure 10.13B), the Coriolis force differences will be in the same sense as before, but the friction will now be greater on the left hand side than on the right, giving a left hand torque opposed to the right hand one of the Coriolis force. Since friction is proportional to the square of the current velocity and Coriolis force only to the velocity, friction becomes increasingly important with increasing velocity. Consequently the net water and sediment circulation round a clockwise ridge is in the anticlockwise sense. Obviously this circulation would be weaker than that for an anticlockwise ridge.

Huthnance (1973) has examined the importance of friction and Coriolis force on an infinite anticlockwise bank and concludes both are important, together with inertia, in producing clockwise circulation. The analysis was carried further in Huthnance (1982a, b) by including sediment movement. In the first paper he considers an infinite length bank with depth uniform tidal currents, but without Coriolis force. Sand transport was assumed, using Bagnold's hypothesis, to be proportional to the third power of the velocity and the effect of the bed slope on the transport was taken into account. He found that there was a maximum bank growth rate when the separation between banks was 250 times the water depth and a minimum timescale of 20 years was required for their growth. For a limited sand supply the banks narrow to about one fifth their

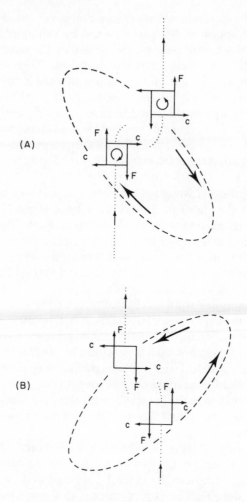

(A)

(B)

Figure 10.13 Forces acting on a water element passing over (**a**) an anticlockwise ridge; (**b**) a clockwise ridge. Dotted lines, the current stream-lines; C Coriolis force relative to the mean; F Friction relative to the mean; Thick arrows resultant sediment circulation. After Zimmerman (1981)

separation. The orientation of the sand bank tended to stabilize at around 27.8°, but with a broad maximum that may make the final orientation susceptible to outside influences such as the coastline. Comparison with the Norfolk banks shows that the observed steeper slope on their northern sides was due to sand transport on the stronger ebb tide. The difference between on-bank and off-bank transport depended strongly on the exponent of velocity in the sediment transport formula.

Using very similar assumptions Huthnance (1982b) examined the stability of elliptical banks and found that the bank tends to narrow and lengthen provided the angle of inclination of the bank to the current was between 18°–48°. It also tended to rotate towards an inclination of about 27°, which was stable and close to the inclination of maximum extension rate and volume growth rate. The influence of waves was also considered by inclusion of Bijkers' stress enhancement factor. The results showed that the bank was lower and flatter topped with waves present. Bank height was also affected by limited sand supply since any increase in bank height is at the expense of bank width, so that the sides tend to steepen until the increased downslope transport inhibits further growth. An overall bedslope in the downstream direction was found to prevent the formation of stationary banks.

Zimmerman (1981) has reviewed the vorticity effects around headlands and over banks, and considered the length scales of the features. The length scale ratio

$$\lambda = \frac{\pi \times \text{ tidal excursion}}{\text{ridge length}}$$

was derived as a fundamental parameter. Comparison with the dimensions of the ridges quoted by Off gave $\lambda = 1.9 \pm 1.2$ and this was the range where the maximum response of the residual circulation was expected. Zimmerman applied the analysis to relict linear ridges to the north west of the Dogger Bank in the North Sea, formed about 7000 BC when the sea level was 40 to 45 m below present. According to the ridge length (33 m), the tidal current speeds, using the above scaling factor, must have been at least twice their present value of $\sim 0.5\,\text{s}^{-1}$.

Both the length scale and the inclination of linear ridges to the tidal current are such that maximum residual circulation is produced round the ridge. The circulation must be necessary to maintain the ridge shape, and possibly to minimize the loss of sand in the direction of net flow.

Headland Banks

In areas where the current strengths exceed about $0.5\,\text{m s}^{-1}$ sand banks are likely to appear on one side or other of the headland on the coast. In some cases a bank occurs on both sides of the promontory. The presence of the banks also appears to be aided by a fairly rapid deepening of water away from the coast; they are less evident off coasts with a low offshore slope. The banks are only a few kilometres in length and with an elongated pear-shaped form, the broad end being directed towards the tip of the headland. The shallower part is nearest the headland and towards the pointed end the crestal depth gradually increases. The side of the bank facing the sea has a steeper slope than that on the landward side, and the bank is separated from the headland by a deep narrow channel. Within this channel there is a very marked asymmetry of the tidal current giving a residual flow towards the headland. On the outer side of the bank there is an

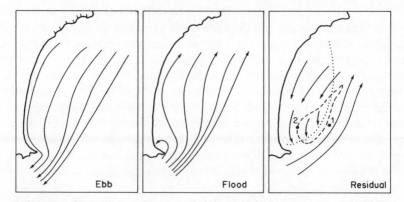

Figure 10.14 Schematic representation of the stream-lines of the maximum ebb, flood and of the residual currents around Start Point and over the Skerries Bank. Positions 1 and 2 show the positions of the current measurements in Figure 10.15. Dotted line, zero residual flow. Dashed line, outline of banks. After Pingree and Maddock (1979), and Acton and Dyer (1975)

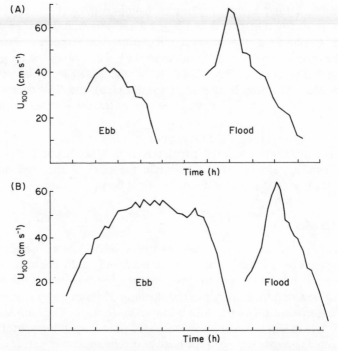

Figure 10.15 Example of current measurements obtained on either side of a banner bank, Skerries Bank, S. Devon. The measurements were not obtained at the same time. (A) Outside the bank at a flood dominated site Station 1. (B) Inside the bank at an ebb dominated site Station 2. The positions of the measurements are shown in Figure 10.14

asymmetry in the opposite sense and a residual away from the headland. Consequently an eddy in the residual water circulation is inferred. Figure 10.14 shows the flow around Start Point, South Devon and over the Skerries Bank formed in its lee. The stream-lines of the ebb and flood currents are not coincident, and the residual flow shows an eddy, with the line dividing ebb and flood dominance lying along the axis of the Bank. Pingree and Maddock (1979) have shown, using a mathematical model, that though the mean bed stress tends to be directed around the Skerries Bank in an anticlockwise sense, the maximum stress is generally northwards, particularly over its southern end. This is shown by comparison between the results of current measurements made on either side of the bank (Figure 10.15). On the inshore side, the short sharp flood stage gives a higher stress than the longer flatter ebb, but the asymmetry is sufficient to ensure that the mean sediment transport is in the ebb direction. On the outside of the bank, the ebb current is shorter than the flood and is less strong. The asymmetry of the tidal cycle results from the current turning earlier close inshore towards the end of the flood tide, and flowing anticlockwise around the tip of the bank. Sand wave facing directions suggest a residual circulation of sediment corresponding to that of the water. However, consideration of the fluid dynamics of the situation shows that the residual eddies in the water flow are created by the headland, and that the banks then arise from the eddies, rather than vice versa.

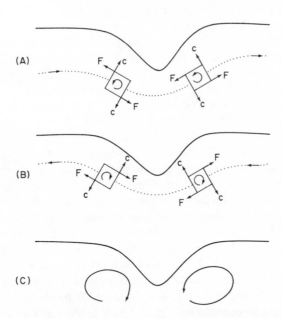

Figure 10.16 Illustration of the mechanism of vorticity generation by tidal flow round a headland. (A) Flood tide. (B) Ebb tide. (C) Residual current eddies. C Coriolis force. F Frictional force

Let us consider a tidal flow in the northern hemisphere past the idealized headland shown in Figure 10.16. Offshore the tidal current will be virtually rectilinear, but as the coastline is approached the currents become perturbed by the headland, they tend to accelerate as the headland is approached and decelerate after they pass it. Near the coast during the flood tide, as the current streams towards the east, it is deflected by the coast and continuity requires an acceleration. Because of the higher velocity the laterally-directed Coriolis force increases towards the headland. This is balanced by a slight increase in the water slope transverse to the current, and this leads to the water level close to the tip of the headland becoming relatively depressed. In addition the curvature of the stream-lines around the tip of the headland produces an outwardly directed centrifugal force, also balanced by an increase in the offshore water slope.

The forces on a parcel of water in the flow can be considered in the same way as was done for linear sand banks. On the upstream side of the promontory Coriolis and centrifugal forces increase, and create clockwise vorticity, but their sum is more than exceeded by the anticlockwise vorticity caused by the frictional couple due to the velocity rapidly increasing towards the coast. In the lee of the headland the frictional couple is reduced by the expansion of the flow, but the Coriolis and centrifugal forces are now decreasing because of the diminishing current speed, so that they add to anticlockwise vorticity. Fluid inertia causes an advection of the vorticity downstream, leading to a net anticlockwise vorticity on the east side of the headland.

On the westward-flowing stream the reverse situation occurs and, in this case, clockwise vorticity is generated and advected to the western side of the headland. Again the water level at the headland is depressed and there is a mean lowering of sea level. In the vicinity of Portland Bill on the south coast of England, Pingree and Maddock (1977) have estimated a lowering of mean sea level of 15 cm. However, on the westerly flow, Coriolis force and centrifugal force will be opposing rather than enhancing each other, and it is to be expected that the westerly eddy would be rather weaker than that to the east.

The result of the vorticity generation is the production of a pair of eddies in the mean circulation, one on either side of the headland. There is a residual circulation of water towards the headland close inshore, and a strong residual flow seawards from the tip of the headland, known as the headland flow. The relative strengths of the eddies are likely to be sensitive to the orientation of the coastline on each side of the headland with respect to the flow direction. Consequently a symmetrical headland such as Portland Bill can produce an eddy on either side, but an asymmetrical one such as Start Point, at a step-like widening of the English Channel, produces only one (Pingree and Maddock, 1979).

The influence of the general offshore slope on the production of the residual eddies is apparent by considering Figure 10.16. If the water shallows gradually towards the shore, the current strengths will be considerably smaller, causing the current velocities round the tip of the headland to be relatively less. Consequently the frictional torques will be diminished and the eddies will be less well developed.

The processes causing the generation of the residual eddies have been discussed by Pingree and Maddock (1977, 1979) Pingree (1978), Heathershaw and Hammond (1980), Zimmerman (1981) and Robinson (1981). As can be seen from the above discussion, the eddies are generated even when the bank is absent, but we can now consider how the presence of the eddy may cause a bank to form. We have already stated that there is a balance between the Coriolis and centrifugal forces, and the surface water slope. Of course the current velocity is greater at the surface than at the bottom and consequently both the Coriolis and the centrifugal forces are largest at the surface and decrease towards the bed. In fact centrifugal force decreases faster as it depends on the square of the velocity. The water slope gives an opposing force that is constant with depth. Since a balance is only maintained in terms of the depth average, in the anti-clockwise eddy there is a net outward force at the surface and a net inward force near the bed. These slight inequalities drive a secondary circulation directed towards the centre of the eddy near the bed and away from it on the surface with an upwelling in the centre (Pingree and Maddock, 1979). The near-bed convergence causes accumulation of sediment, in much the same way as convergence beneath helical vortices causes sand ribbons. Eventually as the bank develops, the tidal flow and the secondary circulation become modified by its presence until an equilibrium is created, so that the net sediment transport is around the bank rather than having a significant component towards its crest. Soulsby (1981) has reported current measurements made on the Skerries Bank, in the lee of Start Point (Position 1, Figure 10.14). Near the bed the residual current ran nearly parallel to the isobaths, but the surface current flowed approximately across the bank. However turbulence measurements made 65 cm above the bed gave a mean Reynolds stress with an azimuth directed 6° anticlockwise of the near-bed current, i.e. slightly towards the bank. Consequently fine sediment in suspension was more likely to be carried towards the bank than any coarser material in suspension or in bedload motion. Thus an eventual sorting of sediment may result once the bank is formed, with finer material near the crest.

In the clockwise eddy Coriolis and centrifugal forces act in opposite directions and if Coriolis is the greater then a bottom divergence could result from the presence of a net inward surface force and a net outward directed bottom force. An approximate prediction of whether this will occur can be obtained by consideration of the Rossby number (Pingree, 1978). The Rossby number R_0 compares the relative importance of Coriolis and centrifugal effects. Thus

$$R_0 = \frac{u^2/R}{fu} = \frac{u}{fR}$$

where R is the radius of curvature and f the Coriolis parameter.

When $R_0 \gg 1$, both clockwise and anticlockwise eddies produce convergence of water near the bottom and conditions suitable for bank formation. If $R_0 \ll 1$, Coriolis force dominates. Then for the anticlockwise eddy in the northern hemisphere the near-bed flow will be divergent. For $R_0 \sim 1$, the anticlockwise eddy will have convergent bottom flow, but the clockwise eddy should have no tendency for bank formation since the centrifugal force will just balance the Coriolis force.

Portland Bill has an $R_0 \sim 1$ and one bank is present in the anticlockwise eddy to the east of the headland. On the other hand Lundy Island has an $R_0 \sim 5$ and four banks are present, two on either side of the island (Pingree, 1978). Scarweather Bank in the Bristol Channel, formed in the clockwise eddy west of Porthcawl Point, has an R_0 slightly greater than unity (Heathershaw and Hammond, 1980), but no bank can be found on the eastern side of the point because of the constriction of the channel. Current measurements at a few points around the bank demonstrated the presence of a mean near-bed convergence in both the water and sediment circulations.

Alternatively, by analogy with solid rotation, Pingree (1978) has suggested a criterion for sand bank formation based on the ratio between the tidal vorticity $\bar{\omega}$ and Coriolis force, where

$$\bar{\omega} = \frac{\delta \bar{v}}{\delta x} - \frac{\delta \bar{u}}{\delta y}$$

The overbar denotes a tidal mean value. The vorticity can thus be calculated from the spatial velocity field. When $\bar{\omega}/f > 0$ sand banks form on the right hand side of the headland in the northern hemisphere, looking from the sea, and when $\bar{\omega}/f < -2$ they form on the left hand side. To the east of Portland Bill $\bar{\omega}/f$ is positive, but to the west the quotient is more negative than -2 over only a small area. This supports the above analysis in terms of Rossby number.

Both linear sand banks and the headland associated banks are created beneath eddies that result from the transfer of vorticity from the tidal component of vorticity into the mean circulation. Again the presence of the banks and the circulation of sediment provides a good means of sorting the sediment, as well as reducing the rate of loss of material down the regional transport path.

As can be seen in this chapter the sediment moves along the transport path in response to the fluid forces, but it does so in a manner that minimizes the overall transport rate. In furrows and sand ribbons the sediment is driven towards the zones of lowest bed stress. When megaripples and sand waves develop, the form drag reduces the friction available at the bed, and consequently the transport rate. The circulation round sand banks also causes the regional transport rates to be low. These effects could be thought of as an application of Le Chateliers' principle. This states that every change of one of the factors of an equilibrium brings about an adjustment which tends to minimize the original change.

The continual movement on the transport paths enables the finer sediment to be winnowed out, and the coarser material to be left behind, so that, at each stage the sediment becomes well sorted. At the moment our knowledge of the dynamics of sediment transport over bedforms in oscillatory flow, especially where the sediments are poorly sorted, is inadequate for the quantification of regional sediment transport patterns.

CHAPTER 11

Beach Processes

SHALLOW WATER WAVE EFFECTS

In the nearshore zone the decreasing water depth affects the wave motion, eventually leading to the waves breaking and running up onto the beach. Four zones can be distinguished (Figure 11.1):

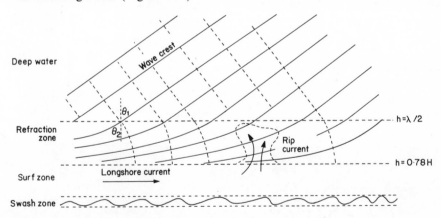

Figure 11.1 Zonation of the nearshore region caused by wave refraction, breaking and run-up onto the beach. Dashed lines are the wave orthogonals

1. Deep water: here the water depth is large compared with the wavelength of the waves, the wave crests are straight and the velocity of propagation c and angle of incidence θ, relative to the shore, are constant.
2. Refraction zone: the waves feel the bottom, their wavelength and velocity vary as they progress, though their period remains constant. The wave crests are progressively bent to parallel the shoreline, and the waves steepen and eventually break.

293

3. Surf zone: this is the zone between the breaker point and the shore. The breaker line is not a fixed line since the position of breaking varies with wave height, higher waves breaking farther down the shore than smaller ones. Within the surf zone measurements have shown that there is a well-defined longshore current.
4. Swash zone: this is limited by the highest point on the beach that the breaking waves run up to, and the lowest point to which the water recedes between waves.

In the alongshore direction there are water circulation cells that are often associated with the larger scale topographic features of the beach. At intervals along the beach strong rip currents may penetrate from within the surf zone to well outside the breaker line. They can reach a maximum distance offshore of about three surf zone widths and attain speeds of $2 \, \text{m s}^{-1}$. The rip currents are fed by the longshore currents within the surf zone and end in an expanded rip-head offshore where the currents fairly rapidly diminish. The discharge of water offshore in the rip currents is balanced by an onshore transport through the rest of the breaker zone, and the longshore currents must be driven by an alongshore gradient of the mean water level. Consequently the speed and length of the rip currents are related to the offshore wave height, and observations have shown that the rip currents are located in areas of locally lower breaking waves.

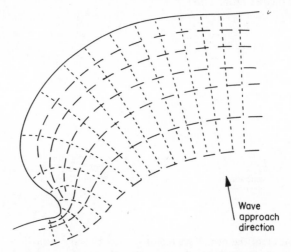

Wave approach direction

Figure 11.2 Refraction of waves in a ζ curve bay. Wave orthogonals (dotted lines) show normal approach to beach throughout the bay, but concentration of energy on the headland

The nearshore water circulation cells affect the sediment circulation and, under conditions of steady wave input, the beach shape adjusts so that the intensity of the longshore currents and the rip currents are minimized. When this happens on a large scale the log spiral or zeta curve (ζ) bay can be formed. These bays develop particularly in situations where a single headland protects one end of the beach from a dominant wave direction. A combination of wave refraction and diffraction causes a tight curvature for the beach close to the headland, but it gradually turns to face the prevailing wave direction further away (Figure 11.2). The prevailing waves then tend to approach the beach normally over its whole length and this minimizes the longshore transport and maintains equilibrium.

A crucial factor in the shape and equilibrium of beaches is therefore the intensity of the longshore currents. These are affected by wave refraction, wave set-up, edge waves and oblique wave approach.

Wave Refraction

When waves travel into shallow water they are subject to refraction, because their speed of travel diminishes with water depth. As the wave crest approaches obliquely towards the shore (Figure 11.1), the nearshore end will travel slower than its offshore end. Consequently the crest bends round to become progressively more nearly parallel to the depth contours and to the shore, with the wave orthogonals becoming more normal to the shore. The process is very similar to the bending of light rays in optics and can also be represented by Snell's Law

$$\frac{\sin \theta_1}{c_1} = \frac{\sin \theta_2}{c_2} = \text{constant} \tag{11.1}$$

where c_1 and c_2 are the wave velocities at two different depths, values of which can be calculated from Equation 3.

As the waves are refracted the wave orthogonals may either spread out or converge. For the simple straight beach with oblique incidence shown in Figure 11.1, the spacing of the orthogonals at the shoreline is greater than that along the wave crest offshore. Providing there are no energy losses the energy flux in the shoaling waves remains equal to its deep water value and as the wave approaches the shore its energy E, and thus its height, increases. The wave energy flux per unit width $P = Ec_g = ECn$, where c_g is the group velocity, c the phase velocity, and n equals unity in shallow water and a half in deep water. Waves lose energy by bottom friction, breaking and turbulence. Since the wave energy flux between orthogonals remains essentially constant up to break point, it is easy to see that the wave energy per unit beach length is less for oblique approach than for the same waves approaching directly from offshore. Also the wave height at the beach would be greater for the normal approach case.

With more complex offshore topography it is possible to have focussing of wave energy, or the converse, at different points along the coast. A convex underwater profile, a ridge, leads to convergence, and a concave bay causes divergence. As can be seen from Figure 11.2 wave heights will normally be greatest on headlands and least in bays.

Wave Breaking

For the Stokes wave in deep water Michell showed in 1893 that there is a limiting crest angle of 120°. This gives a theoretical limiting wave steepness in deep water of

$$(H/\lambda)_B = 0.142 \tag{11.2}$$

where the subscript B denotes the value at breaking.

In shallow water breaking waves act as solitary waves, and their maximum amplitude is controlled by the depth. Obviously the point of collapse occurs when the orbital velocity at the crest exceeds the phase speed of the wave. This should occur when $H/h = 1$. Observations give the ratio $(H/h)_B$ varying between 0.73 and 1.03, though McCowan's results of $(H/h)_B = 0.78$ is the most widely used. However Tucker *et al.* (1983), from a study of the effect of an offshore bank on wave height on its nearshore side, found a ratio $(H/h)_B$ of 0.48 and 0.53 at high tide and low tide respectively.

Breaking over the entire range of depth is covered by the formula

$$(H/\lambda)_B = 0.14 \tanh (2\pi h_B/\lambda_B) \tag{11.3}$$

in which λ_B is the wavelength of the breaking wave from linear theory. Equation 11.3 gives a limiting steepness in deep water of 0.14 and in shallow water of 0.88.

There are three types of breaking wave: spilling, plunging and surging (Figure 11.3). With spilling breakers, the spilling results from the crest portion of the wave moving slightly faster than the rest of the wave. This cascades down the front face of the wave and in deep water causes the amplitude to diminish

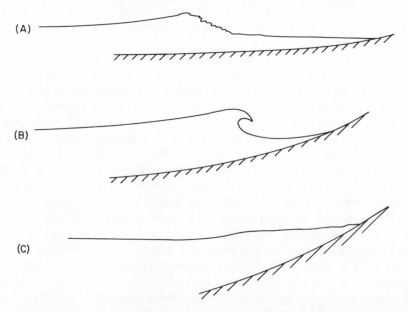

Figure 11.3 Types of breaking waves. (A) Spilling; (B) Plunging; (C) Surging. Not to scale

slightly so that the wave slows down and resumes its smooth profile. With plunging waves the whole front face of the wave steepens until vertical, and the crest curls over and falls onto the base of the wave. The entrapped air erupts from the back of the wave as a whole wave collapses. With surging wave the front face and crest of the wave remain fairly smooth and the wave rushes up the beach with only minor amounts of foam.

There is a continuous sequence of breaking wave types, and the type is a function of the deep water wave steepness and beach slope. Spilling waves occur with steep waves on gently inclined beaches, plunging waves are of intermediate steepness on steeper beaches, and surging is of low amplitude waves on steep beaches. Galvin (1972) has proposed the parameter $B_B = H_B/g \tan \beta \, T^2$ as distinguishing the different wave breaker types, where $\tan \beta$ is the beach slope. The surge–plunge limit was given as $B_B = 0.003$, and the plunge-spill limit as $B_B = 0.068$. Alternatively for $\tan \beta < 0.1$ results can be represented by

$$(H/h)_B = 0.72 (1 + 6.4 \tan \beta) \tag{11.4}$$

This gives values of the height to depth ratio between 0.72–1.18 which corresponds well with that stated above.

Gaughan and Komar (1975) have examined the relationship between the deep water wave steepness, the beach slope and the occurrence of plunging and spilling breakers. Their results are shown in Figure 11.4.

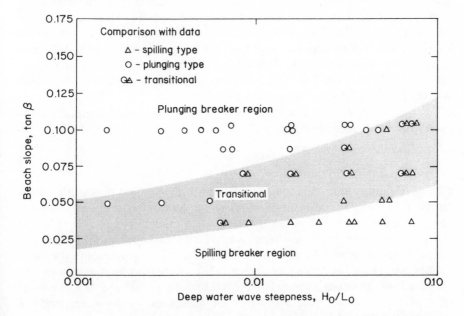

Figure 11.4 Relationship between breaker type, deep water wave steepness and beach slope. *From Gaughan and Komar, 1975, J. Geophys. Res., **80**, 2991–2996. Reproduced by permission of American Geophysical Union*

Upon breaking some of the wave energy is reflected back out to sea by the beach. In general steep beaches are reflective and low slopes are dissipative, though the effect also depends on the wave characteristics. A simple dimensionless parameter Ω has been used to express the reflective nature of the beach

$$\Omega = H_B/w_sT$$

where w_s is the settling velocity of the beach sediment and T the wave period. A value of $\Omega < 1$ denotes a reflective beach, and $\Omega > 6$ a dissipative beach. Between these values the intermediate beaches show strongly barred profiles (Wright and Short, 1984).

Wave Set-up

As the deep water waves encounter a sloping beach and break they cause a change in the level of the mean water surface. Initially near the breaker line there is a set down, a lowering of the mean surface. This is followed further inshore by an almost linear increase in elevation, the wave set-up, which can reach the order of a metre above still water level at the edge of the swash zone (Figure 11.5). The slope of the water surface is proportional to the beach slope, and the seaward pressure gradient of the sloping water surface is balanced by the gradient of the incoming momentum. The surface waves possess momentum which is directed in the direction of propagation and is proportional to the square of the wave amplitude. If a wave train is reflected from an obstacle its momentum must be reversed. Conservation of momentum then requires that there is a force on the obstacle equal to the rate of change of wave momentum. This force is a manifestation of what Longuet-Higgins and Stewart (1964) have termed the radiation stress. The change of momentum is by definition equivalent to the stress. The radiation stress may thus be defined as the excess flow of momentum due to the presence of waves.

For waves approaching the shore with crests parallel to the shore, the radiation stress in the offshore direction S_{xx} is

$$S_{xx} = \frac{3}{2} E = \frac{3}{16} \rho gH^2 \tag{11.5}$$

Part of this momentum flux may be reflected to the sea and part will be dissipated by friction. The rest causes wave set-up. As can be seen, wave set-up is likely to be greater on dissipative beaches.

The longer waves break in deeper water than smaller waves and the set-up starts further seaward for large waves. Thus, because the gradients of the water surface are about the same, the larger breakers create a higher set-up. The magnitude of the set-up is about $H/6$ for sinusoidal waves. However Symonds *et al.* (1982) have shown that, because of the groupiness in the waves, and the consequent intermittent larger waves, in the natural situation the set-up is greater, reaching $H/3$. Consequently when there is an alongshore variation in waveheight, there is also a variation in the mean water level within the surf zone caused by the set-up. The alongshore gradients of water level can then drive longshore currents away from the positions of maximum set-up, and largest wave height (Bowen, 1969a; Bowen and Inman, 1969). Rip currents therefore are located in areas of lowest wave height. The differences in wave height along the beach can be caused by wave refraction and by edge waves.

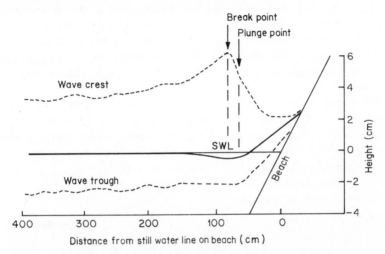

Figure 11.5 Result of an experiment showing wave set-up. *From Bowen, 1969a, J. Geophys. Res., 74, 5467–5478. Reproduced by permission of American Geophysical Union.* Dotted lines, envelope of wave height. Thick line, mean water level

BEACH CUSPS AND EDGE WAVES

Cuspate and crescentic topographic features with a variety of wavelengths occur on beaches. They have been found with wavelengths of as little as 10 cm on lakeshores, and range up to kilometres on open beaches. The nomenclature of the features is not clear. The smaller features are generally known as beach cusps, but the longer ones have been termed giant cusps, and rhythmic topography, amongst other terms.

Beach cusps are concave seaward-facing features which are generally extremely regular in wavelength. The horns of the cusps are normally coarser than the cusp valleys and can have elevations of tens of centimetres.

Typical beach cusps are 10–100 m wavelength, and they fall into two categories: those formed on the beachface by the swash and backwash, and those formed in the surf zone by the nearshore circulation system. With the former type there is a distinction between those generated on beaches where wave breaking and dissipation are important, i.e. the lower slope beaches, and those formed on the steeper more reflective beaches. The beach cusps are always best developed near high water mark where they do not get obliterated by the chang-

300

ing water level, and they are most persistent during neap tides when the change in water level is least.

The water circulation formed by the wave interaction with the cusps appears to be different for high and lower slope beach cusps. On high slope, reflective beaches the breaking wave is split by the horns of the cusp, and flows around towards the head of the cusp before flowing seawards as the backwash in the centre of cusp valley where it impedes the progression of the next wave (Figure 11.6A). This action would tend to shorten the horns of the cusp, and the backwash carries sediment seaward to be deposited where the backwash interacts with the next breaking wave. The overall result is a tendency for the beach slope to be reduced.

With lower slope beaches the spilling wave is refracted by the cusp. The wave travels fastest up the cusp valley, spreads out at the head of the valley and travels seawards along the sides of the cusp before dying away, or flowing towards the centre of the valley (Figure 11.6B). The swash flowing up the valley carries sediment landwards until an equilibrium slope is produced, and the seawards flow along the horns causes them to lengthen. Consequently the tendency is for the beach slope to steepen.

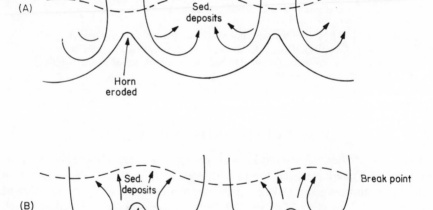

Figure 11.6 Schematic diagram showing the circulation of the swash and backwash during the breaking of a wave on a beach with cusps. (A) Steep beach; (B) Lower slope beach

Obviously the equilibrium situation is for the slopes of the cusps to be such that the water circulation is weak and creates no net sediment movement. The two situations described above will be on either side of this equilibrium. Cusps are best developed on steep coarse beaches exposed to long period swell. A description of the development of beach cusps in a wave basin can be found in Guza and Inman (1975).

Just seaward of the cusps there is often a crescentic bar which has the same wavelength as the cusp, and with the points of the crescents coinciding with the horns of the cusp.

Though several theories have been put forward for the generation of beach cusps, the theory of their generation by edge waves is the most satisfactory. Edge waves are standing waves which are produced parallel to the shore, with crests normal to the shoreline, by the interaction of two wave trains, one of which could be reflected from the beach or from headlands. The standing wave produces nodes and antinodes of water surface oscillation spaced regularly along the beach (Figure 11.7). The edge waves have a constant phase relationship with the incoming waves, and the standing wave pattern is fixed. For incoming and edge waves of the same period, the incoming breaker will coincide with the standing wave at its maximum height at alternate antinodes, and produce slightly higher breaking waves (Figure 11.7). At the nodes in between, at the same time, the breaker will coincide with the minimum standing wave elevation producing slightly lower waves. Since the periods are the same, the high and low breakers persist in the same place, and rip currents can develop where the waves are lowest.

Edge wave motions can be theoretically predicted for a variety of beach profiles. For a linear profile the amplitude of the surface oscillation at the antinodes is maximum at the shore and decreases exponentially offshore, being negligible outside the breaker line. It appears that the most easily excited edge wave and the one with the largest amplitude is the subharmonic of the incident wave, i.e. the edge wave has a period twice that of the incident wave (Inman and Guza, 1982). The amplitude of the subharmonic wave is at least an order of magnitude greater than that of synchronous waves. Experiments show that subharmonic edge wave generation occurs on non-erodable plane laboratory beaches only when the incident waves are strongly reflected at the beach (Guza and Inman, 1975), and this is supported by the field measurements of Huntley and Bowen (1975a). In the field observations of wave run-up on steep beaches have shown significant energy at subharmonic frequencies.

As is shown in Figure 11.8, the cusp spacing λ_c produced by subharmonic waves will be half the edge wave wavelength λ_e. Inman and Guza (1982) give λ_c as

$$\lambda_c = \frac{\lambda_e}{2} = \frac{g}{\pi} T_i^2 \tan \beta$$

where T_i is the incoming wave period. This predicts a maximum swash cusp wavelength of about 150 m for long period waves on the steepest beach. The maximum incident wave amplitude which leads to subharmonic excitation is taken by Guza and Inman (1975) as

$$a_i = 2g \tan^2 \beta \left(\frac{2\pi}{T_i} \right)^{-2}$$

302

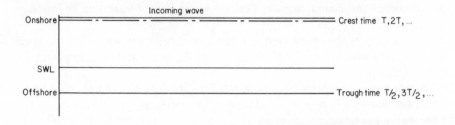

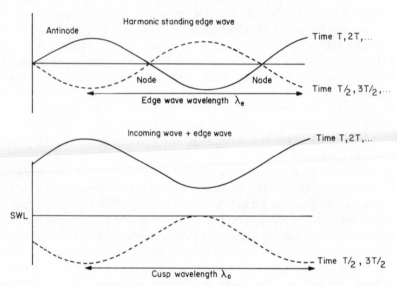

Figure 11.7 Interaction of an incoming wave with a standing edge wave of the same period T, to produce a cusp of the same wavelength as the edge wave. *After Bowen, 1969a, J. Geophys. Res., 74, 5467–5478. Reproduced by permission of American Geophysical Union*

and this corresponds to the transition between surging and plunging breakers. However, there is no theory that predicts which frequency of an incident spectrum of waves will generate the largest edge waves. Consequently comparison between cusp wavelength and edge waves calculated from an incident wave period show large scatter. However edge waves are probably only required for the initiation of beach cusps. Once the bedform has been initiated the incident waves by themselves may allow the cusp to develop to its maximum amplitude and the presence of the cusps may interact with the edge waves. There is evidence that as cusps grow, the subharmonic edge waves are reduced in amplitude. For fairly simple conditions Inman and Guza (1982) suggest that edge waves can only exist if the cusp height η_c

$$\eta_c \leqslant 0.13 \lambda_c \tan \beta$$

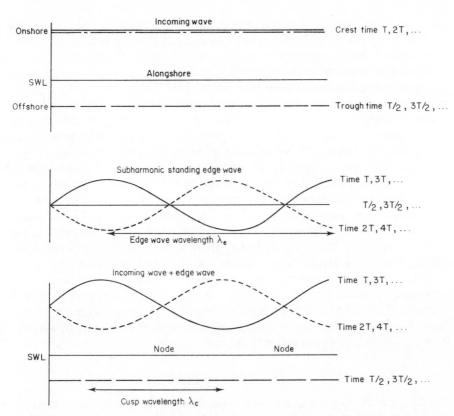

Figure 11.8 Interaction of an incoming wave of period T with a standing edge wave of twice the period, to produce a cusp of half the wavelength of the edge wave

Huntley and Bowen (1975a) have observed subharmonic edge waves on a steep beach which had a calculated amplitude of 27 cm at the shoreline, and a wavelength of 34 ± 6 m. No cusps were observed at this time, but this was thought to be the result of a steady longshore current of about 10 cm s^{-1} caused by oblique wave incidence.

With small longshore currents gradual progressive movement of beach cusps has been observed, but they are unlikely to survive in large currents. Consequently cusps are most likely to occur when a regular swell is approaching normally onto a steep reflective beach.

The most conclusive proof of a connection between edge waves and beach cusps has been reported by Huntley and Bowen (1978). In observations on a Nova Scotia beach the incident wave period was 6.9 s, and spectral analysis of velocity measurements showed a peak at the subharmonic frequency, a period of about 14 s. The phase relationship between the onshore and longshore fluctuations in velocity (180°) and between onshore and vertical (0°) show the peak to be due to a standing subharmonic edge wave. The predicted longshore wavelength for beach cusps was 12 m. Towards the end of their measurements cusps began to appear and within an hour appeared to reach an equilibrium size with a mean spacing of 12.7 m. They also concluded that a reasonably steady shoreline position was required to allow cusps to form.

The larger cuspate forms are generally considered to be related to the near-shore circulation cells associated with rip currents. The position of the rip currents coincides with the embayment, and longshore transport carries sediment towards the rip current whrch then takes it offshore. However considerable complexities are apparent with oblique wave approach, or nearshore tidal currents. Rip currents may in some instances result from the interaction of synchronous edge waves with incident waves, producing longshore variations in wave heights and wave set-up. But often the rip current spacing is too large for this to be a plausible explanation. Holman and Bowen (1982) consider standing edge waves to be a special case, and have considered the more general situation of edge waves with different modes and different amplitudes propagating in the same or opposite directions. Providing there is some coherence between the modes, the waves become 'phase locked' and a regular longshore pattern will result. For all other cases than a standing wave, a series of long wavelength rhythmic crescentic bars were formed which bear strong resemblance to natural features. Thus edge waves can explain many regular spaced sedimentary features found along beaches, and they can be produced by forcing by incident wave and reflection of waves from headlands.

OBLIQUE WAVES

So far we have only considered waves incident on the beach with their crests parallel to it. When the waves approach at an angle then the lateral thrust of the radiation stress becomes

$$S_{xy} = E n \sin \theta \cos \theta \qquad (11.6)$$

The $\cos \theta$ factor takes account of the change in the length of the wave crests due to refraction, and the $\sin \theta$ the changing direction of the waves. It is difficult, however, to measure the wave direction. S_{xy} is the flux towards the shoreline of the momentum directed along the beach. It remains nearly constant during the shoaling of waves over a uniform slope, and is only dissipated when the waves reach nearshore. Nevertheless Bowen (1969b) shows that this provides a forcing within the surf zone that tends to produce a longshore current.

Longshore Currents

Within the surf zone the longshore currents can be produced by wave set-up, edge waves and by oblique-wave approach. The normal situation is to have two, or more, of these generating mechanisms working at the same time. It is thus difficult to separate them.

Let us consider the oblique wave approach in isolation to start with. Bowen (1969b) predicted the form for the velocity profile in the longshore current with distance from the shore, based on the radiation stress concept. The magnitude of the velocity, however, depended on the horizontal mixing assumed to take place across the breaker line. Longuet-Higgins (1970) extended this approach and calibrated the mean value against experimental data. This gave the representative curves shown in Figure 11.9. Assuming the radiation stress produces the longshore current and that it is dissipated by the associated bottom friction, the mean longshore current is given by

$$\bar{v}_l = \frac{5}{8} \pi \frac{\tan \beta}{C_f} u_m \sin \theta_B \tag{11.7}$$

where θ_B is the wave direction at the break point, C_f is the bottom drag coefficient, and u_m is the maximum horizontal orbital velocity at the break point, with $u_m = \left(\dfrac{2 E_B}{H_B}\right)^{\frac{1}{2}}$.

From sand transport studies Komar and Inman (1970) found that

$$\bar{v}_l = 2.7 \, u_m \sin \theta_B \tag{11.8}$$

Comparisons between Equations 11.7 and 11.8 suggest that $\tan \beta/C_f$ is a constant. However, there is discussion about whether value of C_f is greater with steeper beach slopes, or whether the constancy might be due to increased horizontal mixing on steeper beaches.

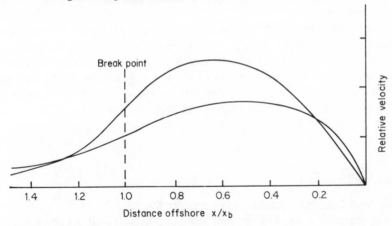

Figure 11.9 Distribution of longshore current within and just outside the surf zone. Curves show approximate limits of experimental data. *After Longuet-Higgins, 1972, J. Geophys. Res., 75, 6790–6801. Copyright by the American Geophysical Union*

Komar (1975) reviewed a large amount of field and laboratory data and found that a better relationship was

$$\bar{v}_l = 2.7 \, u_m \sin \theta_B \cos \theta_B \tag{11.9a}$$

which gives a maximum longshore current when $\theta = 45°$.

This obviously corresponds more closely to Equation 11.6. Equation 11.9 can also be written as

$$\bar{v}_1 = 1.17\,(gH_B)^{\frac{1}{2}}\sin\theta_B\cos\theta_B \tag{11.9b}$$

Komar (1975) also added the oblique wave component approach to a component caused by the longshore variation in wave height and obtained

$$\bar{v}_1 = 2.7\,u_m\sin\theta_B\cos\theta_B -$$

$$\frac{\pi\sqrt{2}}{C_f\gamma_B{}^3}\left(1 + \frac{3\gamma_B{}^2}{8} - \frac{\gamma_B{}^2}{4}\cos^2\theta_B\right)u_m\frac{\delta H_B}{\delta y} \tag{11.10}$$

where γ_B is the ratio of wave height to water depth at the breaker point. Equation 11.10 indicates that the longshore thrust due to the oblique wave approach can be either enhanced or reduced by the wave height variation, depending on the size and magnitude of $\delta H_B/\delta y$. As the breaker height can vary in a fairly periodic way along the beach, the longshore current strength will also vary, converging on a point somewhat downcoast from the position of the lowest waves, and diverging from near the position of highest waves. The displacement downcoast will depend on the strength of the longshore current produced by the oblique wave approach. The points of convergence will then fix the position of the rip currents.

So far we have not considered the effect of tidal currents on the breaker zone. As far as the longshore current is concerned a further time dependent term could be included in Equation 11.10 and this would cause a continual change in the magnitude of the longshore current, and adjustment of the rip current systems. However the breaker heights may also be related to tide level. Because the beach profile is normally steeper near high tide mark than at low tide, the maximum breaker height is likely at around high water, other things being equal.

Longshore Transport

The term longshore transport is normally reserved for the longshore movement of sediment caused by the longshore currents. It is also known as littoral drift, or littoral transport.

Engineering approaches have attempted to relate the littoral drift to the 'longshore component of the wave power' P_1. The wave power or wave energy flux $P = (E\,cn)$ and the longshore component is written

$$P_1 = (E\,cn)_B\sin\theta_B\cos\theta_B \tag{11.11}$$

This is obviously the product of the lateral thrust of the radiation stress evaluated at the breaker point (Equation 11.6), and the wave velocity. Longuet-Higgins (1972) has objected to the term longshore component, arguing that the cosine factor should not be included in anything termed a longshore component, since its function is to convert the power per unit crest length into a power per unit shore length. A good alternative term has not yet been universally accepted, though it is often called the wave thrust.

A number of results gives the longshore volume transport rate of sand S_l, as $S_l \propto P_l^m$ where m is 0.8 to 1.0. However, the dimensions of S_l and P_l are different so that the value of the constant depends on the units in which each is measured. For instance Galvin and Vitale (1976) quote $S_l = 7500\ P_l$ when S_l is in $yd^3 y^{-1}$ and P_l is in ft–lb $s^{-1} ft^{-1}$. However, Inman and Bagnold (1963) point out that the transport rate of sediment should be expressed as the immersed weight transport rate given by

$$I_l = (\rho_s - \rho)\, g\, n\, S_l \qquad (11.12)$$

where n is the packing fraction (~ 0.6). As this approach expresses the sand transport in terms of an immersed weight rather than a volume, L_l and P_l then have the same units. Consequently

$$I_l = K\, P_l \qquad (11.13)$$

Equation 11.13 holds solely for conditions of oblique wave approach.

For several sets of data Komar and Inman (1970) found that $L_l = 0.77\ P_l$, a result not showing any dependence on beach slope, as might be expected from Equation 11.7 with tan β/C_f constant. Komar (1971) showed that Equation 11.13 applies to the swash zone as well as the breaker zone despite the fact that within the swash zone the sediment particles undergo a saw-toothed pattern of movement; the swash taking the particles obliquely up the beach and flowing back normal to the beach slope under gravity in the backwash.

Bagnold (1963) extended his concept of work done by water in moving sediment particles to include wave effects. He assumed that a mean mass of sediment is supported over a unit bed area by the mean velocity $\bar{u}_0$ of the to-and-fro motion. The immersed weight of the sediment will be $q/\bar{u}_0$. This can be written as $K'W/\bar{u}_0$. Since the sediment is already supported by the wave action, no additional tangential stress is needed to transport the sediment at a velocity U_θ by a steady current of velocity U_θ in any direction θ. Assuming that $u_\theta/\overline{U}_0 \approx u_\theta/u_0$ where u_0 is the orbital velocity measured at the same height as u_θ, then the measured weight transport rate should be

$$q = K'W \frac{u_\theta}{u_0} \qquad (11.14)$$

In other words the available power from the wave motion W supports the sediment and W/u_0 is the stress exerted by the waves. Inman and Bagnold (1963) applied this model to the surf zone, assuming that a portion of the wave energy flux $(E\, cn)_B \cos \theta_B$ is dissipated in placing the sand in motion. The mean stress applied to the beach is then proportional to $(E\, cn)_B \cos \theta_B / u_0$. Consequently

$$I_l = K'(E\, cn)_B \cos \theta_B\ \frac{\overline{v}_l}{u_0} \qquad (11.15)$$

308

Komar and Inman (1970) found for several sets of data that $K' = 0.28$ when u_0 was assumed to be equal to $u_m = (2 E_B/\rho h_B)^{\frac{1}{2}}$.

Komar (1971) has used the approach of Equation 11.15 to determine the variation of sand transport across the surf zone. Basically the distribution is proportional to the product of the local stress exerted by the waves, and the longshore current velocity. The maximum sand transport rate therefore occurs to seaward of the maximum in the longshore current (Figure 11.10), the average transport across the surf zone being given by Equation 11.15.

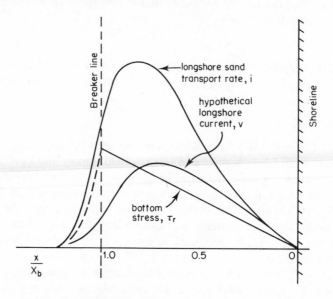

Figure 11.10 Diagrammatic distribution of bottom stress, longshore current and longshore transport rate across the surf zone. *From Komar, 1971,* J. Geophys. Res., *76, 713–721. Reproduced by permission of American Geophysical Union*

BEACH PROFILES

The slope of the beach depends on the grain size of the sediment and on the incident wave steepness. However beach slope is difficult to define since they are seldom planar and more than one slope can be present across the beach face. Most consistent relationships are found for the swash–backwash zone at the top of the beach. The incoming swash carries a number of particles with it up the slope against the gravitational forces. However, because of the permeability of the beach, a certain proportion of the swash flows into the beach, and velocity of the backwash is reduced. Consequently, despite the favourable slope the backwash can carry fewer particles seaward. Eventually an equilibrium beach slope is formed when as many grains are carried upbeach on the swash as are swept back with the backwash. The velocity of the swash increases with the wave steepness, and the amount of percolation rises with the sediment grain size, since permeability increases with grain size. The result is for a lower beach slope to form under steep plunging waves, and a higher slope under lower steepness spilling waves. Figure 11.11 shows a series of results for beach slope, against sediment size and wave steepness summarized by Carter et al. (1973). Since low waves lead to an increase in beach slope, they are often termed constructive waves. Alternatively, steep waves are destructive.

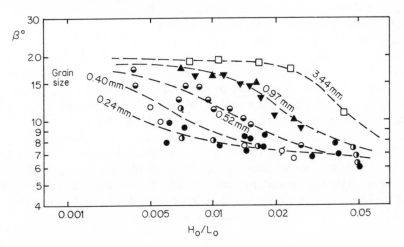

Figure 11.11 Relationship between beach slope, grain size and offshore wave steepness. *From Carter* et al., *1973,* Proc. Amer. Soc. Civil Eng., *99, WW2, 165–184. Reproduced by permission of American Society of Civil Engineers*

Bagnold (1963) proposed a formulation for beach slope by assuming a zero net transport at equilibrium, and a balance of forces induced by asymmetry in the onshore–offshore velocities

$$\tan \beta \ = \ \tan \varphi \left(\frac{1 - \varepsilon}{1 + \varepsilon} \right) \tag{11.16}$$

where φ is the angle of internal friction, and ε is the ratio of the energy dissipation in the offshore flow to that in the onshore flow. Inman and Frautschy (1966) proposed that ε should be proportional to the ratio $(u_m$ offshore$)^3/(u_m$ onshore$)^3$. Greenwood and Mittler (1984) found that this agreed fairly well with measurements, when the observed slope of an underwater bar was compared with that calculated from orbital velocities during a storm.

However Jago and Hardisty (1984) suggest

$$\tan \beta \ = \ \tan \varphi \left(\frac{1 - V^2}{1 - V^2} \right) \tag{11.17}$$

where $V = (\bar{u}$ onshore$/\bar{u}$ offshore$)$. They also found good agreement between measured beach slopes and that calculated from measured orbital velocities. At the moment no comparison has been made to determine which of the above approaches is the most reliable. However there are also difficulties in knowing when the beach slope is in equilibrium.

Using data from twenty-seven beaches, King (1972) obtained the empirical formula

$$\text{Beach slope} = 407.71 + 4.2 \, \phi D - 0.17 \log E \qquad (11.18)$$

where E is the wave energy, and D the grain size in phi units.

Extensive descriptions of beach profiles and their relationships to storms etc. can be found in King (1972) and Komar (1976).

In situations where the tidal rise and fall creates a wide intertidal area, the beach slope is affected by the passage of the surf and swash zones across the beach. Also the water table in the beach, arising from the underlying rocks and percolation from the upper part of the beach, will intersect the beach face. Where this occurs there is generally an abrupt decrease in beach slope. Below this point the mean flow is out of the beach during the tidal cycle and this increases the potential mobility of the sediment resulting in a lower slope.

The differences between the velocities of swash and backwash on the upper part of the beach, due to percolation, causes an effective sorting of the beach sediments. The coarser material tends to accumulate near the high tide level, and the finer grains are moved to the lower part of the beach. Consequently the mean beach slope decreases towards low tide mark.

Onshore–Offshore Transport

In order to maintain an equilibrium beach profile, considerable onshore–offshore sediment transport is necessary under conditions of varying incident wave steepness. Low steepness spilling waves lead to a steepening of the beach, and movement of material from the lower beach to the upper beach. Steeper waves cause the reverse movement. Under these conditions the approach defined by Equation 11.16 can only hold at one or two restricted positions.

Laboratory experiments have shown that seaward of the breaker line there is a neutral line, or null point, where, though there is an oscillation of the sediment particles, there is an equilibrium between the force of gravity and the wave-induced forces. Seawards of the null point there is an offshore movement of sediment and landward of it movement towards the shore. The position of the null point varies with the wave steepness, for increased steepness the null point moving shoreward.

Prediction of the position of the neutral line has been tackled by many authors, including Ippen and Eagleson (1955), Miller and Zeigler (1958) and Eagleson and Dean (1961). Ippen and Eagleson (1955) showed that inside the null point the shoreward directed grain velocities increased to a maximum just seaward of the breaker line (Figure 11.12). This would lead to an accumulation of sediment just shoreward of this maximum where the velocities diminish, and leads to the formation of a break point bar. The null point would be an area of net erosion with material moving both seawards and landward from it. Ippen and Eagleson's measurements were for a single beach slope of 1 : 15, but they found the neutral line occurred where

$$\left(\frac{H}{h}\right)^2 \left(\frac{\lambda}{H}\right) \left(\frac{c}{w_s}\right) = 11.6$$

where H, λ and c are the local height, wavelength and phase velocity of the waves, h is the water depth and w_s the grain settling velocity. This result implies that each grain size has a different null point, the one for fine grains being further seaward than that for coarser grains, which could lead to a sorting of the different size fractions.

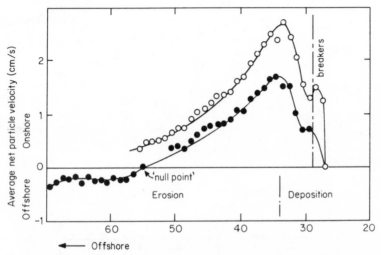

Figure 11.12 Results of laboratory experiments on the movement of two sizes of grains under the same wave conditions. ○ 2 mm plastic grains; ● 3.17 mm plastic grains. After Ippen and Eagleson (1955)

The concept of a neutral line has not been verified in the field. This may be mainly because the incoming waves are not as regular as those in a laboratory flume, and if a neutral line exists it would be very variable in its position.

An alternative explanation for the neutral line has been proposed by Wells (1967). He has suggested that onshore or offshore sediment movement could be caused by slight asymmetries in the orbital velocities, and these asymmetries would be manifest in the skewness of the probability distribution of the horizontal water velocities. When a spectrum of waves approaches the shore the waves become more peaked, and the troughs longer and flatter. Wells showed, for a particular form of the deep water spectrum, that this causes a change in skewness of the velocities, with a negative skewness for a depth to wavelength ratio greater than about 0.09, and he predicted that this would cause offshore sediment movement. Inside this point the skewness was positive, leading to an

312

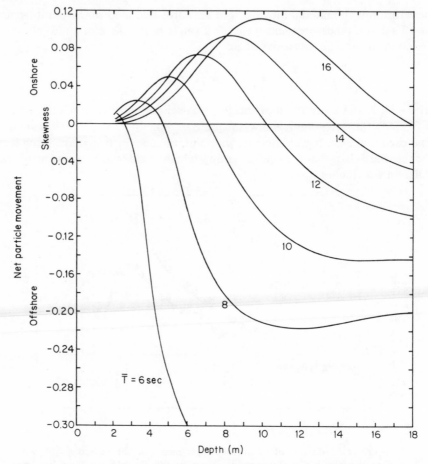

Figure 11.13 Theoretical curves of wave velocity skewness in shallowing water and associated particle movement. *From Wells, 1967, J. Geophys. Res., 72, 497–504. Reproduced by permission of American Geophysical Union.* Compare the effect of a positively skewed wave with that of an asymmetrical tide in intertidal areas (Figure 9.18)

onshore sediment movement. The neutral line may thus be considered as a position of zero skewness in the wave velocities and the direction of net particle movement is determined by the sign of the skewness. As is shown in Figure 11.13 the neutral line moves offshore with increased wave period.

Some support for this explanation has been demonstrated by Huntley and Bowen (1975b), who found positive skewness on a low slope beach under constructive waves, and negative skewness on a steeper beach with steeper waves. Additionally Cook and Gorsline (1972) have observed asymmetries in the magnitudes and durations of velocities on several beaches which were associated with both offshore and onshore sand transport. The best correlation with sand transport was achieved by considering the onshore–offshore velocity ratios for those waves causing sediment movement. They concluded that onshore velocities predominated during summer periods when long period swells reached the coast. Presumably as the slopes approach equilibrium values the skewness of the water velocity distribution approaches a mean value compatible with the requirements of Equations 11.16 or 11.17

Longshore Bars

A common feature of onshore–offshore transport is the presence of longshore bars. They are typical of sandy beaches, but there is a distinct difference between those formed in areas of high and low tidal range.

Typical of those found in high tidal range areas are those present off the Lancashire coast of England and described by Parker (1976). These intertidal features are called ridge and runnel. The ridges are broad flat sandy hummocks, generally with about 0.5–1 m of relief and trending roughly parallel to the shoreline. They are separated by wide runnels which are often slightly finer grained and with thin muddy layers within the sediment. During flooding and ebbing tide, flows parallel to the beach occur along the runnels, with occasional breakthrough from one runnel to the next. The ridges can show steep faces to either seaward or landward, and both offshore and onshore migration have been observed. They do not seem to be greatly affected by storms, however. The wavelength of the ridge–runnel system is 300–500 m. The origin of the features is obscure, but King (1972) has observed that the ridges are an attempt by the waves to produce an equilibrium swash slope gradient on a beach that is considerably flatter than the equilibrium gradient. Of course this is most likely when the waves are attacking a narrow zone of a wide beach. King (1972) has correlated the positions of the ridges with the heights of mean high water neaps, and mean low water neap and spring tides. Because the water level spends a greater amount of time near high water and low water marks on any tide, the monthly variation causes a bimodal distribution of the 'tidal duration factor', a factor expressing the duration of exposure of the beach to wave action. The implication of this for the development of the beach face has been discussed by Carr and Graff (1982). Obviously the wave dominated beach processes will be most active where the tidal duration causes longest exposure, and constructive waves could create the ridges at those levels.

The term ridge and runnel has also been applied to subtidal features on low tidal range beaches. The implications of this for features which may bear a morphological similarity, but which may arise from different processes, has been stressed by Orford and Wright (1978). Longshore bars may be a better name for the subtidal features.

Longshore bars can be seen seaward of the break point bar on beaches with only a low tidal range, such as those in the Mediterranean. These are very regular long crested features running parallel to the shoreline, but in this case with wavelengths of only a few tens of metres. The distance between the bars and their heights and widths increase offshore. Lau and Travis (1973) and Carter *et al.* (1973) have shown that these bars are formed by partial standing waves created by interaction between the incoming waves and their reflection from the beach.

The incoming and reflected waves will interfere with each other and the surface will have nodal lines at which there is a minimum elevation change at intervals of half of the basic wavelength, and antinodes between with an amplitude greater than that of the primary wave. With total reflection the amplitude of the reflected wave equals that of the incoming wave and the elevation change at the antinodes would be twice the basic wave amplitude.

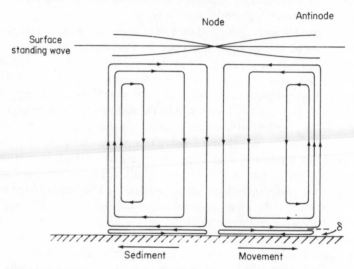

Figure 11.14 The mean water circulation beneath a surface standing wave. δ is the boundary layer thickness. When ripples are present circulation within the boundary layer will be disrupted, and mean sediment movement will be towards the antinodes. *After Liu and Davies, 1977, J. Fluid Mech., 81, 63–84. Reproduced by permission of Cambridge University Press*

Beneath standing waves there is a circulation similar to that shown in Figure 11.14. Within the body of the fluid close to the bed, but outside the boundary layer, there is a residual flow from the nodes to the antinodes of the surface elevation. In the boundary layer the opposite occurs. The boundary layer thickness is $\delta = (vT/2\pi)^{\frac{1}{2}}$, v being the kinematic viscosity and T the wave period. For a typical wave period of 8 s therefore $\delta \sim 0.1$ cm. Noda (1968) has confirmed the existence of these residual flows in a laboratory experiment using dye outside the boundary layer, and vinyl pellets within it. The pellets accumulated at the wave nodes and the dye travelled to the antinodes. Consequently Noda proposed that the direction of travel will depend on the ratio of the sediment diameter to the boundary layer thickness, and on whether the sediment is lifted into suspension. This mechanism could form a means of separating coarse and fine sediment to the nodes and antinodes respectively. However, the boundary layer circulation is likely to be broken up by the presence of ripples, and it can be presumed that sand would move preferentially towards the antinodes.

Lau and Travis (1973) and Carter *et al.* (1973) have shown that in partial standing wave systems the near-bed circulation and sediment transport is towards the shore when the reflection coefficient of the beach is less than 0.41, and only undergoes reversal between antinodes and nodes when this degree of reflection is exceeded. They therefore conclude that transverse bars should only form on beaches steeper than about 17°. This appears to be a rather stringent limit for the formation of longshore bars because even though the velocities may not undergo a reversal, there may be a sufficient perturbation in the onshore flow to cause temporary accumulation under the antinodes. The resulting bedforms would be asymmetrical in this case, with the steep face towards the shore.

Heathershaw (1982b) and Heathershaw and Davies (1985) have reported laboratory measurements suggesting the generation of bars by reflection of waves, though in this case the reflection was caused by the surface waves impingeing upon an undulating sea bed. Consequently the reflected waves could well be produced initially by the breakpoint bar, and the bars progresssively spread seawards with a steadily increasing degree of reflection enhancing their formation.

The form and profile of beaches are responsive to the changing wave climate and in some situations may not readily achieve equilibrium. Most of the detailed work on beach dynamics has been carried out on American beaches where the tidal range is low, the beach material generally uniform, and the beaches are long. In comparison the studies carried out on tidal beaches, with mixed sediment, and interrupted by headlands are less comprehensive. Even for the former situation the sand transport and rip current dynamics are not well known.

This chapter includes a brief discussion of only some of the complex processes and features exhibited by studies of beach processes. Much more extensive coverage is to be found especially in King (1972) and Komar (1976, 1983).

References

Abbott, J. E. and Francis, J. R. D. 1977. Saltation and suspension trajectories of solid grains in a water stream. *Phil. Trans. R. Soc. London*, **A284**, 225–254.

Abou-Seida, M. M. 1965. Bedload function due to wave action. *Hydraulic Eng. Lab. Univ. California, Report HFL–2–11*, 78 pp.

Ackers, P. and White, W. R. 1973. Sediment Transport: new approach and analysis. *J. Hydraul. Div. ASCE*, **99**, HY11, 2041–2060.

Acton, J. R., and Dyer, C. M. 1975. Mapping of tidal currents near the Skerries Bank. *Quart. J. Geol. Soc. London*, **131**, 63–67.

Adams, C. E. and Weatherby, G. L. 1981. Some effects of suspended sediment stratification on an oceanic bottom boundary layer. *J. Geophys. Res.*, **86**, 4161–4172.

Aksoy, S. 1972. Fluid force acting on a sphere near a solid boundary. *Proc. 15th Cong. Int. Assoc. Hydraul. Res.*, 217–224.

Allen, G. P., Castaing, P. and Klingbiel, A. 1972. Distinction of elementary sand populations in the Gironde estuary (France) by R-mode factor analysis of grain-size data. *Sedimentology*, **19**, 21–35.

Allen, G. P., Salomon, J. C., Bassoulet, P., Du Penhoat, Y. and De Grandpre, C. 1980. Effects of tides on mixing and suspended sediment transport in macrotidal estuaries. *Sedimentary Geol.*, **26**, 69–90.

Allen, G. P., Sauzay, G., Castaing, P. and Jouanneau, J. M. 1977. Transport and deposition of suspended sediment in the Gironde estuary, France. In Wiley, M. (Ed.), *Estuarine Processes, Vol. II*, Academic Press, New York, 63–81.

Allen, J. R. L. 1969. Erosional current marks of weakly cohesive mud beds. *J. Sediment. Petrol.*, **39**, 607–623.

Allen, J. R. L. 1970a. The systematic packing of prolate spheroids with reference to concentration and dilatancy. *Geol. en Mijnbouw*, **49**, 211–220.

Allen, J. R. L. 1970b. The angle of initial yield of haphazard assemblages of equal spheres in bulk. *Geol. en Mijnbouw*, **49**, 13–22.

Allen, J. R. L. 1974. Packing and resistance to compaction of shells. *Sedimentology*, **21**, 71–86.

Allen, J. R. L. 1980. Sand waves: A model of origin and internal structure. *Sedimentary Geol.*, **26**, 281–328.

Allen, J. R. L. 1982. *Sedimentary Structures: Their Character and Physical Basis*. 2 vols, Elsevier, Amsterdam.

316

Allen, J. R. L. and Friend, P. F. 1976a. Changes in intertidal dunes during two spring–neap cycles, Lifeboat Station Bank, Wells-next-the-Sea, Norfolk (England). *Sedimentology*, **23**, 329–346.

Allen, J. R. L. and Friend, P. F. 1976b. Relaxation time of dunes in decelerating aqueous flows. *Quart. J. Geol. Soc. London*, **132**, 17–26.

Allen, J. R. L. and Leeder, M. R. 1980. Criteria for the instability of upper stage plane beds. *Sedimentology*, **27**, 209–217.

Allen, P. A. 1985. Hummocky cross-stratification is not produced purely under progressive gravity waves. *Nature*, **313**, 562–564.

Anderson, F. E. 1972. Resuspension of estuarine sediments by small amplitude waves. *J. Sediment. Petrol.*, **42**, 602–607.

Anderson, F. E. 1983. The northern muddy intertidal: a seasonally changing source of suspended sediments to estuarine waters—a review. *Can. J. Fish Aquat. Sci.*, **40**, Supp. 1, 143–159.

Ariathurai, R. and Arulanandan, K. 1978. Erosion rates of cohesive soils. *J. Hydraul. Div. ASCE*, **104**, HY2, 279–283.

Ariathurai, R. and Krone, R. B. 1976. Finite element model for cohesive sediment transport. *J. Hydraul. Div. ASCE*, **102**, HY3, 323–338.

Avoine, J., Allen, G. P., Nichols, M., Salomon, J. C. and Larsonneur, C. 1981. Suspended-sediment transport in the Seine estuary, France: effect of man-made modifications on estuary-shelf sedimentology. *Marine Geol.*, **40**, 119–137.

Baba, J. and Komar, P. D. 1981. Measurements and analysis of settling velocities of natural quartz sand grains. *J. Sediment. Petrol.*, **51**, 631–640.

Bagnold, R. A. 1937. The size grading of sand by wind. *Proc. R. Soc. London*, **A163**, 250–264.

Bagnold, R. A. 1946. Motion of waves in shallow water. Interaction between waves and sand bottoms. *Proc. R. Soc. London*, **A187**, 1–15.

Bagnold, R. A. 1954. Experiments on a gravity-free dispersion of large solid spheres in a Newtonian fluid under shear. *Proc. R. Soc. London*, **A225**, 49–63.

Bagnold, R. A. 1956. Flow of cohesionless grains in fluids. *Phil. Trans. R. Soc. London*, **A249**, 235–297.

Bagnold, R. A. 1963. Beach and nearshore processes. Part 1, Mechanics of marine sedimentation. In Hill, M. N. (Ed.), *The Sea: Vol. 3*, Wiley–Interscience, New York, 507–528.

Bagnold, R. A. 1966. An approach to the sediment transport problem from general physics. *U.S. Geol. Surv. Prof. Paper*, **422–I**.

Bagnold, R. A. 1973. The nature of saltation and 'bedload' transport in water. *Proc. R. Soc. London*, **A332**, 473–504.

Bagnold, R. A. 1974. Fluid forces on a body in shear-flow; experimental use of 'stationary flow', *Proc. R. Soc. London*, **A340**, 147–171.

Bagnold, R. A. 1980. An empirical correlation of bedload transport rates in flumes and natural rivers. *Proc. R. Soc. London*, **A372**, 453–473.

Bagnold, R. A. and Barndorff-Nielsen, O. 1980. The pattern of natural size distributions. *Sedimentology*, **27**, 199–207.

Bailard, J. A. and Inman, D. L. 1979. A reexamination of Bagnold's granular-fluid model and bedload transport equation. *J. Geophys. Res.*, **84**, 7827–7833.

Bathurst, J. C., Graf, W. H. and Cao, H. H. 1983. Initiation of sediment transport in steep channels with coarse bed material. In Sumer, B. M. and Müller, A. (Eds), *Mechanics of Sediment Transport*, Balkema, Netherlands.

Been, K. and Sills, G. C. 1981. Self weight consolidation of soft soils: an experimental and theoretical study. *Geotechnique*, **31**, 519–535.

Belderson, R. H. and Stride, A. H. 1966. Tidal current fashioning of a basal bed. *Marine Geol.*, **4**, 237–257.

318

Biddle, P. and Miles, J. H. 1972. The nature of contemporary silts in British estuaries. *Sedimentary Geol., 7*, 23–33.

Biggs, R. B. 1970. Sources and distributions of suspended sediment in Northern Chesapeake Bay. *Marine Geol., 9*, 187–201.

Bijker, E. W. 1967. Some considerations about scales for coastal models with movable bed. *Delft Hydraulics Lab. Publ., 50*.

Blackwelder, R. F. and Haritonides, J. H. 1983. Scaling of the bursting frequency in turbulent boundary layers. *J. Fluid Mech., 132*, 87–103.

Blinco, P. H. and Partheniades, E. 1971. Turbulence characteristics in free surface flows over smooth and rough boundaries. *J. Hydraul. Res., 9*, 43–69.

Boersma, J. R. and Terwindt, J. H. J. 1981. Neap–spring tide sequences of intertidal shoal deposits in a mesotidal estuary. *Sedimentology, 28*, 151–170.

Boon, J. D. 1975. Tidal discharge asymmetry in a salt marsh drainage system. *Limnol. Oceanog., 20*, 71–80.

Boothroyd, J. C. and Hubbard, D. K. 1975. Genesis of bedforms in mesotidal estuaries. In Cronin, L. E. (Ed.), *Estuarine Research, Vol. II*, Academic Press, New York, 217–234.

Bowden, K. F. 1962. Measurements of turbulence near the sea bed in a tidal current. *J. Geophys. Res., 67*, 3181–3186.

Bowden, K. F. 1978. Physical problems of the benthic boundary layer. *Geophys. Surv., 3*, 255–296.

Bowden, K. F. and Fairbairn, L. A. 1956. Measurements of turbulent fluctuations and Reynolds stresses in a tidal current. *Proc. R. Soc. London, A237*, 422–438.

Bowden, K. F., Fairbairn, L. A. and Hughes, P. 1959. The distribution of shearing stresses in a tidal current. *Geophys. J. R. astr. Soc., 2*, 288–305.

Bowden, K. F. and Ferguson, S. R. 1980. Variations with height of the turbulence in a tidally-induced bottom boundary layer. In Nihoul, J. C. J. (Ed.), *Marine Turbulence*, Elsevier Science Publishers, Amsterdam, 259–286.

Bowden, K. F. and Howe, M. R. 1963. Observations of turbulence in a tidal channel. *J. Fluid Mech., 17*, 271–284.

Bowen, A. J. 1969a. Rip currents. 1. Theoretical investigations. *J. Geophys. Res., 84*, 5467–5478.

Bowen, A. J. 1969b. The generation of longshore currents on a plane beach. *J. Marine Res., 27*, 206–215.

Bowen, A. J. and Inman, D. L. 1969. Rip currents. 2. Laboratory and field observations. *J. Geophys. Res., 74*, 5479–5490.

Boyadzhiev, L. 1973. On the movement of a spherical particle in vertically oscillating liquid. *J. Fluid Mech., 57*, 545–548.

Bridge, J. S. 1978. Origin of horizontal lamination under turbulent boundary layers. *Sedimentary Geol., 20*, 1–16.

Bridge, J. S. 1981a. Hydraulic interpretation of grain-size distributions using a physical model for bedload transport. *J. Sediment. Petrol., 51*, 1109–1124.

Bridge, J. S. 1981b. A discussion of Bagnold's (1956) bedload transport theory in relation to recent developments in bedload modelling. *Earth Surface Processes, 6*, 187–190.

Bridges, P. H. and Leeder, M. R. 1976. Sedimentary model for intertidal mudflat channels with examples from the Solway Firth, Scotland. *Sedimentology, 23*, 533–552.

Bruun, P. 1978. *Stability of Tidal Inlets: Theory and Engineering*, Elsevier, Amsterdam.

Buller, A. T., Green, C. D. and McManus, J. 1975. Dynamics and sedimentation: the Tay in comparison with other estuaries. In Hails, J. and Carr, A. P. (Eds), *Nearshore Sediment Dynamics and Sedimentation*, John Wiley, Chichester, 201–249.

Cacchione, D. A. and Drake, D. E. 1982. Measurements of storm-generated bottom stresses on the continental shelf. *J. Geophys. Res., 87*, 1952–1960.

Caldwell, D. R. and Chriss, T. M. 1979. The viscous sublayer at the sea floor. *Science,* **205**, 1131–1132.

Carder, K. L., Steward, R. G. and Betzer, P. R. 1982. In-situ holographic measurements of the sizes and settling rates of oceanic particulates. *J. Geophys. Res.,* **87**, 5681–5685.

Carling, P. A. 1983. Threshold of coarse sediment transport in broad and narrow natural streams. *Earth Surface Processes,* **8**, 1–18.

Carr, A. P. and Graff, J. 1982. The total immersion factor and shore platform development: discussion. *Trans. Inst. Brit. Geog. N.S.,* **7**, 240–245.

Carter, T. G., Liu, P.L.-F. and Mei, C. C. 1973. Mass transport by waves and offshore sand bedforms. *J. Wat. Harb. Coastal Eng. Div. ASCE,* **99**, WW2, 165–184.

Cartwright, D. E. 1959. On submarine sand waves and tidal lee waves. *Proc. R. Soc. London,* **A253**, 218–241.

Carver, R. E. 1971. *Procedures in sedimentary petrology,* Wiley–Interscience, New York.

Caston, G. F. 1979. Wreck marks: indicators of net sand transport. *Marine Geol.,* **33**, 193–204.

Caston, G. F. 1981. Potential gain and loss of sand banks in the southern Bight of the North Sea. *Marine Geol.,* **41**, 239–250.

Caston, V. N. D. 1972. Linear sand banks in the southern North Sea. *Sedimentology,* **18**, 63–78.

Chan, K. W., Baird, M. H. I. and Round, G. F. 1972. Behaviour of beds of dense particles in a horizontal oscillatory liquid. *Proc. R. Soc. London,* **A330**, 537–559.

Channon, R. D. 1971. The Bristol fall column for coarse sediment grading. *J. Sediment. Petrol.,* **41**, 867–870.

Channon, R. D. and Hamilton, D. 1971. Sea bottom velocity profiles on the continental shelf south-west of England. *Nature,* **231**, 383–385.

Charnock, H. 1959. Tidal friction from currents near the sea bed. *Geophys. J. R. astr. Soc.,* **2**, 215–221.

Chen, J.-Y., Yun, C.-X. and Xu, H. 1982. The model of development of the Chang Jiang estuary during the last 2000 years. In Kennedy, V. S. (Ed.), *Estuarine Comparisons,* Academic Press, New York, 655–666.

Cheng, E. D. H. and Clyde, C. G. 1972. Instantaneous hydrodynamic lift and drag forces on large roughness elements in turbulent open channel flow. In Shen, H. W. (Ed.), *Sedimentation,* Colorado State Univ. Colorado, US, Chapter 3.

Chepil, W. S. 1958. The use of evenly spaced hemispheres to evaluate aerodynamic forces on a soil surface. *Trans. Amer. Geophys. Union,* **39**, 397–404.

Chepil, W. S. 1959. Equilibrium of soil grains at the threshold of movement by wind. *Soil Sci. Soc. Proc.,* **23**, 422–428.

Chepil, W. S. 1961. The use of spheres to measure lift and drag on wind-eroded soil grains. *Soil Sci. Soc. Proc.* **25**, 343–345.

Chriss, T. M. and Caldwell, D. R. 1982. Evidence for the influence of form drag on bottom boundary layer flow. *J. Geophys. Res.,* **87**, 4148–4154.

Chriss, T. M. and Caldwell, D. R. 1984. Universal similarity and the thickness of the viscous sublayer at the ocean floor. *J. Geophys. Res.,* **89**, 6403–6414.

Clarke, T. L., Lesht, B., Young R. A., Swift, D. J. P. and Freeland G. L. 1982. Sediment resuspension by surface-wave action: an examination of possible mechanisms. *Marine Geol.,* **49**, 43–59.

Cole, P. and Miles, G. V. 1983. Two-dimensional model of mud transport. *J. Hydraul. Eng. ASCE,* **109**, 1, 1–12.

Coleman, N. L. 1967. A theoretical and experimental study of drag and lift forces acting on a sphere resting on a hypothetical streambed. *Proc. 12th Cong. Int. Assoc. Hydraul. Res.,* **3**, 185–192.

Coleman, N. L. 1972. The drag coefficient of a stationary sphere on a boundary of similar spheres. *La Houille Blanche,* No. 1, 17–21.

320

Coleman, N. L. 1977. Extension of the drag coefficient function for a stationary sphere on a boundary of similar spheres. *La Houille Blanche*, No. 4, 325–328.

Coleman, N. L. 1981. Velocity profiles with suspended sediment. *J. Hydraul. Res.*, **19**, 211–229.

Conimos, T. J. and Petersen, D. H. 1977. Suspended-particle transport and circulation in San Francisco Bay: an overview. In Wiley, M. (Ed.), *Estuarine Processes Vol. II*, Academic Press, New York, 89–97.

Cook, D. O. and Gorsline, D. S. 1972. Field observations of sand transport by shoaling waves. *Marine Geol.*, **13**, 31–55.

Corino, E. R. and Brodkey, R. S. 1969. A visual investigation of the wall region in turbulent flow. *J. Fluid Mech.*, **37**, 1–30.

Courtois, G. and Monaco, A. 1969. Radioactive methods for the quantitative determination of coastal drift rate. *Marine Geol.*, **7**, 183–206.

Courtois, G. and Sauzay, G. 1966. Les methodes de bilan des taux de comptage de traceurs radioactifs appliquees a la mesure des debits massiques charriage. *La Houille Blanche*, No. 3, 279–284.

Curray, J. R. 1956. Dimensional grain orientation studies of recent coastal sands. *Bull. Amer. Assoc. Petrol. Geol.*, **40**, 2440–2456.

Dalrymple, R. W. 1984. Morphology and internal structure of sandwaves in the Bay of Fundy. *Sedimentology*, **31**, 365–382.

Dalrymple, R. W., Knight R. J. and Lambiae, J. J. 1978. Bedforms and their hydraulic stability relationships in a tidal environment, Bay of Fundy, Canada. *Nature*, **275**, 100–104.

Davies, A. G. 1977. A mathematical model of sediment in suspension in a uniform reversing tidal flow. *Geophys. J. R. astr. Soc.*, **51**, 503–529.

Davies, A. G. 1980. Field observations of the threshold of sand motion in a transitional wave boundary layer. *Coastal Eng.*, **4**, 23–46.

Davies, A. G. 1984. Field observations of wave-induced motion above the seabed and of resulting sediment movement. *Institute of Oceanographic Sciences Report*, No. 179.

Davies, A. G. 1985. Field observations of the threshold of sediment motion by wave action. *Sedimentology*, **32**, 685–704.

Davies, C. M. 1980. Evidence for the formation and age of a commercial sand deposit in the Bristol Channel. *Estuar. Coastal Mar. Sci.*, **11**, 83–99.

Davies, T. R. H. 1980. Bedform spacing and flow resistance. *J. Hydraul. Div. ASCE*, **106**, HY3, 423–433.

Davies, T. R. H. and Samad, M. F. A. 1978. Fluid dynamic lift on a bed particle. *J. Hydraul. Div. ASCE*, **104**, HY8, 1171–1182.

De Flaun, M. F. and Mayer, L. M. 1983. Relationship between bacteria and grain surfaces in intertidal sediments. *Limnol. Oceanog.*, **28**, 873–881.

Dexter, A. R. and Tanner, D. W. 1971. Packing density of ternary mixtures of spheres. *Nature*, **230**, 177–179.

Dingler, J. R. and Inman, D. L. 1976. Wave formed ripples in seashore sands. *Proc. 15th Coastal Eng. Conf.*, 2109–2126.

Doeglas, D. J. 1946. Interpretation of the results of mechanical analysis. *J. Sediment. Petrol.*, **16**, 19–40.

Doodson, A. T. 1928. The analysis of tidal observations. *Phil. Trans. R. Soc. London*, **A227**, 223–279.

Dorey, A. P., Finch, A. R. and Dyer, K. R. 1975. A miniature transponding pebble for studying gravel movement. *Conf. Instrumentation in Oceanography, Bangor*, 327–332.

Draper, L. 1967. Wave activity at the seabed around Northwestern Europe. *Marine Geol.*, **5**, 133–140.

Dyer, K. R. 1970a. Grain size parameters for sandy gravels. *J. Sediment. Petrol.*, **40**, 616–620.

Dyer, K. R. 1970b. Current velocity profiles in a tidal channel. *Geophys. J. R. astr. Soc.,* **22,** 153–161.

Dyer, K. R. 1970c. Linear erosional furrows in Southampton Water. *Nature,* **225,** 56–58.

Dyer, K. R. 1971. The distribution and movement of sediment in the Solent, southern England. *Marine Geol.,* **11,** 175–187.

Dyer, K. R. 1973. *Estuaries: A Physical Introduction,* John Wiley, London, 140 pp.

Dyer, K. R. 1977. Lateral circulation effects in estuaries. In *Estuaries, Geophysics and the Environment,* National Academy of Sciences, Washington, DC, 22–29.

Dyer, K. R. 1980a. Velocity profiles over a rippled bed and the threshold of movement of sand. *Estuar. Coastal Mar. Sci.,* **10,** 181–199.

Dyer, K. R. 1980b. The Estuary of the Exe in its context. *Essays on the Exe Estuary,* **2,** 1–21. Special Volume, Devonshire Association.

Dyer, K. R. 1982. The initiation of sedimentary furrows by standing internal waves. *Sedimentology,* **29,** 885–889.

Dyer, K. R. and King, H. L. 1975. The residual water flow through the Solent, southern England. *Geophys. J. R. astr. Soc.,* **42,** 97–106.

Eagleson, P. S. and Dean R. G. 1961. Wave-induced motion of bottom sediment particles. *Trans. Amer. Soc. Civil Eng.,* **126,** 1162–1189.

Eckman, J. E., Nowell, A. R. M. and Jumars, P. A. 1981. Sediment destabilization by animal tubes. *J. Marine Res.,* **39,** 361–374.

Einsele, G., Overbeck, R., Schwarz, H. U. and Unsöld, G. 1974. Mass physical properties, sliding and erodibility of experimentally deposited and differently consolidated clayey muds. *Sedimentology,* **21,** 339–372.

Einstein, H. A. 1950. The bedload function for sediment transportation in open channel flows. *Soil Cons. Serv. U.S. Dept. Agric. Tech. Bull.,* No. 1026, 78 pp.

Einstein, H. A. and El-Samni, E.-S. A. 1949. Hydrodynamic forces on a rough wall. *Rev. Modern Physics,* **21,** 520–524.

Einstein, H. A. and Krone, R. B. 1962. Experiments to determine modes of cohesive sediment transport in salt water. *J. Geophys. Res.,* **67,** 1451–1461.

Elliot, W. P. 1958. The growth of the atmospheric internal boundary layer. *Trans. Amer. Geophys. Union,* **39,** 1048–1054.

Emery, K. O. 1938. Rapid method of mechanical analysis of sands. *J. Sediment. Petrol.,* **8,** 105–111.

Engel, P. and Lau, Y. L. 1980. Friction factor for two-dimensional dune roughness. *J. Hyd. Res.,* **18,** 213–225.

Engelund, F. 1970. Instability of erodible beds. *J. Fluid Mech.,* **42,** 225–244.

Engelund, F. and Hansen, E. 1967. *A Monograph on Sediment Transport in Alluvial Streams,* Technisk Vorlag, Copenhagen, 62 pp.

Evans, G. 1965. Intertidal flat sediments and their environment of deposition in the Wash. *Quart. J. Geol. Soc. London,* **121,** 209–241.

Evans, G. and Collins, M. B. 1975. The transportation and deposition of suspended sediment over the intertidal flats of the Wash. In Hails, J. R. and Carr, A. P. (Eds), *Nearshore Sediment Dynamics and Sedimentation,* John Wiley, Chichester, 273–304.

Fenton, J. D. and Abbott, J. E. 1977. Initial movement of grains on a stream bed: The effect of relative protrusion. *Proc. R. Soc. London,* **A352,** 523–527.

Festa, J. F. and Hansen, D. V. 1978. Turbidity maxima in partially mixed estuaries: a two-dimensional numerical model. *Estuar. Coastal Mar. Sci.,* **7,** 347–359.

Fleming, G. 1970. Sediment balance of Clyde estuary. *J. Hydraul. Div. ASCE,* **96,** HY11, 2219–2230.

Flood, R. D. 1981. Distribution, morphology, and origin of sedimentary furrows in cohesive sediments, Southampton Water. *Sedimentology,* **28,** 511–529.

Flood, R. D. 1983. Classification of sedimentary furrows and a model for furrow initiation and evolution. *Geol. Soc. Amer. Bull.,* **94,** 630–639.

322

Flood, R. D. and Hollister, C. D. 1980. Submersible studies of deep sea furrows and transverse ripples in cohesive sediment. *Marine Geol.,* **36,** M1–M9.

Folk, R. L. 1966. A review of grain-size parameters. *Sedimentology,* **6,** 73–93.

Folk, R. L. and Ward, W. C. 1957. Brazos River bar: a study in the significance of grain size parameters. *J. Sediment. Petrol.,* **27,** 3–26.

Francis, J. R. D. 1973. Experiments on the motions of solitary grains along the bed of a water stream. *Proc. R. Soc. London,* **A332,** 443–471.

Fraser, H. J. 1935. Experimental study of the porosity and permeability of clastic sediments. *J. Geol.,* **43,** 910–1010.

Fredsoe, J. 1974. On the development of dunes in erodible channels. *J. Fluid Mech.,* **64,** 1–16.

Fredsoe, J. 1982. Shape and dimensions of stationary dunes in rivers. *J. Hydraul. Div. ASCE,* **108,** HY8, 932–947.

Frey, R. W. and Basan, P. B. 1978. Coastal salt marshes. In Davies, R. A. Jr. (Ed.), *Coastal Sedimentary Environments,* Springer-Verlag, New York, 101–169.

Friedman, G. M. 1961. Distribution between dune, beach, and river sands from textural characteristics. *J. Sediment. Petrol.,* **31,** 514–529.

Frostick, L. E. and McCave, I. N. 1979. Seasonal shifts of sediment within an estuary mediated by algal growth. *Estuar. Coastal Mar. Sci.,* **9,** 569–576.

Fukuda, M. K. and Lick, W. 1980. The entrainment of cohesive sediments in fresh water. *J. Geophys. Res.,* **85,** 2813–2824.

Gadd, P. E., Lavelle, J. W. and Swift, D. J. P. 1978. Estimate of sand transport on the New York shelf using near-bottom current meter observations. *J. Sediment. Petrol.,* **48,** 239–252.

Galvin, C. J. 1972. Wave breaking in shallow water. In Meyer, R. E. (Ed.), *Waves on Beaches,* Academic Press, New York, 413–456.

Galvin, C. J. and Vitale, P. 1976. Longshore transport prediction—SPM 1973 Equation. *Proc. 15th Coastal Eng. Conf.,* 1133–1148.

Garde, R. J. and Sethuraman, S. 1969. Variation of drag coefficient of a sphere rolling along a boundary. *La Houille Blanche,* No. 7, 727–732.

Gaughan, M. K. and Komar, P. D. 1975. The theory of wave propagation in water of gradually varying depth, and the prediction of breaker type and height. *J. Geophys. Res.,* **80,** 2991–2996.

Gelfenbaum, G. 1983. Suspended-sediment response to semidiurnal and fortnightly tidal variations in a mesotidal estuary: Columbia River, USA. *Marine Geol.,* **52,** 39–57.

Gibbs, R. J. 1977. Suspended sediment transport and the turbidity maximum. In *Estuaries, Geophysics and the Environment,* National Academy of Sciences, Washington, DC, 104–109.

Gibbs, R. J. 1983. Coagulation rates of clay minerals and natural sediments. *J. Sediment. Petrol.,* **53,** 1193–1203.

Gibbs, R. J., Matthews, M. D. and Link, D. A. 1971. The relationship between sphere size and settling velocity. *J. Sediment. Petrol.,* **41,** 7–18.

Glen, N. C. 1979. Tidal Measurement. In Dyer, K. R. (Ed.), *Estuarine Hydrography and Sedimentation,* Cambridge University Press, Cambridge, 19–40.

Godin, G. 1972. *The Analysis of Tides,* Liverpool University Press, 264 pp.

Goldberg, E. D. and Griffin, J. J. 1964. Sedimentation rates and mineralogy in the South Atlantic. *J. Geophys. Res.,* **62,** 4293–4309.

Gordon, C. M. 1974. Intermittent momentum transport in a geophysical boundary layer. *Nature,* **248,** 392–394.

Gordon, C. M. 1975. Sediment entrainment and suspension in a tidal flow. *Marine Geol.,* **18,** M57–M64.

Graf, W. H. 1971. *Hydraulics of Sediment Transport,* McGraw-Hill, New York, 513 pp.

Grant, W. D., Boyer, L. F. and Sanford, L. P. 1982. The effects of bioturbation on the initiation of motion of intertidal sands. *J. Marine Res.,* **40,** 659–677.

Grant, W. D. and Madsen, O. S. 1979. Combined wave and current interaction with a rough bottom. *J. Geophys. Res.*, **84**, 1797–1808.

Grant, W. D. and Madsen, O. S. 1982. Movable bed roughness in unsteady oscillatory flow. *J. Geophys. Res.*, **87**, 469–481.

Grass, A. J. 1970. Initial instability of fine bed sand. *J. Hydraul. Div. ASCE*, **96**, HY3, 619–632.

Grass, A. J. 1981. Sediment transport by waves and currents. *SERC London Centre for Marine Technology. Report No. FL29*, 30 pp.

Graton, L. C. and Fraser, H. J. 1935. Systematic packing of spheres with particular relation to porosity and permeability. *J. Geol.*, **43**, 785–909.

Greenwood, B. and Mittler, P. R. 1984. Sediment flux and equilibrium slopes in a barred nearshore. *Marine Geol.*, **60**, 79–98.

Groen, P. 1967. On the residual transport of suspended matter by an alternating tidal current. *Neth. J. Sea Res.*, **3**, 564–574.

Gross, T. F. and Nowell, A. R. M. 1983. Mean flow and turbulence scaling in a tidal boundary layer. *Continental Shelf Res.*, **2**, 109–126.

Gularte, R. C., Kelly, W. E. and Nacci, V. A. 1980. Erosion of cohesive sediments as a rate process. *Ocean. Engng.*, **7**, 539–551.

Gust, G. 1976. Observations on turbulent-drag reduction in a dilute suspension of clay in sea-water. *J. Fluid Mech.*, **75**, 29–47.

Gust, G. and Southard, J. B. 1983. Effects of weak bedload on the universal law of the wall. *J. Geophys. Res.*, **88**, 5939–5952.

Gust, G. and Stolte, S. 1978. Features of a measured wind–wave spectrum at the sea bed. *Marine Geol.*, **27**, M1–M8.

Gust, G. and Walger, E. 1976. The influence of suspended cohesive sediments on boundary-layer structure and erosive activity of turbulent seawater flow. *Marine Geol.*, **22**, 189–206.

Guza, R. T. and Inman, D. L. 1975. Edge waves and beach cusps. *J. Geophys. Res.*, **80**, 2997–3012.

Hallermeier, R. J. 1981. Terminal settling velocity of commonly occurring sand grains. *Sedimentology*, **28**, 859–865.

Hallermeier, R. J. 1982. Oscillatory bedload transport: data review and simple formulation. *Continental Shelf Res.*, **1**, 159–190.

Halliwell, A. R. and O'Connor, B. A. 1966. Suspended sediment in a tidal estuary. *Proc. 10th Coastal Eng. Conf.*, 687–706.

Hamilton, N. and Rees, A. I. 1970. The use of magnetic fabric in palaeocurrent estimation. In Runcorn, S. K. (Ed.), *Palaeogeophysics*, Academic Press, New York, 445–464.

Hammond, F. D. C., Heathershaw, A. D. and Langhorne, D. N. 1984. A comparison between Shields' threshold criterion and the movement of loosely packed gravel in a tidal channel. *Sedimentology*, **31**, 51–62.

Hammond, T. M. and Collins, M. B. 1979. On the threshold of transport of sand-sized sediment under the combined influence of unidirectional and oscillatory flow. *Sedimentology*, **26**, 795–812.

Harbaugh, J. W. and Merriam, D. F. 1968. *Computer Applications in Stratigraphic Analysis*, John Wiley, New York.

Hardisty, J. 1983. An assessment and calibration of formulations for Bagnold's bedload equation. *J. Sediment. Petrol.*, **53**, 1007–1010.

Hardisty, J. and Hamilton, D. 1984. Measurements of sediment transport on the seabed southwest of England. *Geo. Marine Letters*, **4**, 19–23.

Harms, J. C. 1969. Hydraulic significance of some sand ripples. *Geol. Soc. Amer. Bull.*, **80**, 363–396.

Harvey, J. G. 1966. Large sand waves in the Irish Sea. *Marine Geol.*, **4**, 49–55.

Haven, D. S. and Morales-Alamo, R. 1972. Biodeposition as a factor in sedimentation of fine suspended solids in estuaries. In Nelson, B. W. (Ed.), *Environmental Framework of Coastal Plain Estuaries. Geol. Soc. Amer. Mem.,* **133**, 121–130.

Hawkins, A. B. and Sebbage, M. J. 1972. The reversal of sand waves in the Bristol Channel. *Marine Geol.,* **12**, M7–M9.

Hayes, M. O. 1975. Morphology of sand accumulation in estuaries: an introduction to the symposium. In Cronin, L. E. (Ed.), *Estuarine Research Vol.II*, Academic Press, New York, 3–22.

Hayes, M. O. 1978. Impact of hurricanes on sedimentation in estuaries, bays, and lagoons. In Wiley, M. L. (Ed.), *Estuarine Interactions*, Academic Press, New York, 323–346.

Healey, R. G., Pye, K., Stoddart, D. R. and Bayliss-Smith, T. P. 1981. Velocity variations in salt marsh creeks, Norfolk, England. *Estuar. Coastal Shelf Sci.,* **13**, 535–545.

Heathershaw, A. D. 1974. 'Bursting' phenomena in the sea. *Nature,* **248**, 394–395.

Heathershaw, A. D. 1979. The turbulent structure of the bottom boundary layer in a tidal current. *Geophys. J. R. astr. Soc.,* **58**, 395–430.

Heathershaw, A. D. 1981. Comparisons of measured and predicted sediment transport rates in tidal currents. *Marine Geol.,* **42**, 75–104.

Heathershaw, A. D. 1982a. Some observations of currents in shallow water during a storm surge. *Estuar. Coastal Shelf Sci.,* **14**, 635–648.

Heathershaw, A. D. 1982b. Seabed-wave resonance and sand bar growth. *Nature,* **296**, 343–345.

Heathershaw, A. D. and Davies, A. G. 1985. Resonant wave reflection by transverse bedforms and its relation to beaches and offshore bars. *Marine Geol.,* **62**, 321–338.

Heathershaw, A. D. and Hammond, F. D. C. 1979. Swansea Bay (Sker) Project Topic Report 6. Offshore sediment movement and its relation to observed tidal current and wave data. *Institute of Oceanographic Sciences. Report No. 93.*

Heathershaw, A. D. and Hammond F. D. C. 1980. Secondary circulations near sand banks and in coastal embayments. *Deutsche Hydrog. Zeits.,* **33**, 135–151.

Heathershaw, A. D. and Simpson, J. H. 1978. The sampling variability of the Reynolds stress and its relation to boundary shear stress and drag coefficient measurements. *Estuar. Coastal Mar. Sci.,* **6**, 263–274.

Heathershaw, A. D. and Thorne, P. D. 1985. Sea-bed noises reveal role of turbulent bursting phenomenon in sediment transport by tidal currents. *Nature,* **316**, 339–342.

Hinze, J. O. 1959. *Turbulence*, McGraw-Hill, New York, 586 pp.

Hollister, C. D., Flood, R. D., Johnson, D. A., Lonsdale, P. F. and Southard, J. B. 1974. Abyssal furrows and hyperbolic echo traces on the Bahama Outer Ridge. *Geology,* **2**, 395–400.

Holman, R. A. and Bowen, A. J. 1982. Bars, bumps, and holes: models for the generation of complex beach topography. *J. Geophys. Res.,* **87**, 457–468.

Houbolt, J. J. H. C. 1968. Recent sediments in the Southern Bight of the North Sea. *Geol. en Mijnbouw,* **47**, 245–273.

Hulsey, J. D. 1961. Relations of settling velocity of sand size spheres and sample weight. *J. Sediment. Petrol.,* **31**, 101–112.

Hunt, J. N. 1954. The turbulent transport of sediment in open channels. *Proc. R. Soc. London,* **A224**, 322–335.

Huntley, D. A. and Bowen, A. J. 1975a. Field observations of edge waves and discussions of their effect on beach material. *Quart. J. Geol. Soc. London,* **131**, 69–81.

Huntley D. A. and Bowen, A. J. 1975b. Comparison of the hydrodynamics of steep and shallow beaches. In Hails, J. R. and Carr, A. P. (Eds), *Nearshore Sediment Dynamics and Sedimentation*, John Wiley, London, 69–109.

Huntley D. A. and Bowen, A. J. 1978. Beach cusps and edge waves. *Proc. 16th Coastal Eng. Conf.,* 1378–1393.

Huthnance, J. M. 1973. Tidal current asymmetries over the Norfolk sandbanks. *Estuar. Coastal Mar. Sci.,* **1**, 89–99.

Huthnance, J. M. 1982a. On one mechanism forming linear sand banks. *Estuar. Coastal Mar. Sci.,* **14**, 79–99.

Huthnance, J. M. 1982b. On the formation of sand banks of finite extent. *Estuar. Coastal Mar. Sci.,* **15**, 277–299.

Inglis, Sir C. C. and Allen, F. H. 1957. The regimen of the Thames estuary as affected currents, salinities and river flow. *Proc. Inst. Civil Eng.,* **7**, 827–868.

Inglis, Sir C. C. and Kestner, F. J. T. 1958. The long term effects of training walls, reclamation and dredging on estuaries. *Proc. Inst. Civil Eng.,* **9**, 193–216.

Inman, D. L. 1949. Sorting of sediment in light of fluid mechanics. *J. Sediment. Petrol.,* **19**, 51–70.

Inman, D. L. 1952. Measures for describing the size distribution of sediments. *J. Sediment. Petrol.,* **22**, 125–145.

Inman, D. L. and Bagnold, R. A. 1963. Beach and nearshore processes. Part II. Littoral processes. In Hill, M. N. (Ed.), *The Sea: Vol. 3*, Wiley–Interscience, New York, 529–553.

Inman, D. L. and Frautschy, J. D. 1966. Littoral processes and the development of shorelines. *Proc. 10th Coastal Eng. Conf.,* 511–536.

Inman, D. L. and Guza, R. T. 1982. The origin of swash cusps on beaches. *Marine Geol.,* **49**, 133–148.

Ippen, A. T. and Eagleson, P. S. 1955. A study of sediment sorting by waves shoaling on a plane beach. *U.S. Army Corps Eng. Beach Erosion Board Tech. Memo,* **63**.

Ippen, A. T. and Verma, R. P. 1953. The motion of discrete particles along the bed of a turbulent stream. *Proc. Minnesota Int. Hydraul. Conv.*

Itakura, T. and Kishi, T. 1980. Open channel flow with suspended sediments. *J. Hydraul. Div. ASCE,* **106**, HY8, 1325–1343.

Jackson, N. A. 1976. The propagation of modified flow downstream of a change of roughness. *Q. J. R. Metereol. Soc.,* **102**, 775–779.

Jackson, P. S. 1981. On the displacement height in the logarithmic velocity profile. *J. Fluid Mech.,* **111**, 15–25.

Jackson, R. G. 1976. Sedimentological and fluid-dynamic implications of the turbulent bursting phenomenon in geophysical flows. *J. Fluid Mech.,* **77**, 531–560.

Jackson, W. H. 1964. An investigation into silt in suspension in the River Humber. *Dock Harb. Auth.,* **XLV** No. 526.

Jago, C. F. and Hardisty, J. 1984. Sedimentology and morphodynamics of a microtidal beach, Pendine Sands, S. Wales. *Marine Geol.,* **60**, 99–122.

James, A. E. and Williams, D. J. A. 1982. Flocculation and rheology of kaolinite/quartz suspensions, *Rheol. Acta,* **21**, 176–183.

Jones, N. S., Kain, J. M. and Stride, A. H. 1965. The movement of sand waves on Warts Bank, Isle of Man. *Marine Geol.,* **3**, 329–336.

Jonsson, I. G. 1966. Wave boundary layers and friction factors. *Proc. 10th Coastal Eng. Conf.,* 127–148.

Jumars, P. A. and Nowell, A. R. M. 1984. Effects of benthos on sediment transport: difficulties with functional grouping. *Continental Shelf Res.,* **3**, 115–130.

Kachel, N. B. and Sternberg, R. W. 1971. Transport of bedload as ripples during an ebb current. *Marine Geol.,* **10**, 229–244.

Kajiura, K. 1968. A model of the bottom boundary layer in water waves. *Bull. Earthquake Res. Inst.,* **46**, 75–123.

Kalkanis, G. 1964. Transportation of bed material due to wave action. *U.S. Army Corps Eng. Tech. Memo No. 2.*

Kamphuis, J. W. 1974. Determination of sand roughness for fixed beds. *J. Hydraul. Res.,* **12**, 193–203.

326

Kamphuis, J. W. and Hall, K. R. 1983. Cohesive material erosion by unidirectional current. *J. Hydraul. Eng. ASCE*, **109**, No. 1, 49–61.

Kana, T. W. 1978. Surfzone measurements of suspended sediments. *Proc. 16th Coastal Eng. Conf.*, 1725–1743.

Karahan, M. E. and Peterson, A. W. 1980. Visualization of separation over sand waves. *J. Hydraul. Div. ASCE*, **106**, HY8, 1345–1352.

Karl, H. A. 1980. Speculations on processes responsible for mesoscale current lineations on the continental shelf, southern California. *Marine Geol.*, **34**, M9–M18.

Kemp. P. H. and Simons, R. R. 1982. The interaction between waves and a turbulent current: waves propagating with the current. *J. Fluid Mech.*, **116**, 227–250.

Kemp. P. H. and Simons, R. R. 1983. The interaction of waves and a turbulent current: waves propagating against the current. *J. Fluid Mech.*, **130**, 73–89.

Kennedy, J. F. and Locher, F. A. 1972. Sediment suspension by water waves. In Meyer, R. E. (Ed.), *Waves on Beaches*, Academic Press, 249–296.

Kenyon, N. H. 1970. Sand ribbons of European tidal seas. *Marine Geol.*, **9**, 25–39.

Kenyon, N. H. and Stride, A. H. 1968. The crest length and sinuosity of some marine sand waves. *J. Sediment. Petrol.*, **38**, 255–259.

Kenyon, N. H. and Stride, A. H. 1970. The tide-swept continental shelf sediment between the Shetland Isles and France. *Sedimentology*, **14**, 159–173.

Keunen, Ph. H. 1968. Settling convection and grain size analysis. *J. Sediment. Petrol.*, **38**, 817–831.

King, C. A. M. 1972. *Beaches and Coasts*, 2nd ed., Edward Arnold, London, 570 pp.

Kirby, R. and Kelland, N. C. 1972. Adjacent stable and apparently mobile linear sediment ridges in the Southern North Sea. *Nature*, **238**, 111–112.

Kirby, R. and Oele, E. 1975. The geological history of the Sandettie–Fairy Bank area, southern North Sea. *Phil. Trans. R. Soc. London*, **A279**, 257–267.

Kirby, R. and Parker, W. R. 1974. Seabed density measurements related to echo sounder records. *Dock and Harbour Auth.*, **54**, 423–424.

Kirby, R. and Parker, W. R. 1977. Fine sediment studies relevant to dredging practice and control. *Proc. 2nd Int. Symp. Dredging Technology*, B2-15–B2-26.

Kirby, R. and Parker, W. R. 1981. Settled mud deposits in Bridgwater Bay, Bristol Channel. *Institute of Oceanographic Sciences. Report No. 107*, 67 pp.

Kirby, R. and Parker, W. R. 1982. A suspended sediment front in the Severn Estuary. *Nature*, **295**, 396–399.

Kirby, R. and Parker, W. R. 1983. The distribution and behaviour of fine sediment in the Severn Estuary and Inner Bristol Channel. *Can. J. Fish Aquat. Sci.*, **40**, Suppl. 83–95.

Klein, G. de V. 1970. Deposition and dispersal dynamics of intertidal sand bars. *J. Sediment. Petrol.*, **40**, 1095–1127.

Klovan, J. E. 1966. The use of factor analysis in determining depositional environments from grain size distributions. *J. Sediment. Petrol.*, **36**, 115–126.

Knight D. W. and Macdonald, J. A. 1979. Hydraulic resistance of artificial strip roughness. *J. Hydraul. Div. ASCE*, **105**, HY6, 675–690.

Komar, P. D. 1971. The mechanics of sand transport on beaches. *J. Geophys. Res.*, **76**, 713–721.

Komar, P. D. 1974. Oscillatory ripple marks and the evaluation of ancient wave conditions and environments. *J. Sediment. Petrol.*, **44**, 169–180.

Komar, P. D. 1975. Nearshore currents: generation by obliquely incident waves and longshore variations in breaker heights. In Hails, J. R. and Carr, A. P. (Eds), *Nearshore Sediment Dynamics and Sedimentation*, John Wiley, London, 17–45.

Komar, P. D. 1976. *Beach Processes and Sedimentation*, Prentice-Hall, New Jersey, 429 pp.

Komar, P. D. (Ed.) 1983. *CRC Handbook of Coastal Processes and Erosion*, CRC Press, Florida.

Komar, P. D. and Cui, B. 1984. The analysis of grain size measurements by sieving and settling-tube techniques. *J. Sediment. Petrol.*, **54**, 603–614.

Komar, P. D. and Inman, D. L. 1970. Longshore sand transport on beaches. *J. Geophys. Res.*, **75**, 5914–5927.

Komar, P. D. and Li, Z. 1985. A pivoting analysis of the selective entrainment of gravel. *Sedimentology*, (in press).

Komar, P. D. and Miller, M. C. 1973. The threshold of sediment movement under oscillatory water waves. *J. Sediment. Petrol.*, **43**, 1101–1110.

Komar, P. D. and Miller, M. C. 1974. Sediment threshold under oscillatory waves. *Proc. 14th Coastal Eng. Conf.*, 756–775.

Komar, P. D. and Miller, M. C. 1975a. On the comparison between the threshold of sediment motion under waves and unidirectional currents with a discussion of the practical evaluation of the threshold. *J. Sediment. Petrol.*, **45**, 362–367.

Komar, P. D. and Miller, M. C. 1975b. The initiation of oscillatory ripple marks and the development of plane-bed at high shear stresses under waves. *J. Sediment. Petrol.*, **45**, 697–703.

Komar, P. D. and Reimers, C. E. 1978. Grain shape effects on settling rates. *J. Geol.*, **86**, 193–209.

Kranck, K. 1973. Flocculation of suspended sediment in the sea. *Nature*, **246**, 348–350.

Kranck, K. 1981. Particulate matter grain-size characteristics and flocculation in a partially mixed estuary. *Sedimentology*, **28**, 107–114.

Krone, R. B. 1962. *Flume Studies of the Transport of Sediment in Estuarial Shoaling Processes*, Univ. Calif. Hyd. Eng. Lab. and Sanit. Eng. Res. Lab., Berkeley, 110 pp.

Krone, R. B. 1978a. Aggregation of suspended particles in estuaries. In Kjerfve, B. (Ed.), *Estuarine Transport Processes*, Univ. South Carolina Press, 177–190.

Krone, R. B. 1978b. Engineering interest in the benthic boundary layer. In McCave, I. N. (Ed.), *The Benthic Boundary Layer*, Plenum Press, New York, 143–156.

Krone, R. B. 1979. Sedimentation in the San Francisco Bay system. In Conomos, T. J. (Ed.), *San Francisco Bay: The Urbanized Estuary*, Amer. Assoc. Sci., Pacific Div., San Francisco, 85–96.

Krumbein, W. C. 1934. Size frequency distributions of sediments. *J. Sediment. Petrol.*, **4**, 65–77.

Krumbein, W. C. and Sloss, L. L. 1963. *Stratigraphy and Sedimentation*, 2nd. Ed., W. H. Freeman & Co., San Francisco, 660 pp.

Kynch, G. J. 1952. A theory of sedimentation. *Trans. Faraday Soc.*, **48**, 166–176.

Lamb, H. 1932. *Hydrodynamics*, 6th edn. Cambridge University Press, 738 pp.

Langeraar, W. 1966. Sand waves in the North Sea. *Hydrog. Newsletter*, **1**, 243–246.

Langhorne, D. N. 1973. A sandwave field in the Outer Thames Estuary, Great Britain. *Marine Geol.*, **14**, 129–143.

Langhorne, D. N. 1977. Consideration of meteorological conditions when determining tha navigational water depth over a sand wave field. *Int. Hydrog. Rev.*, **LIV**, 17–30.

Langhorne, D. N. 1981. An evaluation of Bagnold's dimensionless coefficient of proportionality using measurements of sandwave movement. *Marine Geol.*, **43**, 49–64.

Langhorne, D. N. 1982a. A study of the dynamics of a marine sandwave. *Sedimentology*, **29**, 571–594.

Langhorne, D. N. 1982b. The stability of the top metre of the seabed—its importance to engineering and navigational projects. *Int. Hydrog. Rev.*, **LIX**, 79–94.

Lau, J. and Travis, B. 1973. Slowly varying Stokes waves and submarine longshore bars. *J. Geophys. Res.*, **78**, 4489–4497.

Laufer, J. 1954. The structure of turbulence in fully developed pipe flow. *Nat. Advisory Comm. Aeronautics Tech. Report*, **1174**.

Lavelle, J. W. and Mofjeld, H. O. 1983. Effects of time-varying viscosity on oscillatory turbulent channel flow. *J. Geophys. Res.*, **88**, 7607–7616.

Lavelle, J. W., Mofjeld, H. O. and Baker, E. T. 1984. An in-situ erosion rate for a fine-grained marine sediment. *J. Geophys. Res.*, **89**, 6543–6552.

Lavelle, J. W. and Thacker, W. C. 1978. Effects of hindered settling on sediment concentration profiles. *J. Hydraul. Res.*, **16**, 347–355.

Lavelle, J. W., Young, R. A., Swift D. J. P. and Clarke, T. L. 1978. Near-bottom sediment concentration and fluid velocity measurements on the inner continental shelf, New York. *J. Geophys. Res.*, **83**, 6052–6062.

Lee, A. J. and Ramster, J. W. (Eds) 1981. *Atlas of the Sea around the British Isles*, Ministry of Agriculture, Fisheries and Food.

Leeder, M. R. 1983. On the dynamics of sediment suspension by residual Reynolds stresses—confirmation of Bagnold's theory. *Sedimentology*, **30**, 485–491.

Lees, B. J. 1979. A new technique for injecting fluorescent sand tracer in sediment transport experiments in a shallow water marine environment. *Marine Geol.*, **33**, M95–M98.

Lees, B. J. 1981a. Relationship between eddy viscosity of seawater and eddy diffusivity of suspended particles. *Geo-Marine Letters*, **1**, 249–254.

Lees, B. J. 1981b. Sediment transport measurements in the Sizewell–Dunwich Banks area, East Anglia, U.K. *Spec. Publs. Int. Ass. Sediment*, **5**, 269–281.

Lees, B. J. 1983. The relationship of sediment transport rates and paths to sandbanks in a tidally dominated area off the coast of East Anglia, U.K. *Sedimentology*, **30**, 461–483.

Lesht, B. M. 1979. Relationship between sediment resuspension and the statistical frequency distribution of bottom shear stress. *Marine Geol.*, **32**, M19–M27.

Lesht, B. M. 1980. Benthic boundary-layer velocity profiles: dependence on averaging period. *J. Physical Ocean.*, **10**, 985–991.

Lettau, H. 1969. Note on aerodynamic roughness parameter estimation on the basis of roughness-element description. *J. Applied Meteorology*, **8**, 828–832.

Liu, A.-K. and Davies, S. H. 1977. Viscous attenuation of mean drift in water waves. *J. Fluid Mech.*, **81**, 63–84.

Liu, H.-K. 1957. Mechanics of ripple formation. *J. Hydraul. Div. ASCE*, **83**, HY2, Paper 1197.

Lofquist, K. E. B. 1978. Sand ripple growth in an oscillatory-flow water tunnel. *U.S. Army Corps Engineers Coastal Eng. Res. Center Tech. Paper No. 78–5*, 101 pp.

Longuet-Higgins, M. S. 1953. Mass transport in water waves. *Phil. Trans. R. Soc. London*, **A245**, 535–581.

Longuet-Higgins, M. S. 1970. Longshore currents generated by obliquely incident sea waves. 2. *J. Geophys. Res.*, **75**, 6790–6801.

Longuet-Higgins, M. S. 1972. Recent progress in the study of longshore currents. In Meyer, R. E. (Ed.), *Waves on Beaches*, Academic Press, New York, 203–248.

Longuet-Higgins, M. S. and Stewart, R. W. 1964. Radiation stress in water waves; a physical discussion, with applications. *Deep-Sea Res.*, **11**, 529–562.

Ludwick, J. C. 1972. Migration of tidal sand waves in Chesapeake Bay entrance. In Swift, D. J. P., Duane, D. B. and Pilkey, O. H. (Eds), *Shelf Sediment Transport: Process and Pattern*, Dowden, Hutchinson and Ross, Stroudsberg, PA, 377–410.

Ludwick, J. C. 1974. Tidal currents and zig-zag sand shoals in a wide estuary entrance. *Geol. Soc. Amer. Bull.*, **85**, 717–726.

Ludwick, J. C. 1975a. Variations in the boundary-drag coefficient in the tidal entrance to Chesapeake Bay, Virginia. *Marine Geol.*, **19**, 19–28.

Ludwick, J. C. 1975b. Tidal currents, sediment transport, and sand banks in Chesapeake Bay entrance, Virginia. In Cronin, L. E. (Ed.), *Estuarine Research, Vol. II*, Academic Press, New York, 365–380.

Ludwick, J. C. and Domurat, G. W. 1982. A deterministic model of the vertical component of sediment motion in a turbulent fluid. *Marine Geol.*, **45**, 1–15.

Madsen, O. S. and Grant, W. D. 1975. The threshold of sediment movement under oscillatory waves: a discussion. *J. Sediment. Petrol.*, **45**, 360–361.

329

Madsen, O. S. and Grant, W. D. 1976. Quantitative description of sediment transport by waves. *Proc. 15th Coastal Eng. Conf., 1093–1112.*

Manohar, M. 1955. Mechanics of bottom sediment movement due to wave action. *U.S. Army Corps Engineers, Beach Erosion Board Tech. Memo No. 75,* 121 pp.

Mantz, P. A. 1973. Cohesionless, fine graded flaked sediment transport by water. *Nature,* **246**, 14–16.

Mason, C. C. and Folk, R. L. 1958. Differentiation of beach, dune and aeolian flat environments by size analysis, Mustang Island, Texas. *J. Sediment. Petrol.,* **28**, 211–226.

Maude, A. D. and Whitmore, R. L. 1958. A generalized theory of sedimentation. *Brit. J. Appl. Phys.,* **9**, 477–482.

McCammon, R. B. 1962. Efficiencies of percentile measures for describing the mean size and sorting of sedimentary particles. *J. Geol.,* **70**, 453–465.

McCave, I. N. 1970. Deposition of fine-grained suspended sediment from tidal currents. *J. Geophys. Res.,* **75**, 4151–4159.

McCave, I. N. 1971a. Wave effectiveness at the sea bed and its relationship to bedforms and deposition of mud. *J. Sediment. Petrol.,* **41**, 89–96.

McCave, I. N. 1971b. Sand waves in the North Sea off the coast of Holland. *Marine Geol.,* **10**, 199–225.

McCave, I. N. 1973. Some boundary-layer characteristics of tidal currents bearing sand in suspension. *Mem. Soc. R. des Sciences de Liège,* **6**, 187–206.

McCave, I. N. 1974. Discussion on Meade, R. H. Net transport of sediment through the mouths of estuaries: seaward or landward? *Mem. Inst. Geol. Bassin Aquitaine,* No. 7, 207–213.

McCave, I. N. 1979a. Suspended sediment. In Dyer, K. R. (Ed.), *Estuarine Hydrography and Sedimentation,* Cambridge University Press, Cambridge, 131–185.

McCave, I. N. 1979b. Tidal currents at the North Hinder lightship, southern North Sea: flow directions and turbulence in relation to maintenance of sand banks. *Marine Geol.,* **31**, 101–114.

McCave, I. N. 1984. Size spectra and aggregation of suspended particles in the deep ocean. *Deep Sea Res.,* **31**, 329–352.

McCave, I. N. and Langhorne, D. N. 1982. Sandwaves and sediment transport around the end of a tidal sand bank. *Sedimentology,* **29**, 95–110.

McCutcheon, S. 1981. Vertical velocity profiles in stratified flows. *J. Hydraul. Div. ASCE,* **107**, HY8, 973–988.

McLean, S. R. 1981. The role of non-uniform roughness in the formation of sand ribbons. *Marine Geol.,* **42**, 49–74.

McLean, S. R. 1983. Turbulence and sediment transport measurements in a North Sea tidal inlet (The Jade). In Sundermann, J. and Lenz, W. (Eds), *North Sea Dynamics,* Springer-Verlag, Berlin, 436–452.

McManus, D. A. 1965. A study of maximum load for small diameter sieves. *J. Sediment. Petrol.,* **35**, 792–796.

McQuivey, R. S. and Richardson, E. V. 1969. Some turbulence measurements in open channel flows. *J. Hydraul. Div. ASCE,* **95**, HY1, 209–223.

Meade, R. H. 1969. Landward transport of bottom sediments in estuaries of the Atlantic Coastal Plain. *J. Sediment. Petrol.,* **39**, 222–234.

Mehta. A. J. 1978. Bed friction characteristics of three tidal entrances. *Coastal Eng.,* **2**, 69–83.

Mehta. A. J. and Christensen, B. A. 1983. Initiation of sand transport over coarse beds in tidal entrances. *Coastal Eng.,* **7**, 61–75.

Mehta. A. J., Parchure, T. M., Dixit, J. G. and Ariathurai, R. 1982. Resuspension potential of deposited cohesive sediment beds. In Kennedy, V. (Ed.), *Estuarine Comparisons,* Academic Press, New York, 591–609.

330

Mehta. A. J. and Partheniades, E. 1975. An investigation of the depositional properties of flocculated fine sediments. *J. Hydraul. Res.*, No. 4, 361–381.

Meyer-Peter, E. and Müller, R. 1948. Formulae for bed-load transport. *Proc. 2nd Cong. Int. Assoc. Hydraul. Res.*, Stockholm.

Middleton, G. V. 1976. Hydraulic interpretation of sand size distributions. *J. Geol.* **84**, 405–426.

Migniot, C. 1968. Étude des propriétés physiques de différents sédiments très fins et de leur compartement sous des actions hydrodynamiques. *La Houille Blanche*, **23**, 591–620.

Miller, M. C. and Komar, P. D. 1980a. A field investigation of the relationship between oscillation ripple spacing and the near-bottom water orbital motions. *J. Sediment. Petrol.*, **50**, 183–191.

Miller, M. C. and Komar, P. D. 1980b. Oscillation sand ripples generated by laboratory apparatus. *J. Sediment. Petrol.*, **50**, 173–182.

Miller, M. C., McCave, I. N. and Komar, P. D. 1977. Threshold of sediment motion under unidirectional currents. *Sedimentology*, **24**, 507–528.

Miller, R. L. and Byrne R. J. 1966. The angle of repose for a single grain on a fixed rough bed. *Sedimentology*, **6**, 303–314.

Miller, R. L. and Zeigler, J. M. 1958. A model relating dynamics and sediment pattern in equilibrium in the region of shoaling waves, breaker zone, and foreshore. *J. Geol.*, **66**, 417–441.

Milliman, J. D. 1980. Sedimentation in the Fraser River and its estuary, Southwestern British Columbia (Canada). *Estuar. Coastal Mar. Sci.*, **10**, 609–633.

Milliman, J. D. and Meade, R. H. 1983. World wide delivery of river sediment to the oceans. *J. Geol.*, **91**, 1–21.

Monin, A. S. and Yaglom, A. M. 1971. *Statistical Fluid Mechanics: Mechanics of Turbulence*, MIT Press, Massachusetts, 769 pp.

Moore, D. G. 1961. Submarine slumps. *J. Sediment. Petrol.*, **31**, 343–357.

Moss, A. J. 1962. The physical nature of common sandy and pebbly deposits, Pt. 1. *Amer. J. Sci.*, **260**, 337–373.

Moss, A. J. 1963. The physical nature of common sandy and pebbly deposits, Pt. 2. **261**, 297–343.

Moss, A. J. 1972. Bed-load sediments. *Sedimentology*, **18**, 159–219.

Müller, A., Gyr, A. and Dracos, T. 1971. Interaction of rotating elements of the boundary layer with grains of the bed: a contribution to the problem of the threshold of sediment transportation. *J. Hydraul. Res.*, **9**, 373–411.

Müller, G. and Förstner, U. 1968. General relationship between suspended sediment concentration and water discharge in the Alpenrhein and some other rivers. *Nature*, **217**, 244–245.

Nalluri, C. and Novak, P. 1973. Turbulence characteristics in a smooth open channel of circular cross-section. *J. Hydraul. Res.*, **11**, 343–368.

Neill, C. R. and Yalin, M. S. 1969. Quantitative definition of beginning of bed movement. *J. Hydraul. Div. ASCE*, **95**, HY1, 585–587.

Nichols, M. 1972. Sediments of the James River Estuary, Virginia. In Nelson, B. W. (Ed.), *Environmental Framework of Coastal Plain Estuaries. Geological Soc. Amer. Mem.* **133**, 169–212.

Nichols, M. M. 1977. Response and recovery of an estuary following a river flood. *J. Sediment. Petrol.*, **47**, 1171–1186.

Nichols, M. and Poor, G. 1967. Sediment transport in a coastal plain estuary. *J. Wat. Harb. Coastal. Eng. Div. ASCE*, **93**, WW4, 83–95.

Nielsen, P. 1981. Dynamics and geometry of wave-generated ripples. *J. Geophys. Res.*, **86**, 6467–6472.

Nielsen, P. 1984. Field measurements of time-averaged suspended sediment concentration under waves. *Coastal Eng.*, **8**, 51–72.

Nixon, S. W. 1980. Between coastal marshes and coastal waters—a review of twenty years of speculation and research on the role of marshes in estuarine productivity and water chemistry. In Hamilton, P. and MacDonald, K. B. (Eds), *Estuarine and Wetland Processes. Plenum Marine Science*, **11**, 437–525.

Noda, H. 1968. A study of mass transport in boundary layers in standing waves. *Proc. 11th Coastal Eng. Conf.*, 227–247.

Nowell, A. R. M. and Church, M. 1979. Turbulent flow in a depth-limited boundary layer. *J. Geophys. Res.*, **84**, 4816–4824.

Nowell, A. R. M., Jumars, P. A. and Eckman, J. E. 1981. Effects of biological activity on the entrainment of marine sediments. *Marine Geol.*, **42**, 133–153.

Nychas, S. G., Hershey, H. C. and Brodkey, R. S. 1973. A visual study of turbulent shear flow. *J. Fluid Mech.*, **61**, 513–540.

Odd, N. V. M. 1982. The feasibility of using mathematical models to predict sediment transport in the Severn Estuary. In *The Severn Barrage*, Thomas Telford Ltd, London, 195–202.

Odd, N. V. M. and Owen, M. W. 1972. A two layer model of mud transport in the Thames estuary. *Proc. Inst. Civil Eng.*, Supp. 9, Paper 75175.

Oertel, G. F. 1972. Sediment transport of estuary entrance shoals and the formation of swash platforms. *J. Sediment. Petrol.*, **42**, 857–863.

Off, T. 1963. Rythmic linear sand bodies caused by tidal currents. *Bull. Amer. Assoc. Petrol. Geol.*, **47**, 324–341.

Offen, G. R. and Kline, S. J. 1975. A proposed model of the bursting process in turbulent boundary layers. *J. Fluid Mech.*, **70**, 209–228.

Officer, C. B. 1976. *Physical Oceanography of Estuaries (and Associated Coastal Waters)*, Wiley–Interscience, New York, 465 pp.

Officer, C. B. 1980. Discussion of the turbidity maximum in partially mixed estuaries. *Estuar. Coastal Mar. Sci.*, **10**, 239–246.

Orford, J. D. and Wright, P. 1978. What's in a name? Descriptive or genetic implications of 'ridge and runnel' topography. *Marine Geol.*, **28**, M1–M8.

Owen, M. W. 1970a. A detailed study of settling velocities of an estuary mud. *Hydraulics Research Station, Report INT 78*.

Owen, M. W. 1970b. Properties of consolidating mud. *Hydraulics Research Station, Report INT 83*.

Owen, M. W. 1971. The effect of turbulence on the settling velocities of silt flocs. *Proc. 14th Cong. Int. Ass. Hydraul. Res.*, 27–32.

Owen, M. W. 1975. Erosion of Avonmouth mud. *Hydraulics Research Station, Report INT 150*.

Owen, M. W. 1976. Problems in the modelling of transport, erosion, and deposition of cohesive sediments. In Goldberg, E. D., McCave, I. N., O'Brien, J. J. and Steele, J. H. (Eds), *The Sea, Vol. 6*, Wiley–Interscience, New York, 515–537.

Owen, M. W. and Thorn, M. F. C. 1978. Effect of waves on sand transport by currents. *Proc. 16th Coastal Eng. Conf.*, 1675–1687.

Owen, P. R. 1964. Saltation of uniform grains in air. *J. Fluid Mech.*, **20**, 225–242.

Pantin, H. M., Hamilton, D. and Evans, C. D. R. 1981. Secondary flow caused by differential roughness, Langmuir circulations, and their effect on the developments of sand ribbons. *Geo-Marine Letters*, **1**, 255–260.

Parker, W. R. 1976. Sediment mobility and erosion on a multibarred foreshore (Southwest Lancashire, U.K.). In Hails, J. R. and Carr, A. P. (Eds), *Nearshore Sediment Dynamics and Sedimentation*, John Wiley, Chichester, 151–179.

Parker, W. R. and Kirby, R. 1977. Fine sediment studies relevant to dredging practice and control. *Proc. 2nd Int. Symp. Dredging Tech.*, B2-15–B2-26.

Parker, W. R. and Kirby, R. 1982a. Time dependent properties of cohesive sediment relevant to sedimentation management—European experience. In Kennedy, V. (Ed.), *Estuarine Comparisons*, Academic Press, New York.

332

Parker, W. R. and Kirby, R. 1982b. Sources and transport patterns of sediment in the inner Bristol Channel and Severn Estuary. In *Severn Barrage*, Thomas Telford Ltd, London, 181–194.

Partheniades, E. 1965. Erosion and deposition of cohesive soils. *J. Hydraul. Div. ASCE,* **91**, HY1, 105–139.

Passega, R. 1957. Texture as a characteristic of clastic deposition. *Bull. Amer. Assoc. Petrol. Geol.,* **41**, 1952–1984.

Passega, R. 1964. Grain size representation by CM patterns as a geological tool. *J. Sediment. Petrol.,* **34**, 830–847.

Peirce, T. J., Jarman, R. T. and de Turville, C. M. 1970. An experimental study of silt scouring. *Proc. Inst. Civil Eng.,* **45**, 231–243.

Peirce, T. J. and Williams, D. J. A. 1966. Experiments on certain aspects of sedimentation of estuarine muds. *Proc. Inst. Civil Eng.,* **34**, 391–402.

Pethick, J. S. 1980. Velocity surges and asymmetry in tidal channels. *Estuar. Coastal Mar. Sci.,* **11**, 331–345.

Pingree, R. D. 1978. The formation of the Shambles and other banks by tidal stirring of the seas. *J. Mar. Biol. Ass. U.K.,* **58**, 211–226.

Pingree, R. D. and Maddock, L. 1977. Tidal eddies and coastal discharge. *J. Mar. Biol. Ass. U.K.,* **57**, 869–875.

Pingree, R. D. and Maddock, L. 1979. The tidal physics of headland flows and offshore tidal bank formation. *Marine Geol.,* **32**, 269–289.

Postma, H. 1961. Transport and accumulation of suspended matter in the Dutch Wadden Sea. *Neth. J. Sea Res.,* **1**, 148–190.

Postma, H. 1967. Sediment transport and sedimentation in the estuarine environment. In Lauff, G. H. (Ed.), *Estuaries, Amer. Assoc. Adv. Sci. Pub. No. 83,* 158–179.

Price, W. A. and Kendrick, M. P. 1963. Field and model investigation into the reasons for siltation in the Mersey estuary. *Proc. Inst. Civil Eng.,* **24**, 473–517.

Pugh, D. T. 1982. Estimating extreme currents by combining tidal and surge probabilities. *Ocean Engng.,* **9**, 361–372.

Rana, S. A., Simons, D. B. and Mahmood, K. 1973. Analysis of sediment sorting in alluvial channels. *J. Hydraul. Div. ASCE,* **99**, HY11, 1967–1980.

Raudkivi, A. J. 1963. Study of sediment ripple formation. *J. Hydraul. Div. ASCE,* **89**, HY6, 15–33.

Raudkivi, A. J. 1967. *Loose Boundary Hydraulics*, Pergamon Press, Oxford, 331 pp.

Raudkivi, A. J. and Hutchinson, D. L. 1974. Erosion of kaolinite clay by flowing water. *Proc. R. Soc. London,* **A337**, 537–554.

Rees, A. I. 1965. The use of anisotropy of magnetic susceptibility in the estimation of sedimentary fabric. *Sedimentology,* **4**, 257–271.

Rees, A. I. 1966 Some flume experiments with a fine silt. *Sedimentology,* **6**, 209–240.

Reineck. H.-E. and Singh. I. B. 1980. *Depositional Sedimentary Environments, with reference to Terrigenous clastics*, Springer-Verlag, Berlin, 2nd Ed.

Rhoads, D. C. and Young, D. K. 1970. The influence of deposit-feeding organisms on sedimentary stability and community trophic structure. *J. Marine Res.,* **28**, 150–178.

Rhoads, D. C., Yingst, J. Y. and Ullman, W. J. 1978. Sea floor stability in Long Island Sound. Part I. Temporal changes in erodibility of fine-grained sediments. In Wiley, M. L. (Ed.), *Estuarine Interactions*, Academic Press, New York.

Richards, K. J. 1980. The formation of ripples and dunes on an erodible bed. *J. Fluid Mech.,* **99**, 597–618.

Richards, K. J. and Taylor, P. A. 1981. A numerical model of flow over sand waves in water of finite depth. *Geophys. J. R. astr. Soc.,* **65**, 103–128.

Rifai, M. F. and Smith, K. V. H. 1971. Flow over triangular elements simulating dunes. *J. Hydraul. Div. ASCE,* **97**, HY7, 963–976.

Rigler, J. K., Collins, M. B. and Williams, S. J. 1981. A high precision digital recording sedimentation tower for sands. *J. Sediment. Petrol.,* **51**, 642–644.

333

Robinson, I. S. 1981. Tidal vorticity and residual circulation. *Deep-Sea Res.,* **28,** 195–212.
Russel, R. J. and Russel, R. D. 1939. Mississippi river delta sedimentation. In Trask, P. D. (Ed.), *Recent Marine Sediments,* Dover Pubs., New York, 153–177.
Saffman, P. G. and Turner, J. S. 1956. On the collision of drops in turbulent clouds. *J. Fluid Mech.,* **1,** 16–30.
Salsman, G. G., Tolbert, W. H. and Villars, R. G. 1966. Sand ridge migration in St. Andrews Bay, Florida. *Marine Geol.,* **4,** 11–19.
Schlee, J. 1966. A modified Woods Hole rapid sediment analyser. *J. Sediment. Petrol.,* **36,** 404–413.
Schlichtling, H. 1968. *Boundary Layer Theory,* (6th ed), McGraw-Hill, New York, 647 pp.
Schubel, J. R. 1969. Distribution and transport of suspended sediment in Upper Chesapeake Bay. *Tech. Rept. 60, Ref. 69–13,* Chesapeake Bay Institute, Johns Hopkins University.
Schubel, J. R. 1971. Tidal variation of the size distribution of suspended sediment at a station in the Chesapeake Bay turbidity maximum. *Neth. J. Sea Res.,* **5,** 252–266.
Schubel, J. R. 1972. Distribution and transportation of suspended sediment in Upper Chesapeake Bay. In Nelson, B. W. (Ed.), *Environmental Framework of Coastal Plain Estuaries. Geol. Soc. Amer. Mem.,* **133,** 151–167.
Schubel, J. R. 1974. Effects of tropical storm Agnes on the suspended solids of the Northern Chesapeake Bay. In Gibbs, R. J. (Ed.), *Suspended Solids in Water. Plenum Marine Science,* **4,** 113–132.
Schubel, J. R. and Carter, H. H. 1977. Suspended sediment budget for Chesapeake Bay. In Wiley, M. (Ed.), *Estuarine Processes, Vol. II,* Academic Press, New York, 48–62.
Schubel, J. R. and Hirschberg, D. J. 1978. Estuarine graveyards, climatic change, and the importance of the estuarine environment. In Wiley, M. (Ed.), *Estuarine Interactions,* Academic Press, New York, 285–303.
Schubel, J. R. and Schiemer, E. W. 1973. The cause of the acoustically impenetrable, or turbid, character of Chesapeake Bay sediment. *Mar. Geophys. Res.,* **2,** 61–71.
Scott, G. D. 1960. Packing of equal spheres. *Nature,* **188,** 908–909.
Seward-Thompson, B. L. and Hails, J. R. 1973. An appraisal of the computation of statistical parameters in grain size analysis. *Sedimentology,* **20,** 161–169.
Shaw, T. L. 1980. Tides, currents and waves. In Shaw, T. L. (Ed.), *An Environmental Appraisal of the Severn Barrage,* Pitman Publishing Ltd., 1–34.
Sheldon, R. W. 1968. Sedimentation in the estuary of the River Crouch, Essex, England. *Limnol. Oceanog.,* **13,** 72–83.
Sheng, Y. P. and Lick, W. 1979. The transport and resuspension of sediments in a shallow lake. *J. Geophys. Res.,* **84,** 1809–1826.
Shepard, F. P. 1954. Nomenclature based on sand–silt–clay ratios. *J. Sediment. Petrol.,* **24,** 151–158.
Siegenthaler, C. 1981. Tidal cross-strata and the sediment transport rate problem: a geologist's approach. *Marine Geol.,* **45,** 227–240.
Simons, D. B., Richardson, E. V. and Albertson, M. L. 1961. Flume studies using medium sand (0.45 mm). *U.S. Geol. Surv. Water Supply Paper 1498–A.*
Skempton, A. W. 1970. The consolidation of clays by gravitational compaction. *Quart. J. Geol. Soc. London,* **125,** 373–408.
Sleath, J. F. A. 1970. Velocity measurements close to the bed in a wave tank. *J. Fluid Mech.,* **42,** 111–123.
Sleath, J. F. A. 1974. Stability of laminar flow at seabed. *J. Wat. Harb. Coastal Eng. Div. ASCE,* **100,** WW2, 105–122.
Sleath, J. F. A. 1975. A contribution to the study of vortex ripples. *J. Hydraul. Res.,* **13,** 315–328.
Sleath, J. F. A. 1976. On rolling grain ripples. *J. Hydraul. Res.,* **14,** 69–81.

334

Sleath, J. F. A. 1982. The suspension of sand by waves. *J. Hydraul. Res.*, **20**, 439–452.

Smith, J. D. 1969. Geomorphology of a sand ridge. *J. Geol.*, **77**, 39–55.

Smith, J. D. 1970. Stability of a sand bed subjected to a shear flow of low Froude number. *J. Geophys. Res.*, **75**, 5928–5940.

Smith, J. D. 1977. Modeling of sediment transport on continental shelves. In Goldberg, E. D., McCave I. N., O'Brien, J. J. and Steele, J. H. (Eds), *The Sea, Vol. 6*, Wiley–Interscience, New York.

Smith, J. D. and McLean, S. R. 1977. Spatially averaged flow over a wavy surface. *J. Geophys. Res.*, **82**, 1735–1746.

Song, W., Yoo, D. and Dyer, K. R. 1983. Sediment distribution, circulation and provenance in a macrotidal bay: Garolim Bay, Korea. *Marine Geol.*, **52**, 121–140.

Soulsby, R. L. 1977. Similarity scaling of turbulence spectra in marine and atmospheric boundary layers. *J. Physical Ocean.*, **7**, 934–937.

Soulsby, R. L. 1980. Selecting record length and digitization rate for near-bed turbulence measurements. *J. Physical Ocean.*, **10**, 208–219.

Soulsby, R. L. 1981. Measurements of the Reynolds stress components close to a marine sand bank. *Marine Geol.*, **42**, 35–47.

Soulsby, R. L. 1983. The bottom boundary layer of shelf seas. In Johns, B. (Ed.), *Physical Oceanography of Coastal and Shelf Seas*, Elsevier Science Publishers, Amsterdam, Chapter 5.

Soulsby, R. L. and Dyer, K. R. 1981. The form of the near-bed velocity profile in a tidally accelerating flow. *J. Geophys. Res.*, **86**, 8067–8074.

Soulsby, R. L., Salkield, A. P. and LeGood, G. P. 1984. Measurements of the turbulence characteristics of sand suspended by a tidal current. *Continental Shelf Res.*, **3**, 439–454.

Soulsby, R. L. and Wainwright, B. L. S. A. 1985. A criterion for the effect of suspended sediment on near-bottom velocity profiles. (In preparation)

Southard, J. B. and Boguchwal, L. A. 1973. Flume experiments on the transition from ripples to lower flat bed with increasing sand size. *J. Sediment. Petrol.*, **43**, 1114–1121.

Southard, J. B. and Dingler, J. R. 1971. Flume study of ripple propagation behind mounds on flat sand beds. *Sedimentology*, **16**, 251–263.

Spielman, L. A. 1978. Hydrodynamic aspects of flocculation. In Ives, K. J. (Ed.), *The Scientific Basis of Flocculation*, Sijthoff & Noordhoff. Alphen aan den Rijn. Neth., 63–88.

Stanley, D. J. and Swift D. J. P. 1976. *Marine Sediment Transport and Environmental Management*, Wiley–Interscience, New York, 602 pp.

Statham, I. 1974. The relationship of porosity and angle of repose to mixture proportions in assemblages of different sized materials. *Sedimentology*, **21**, 149–162.

Stearns, C. R. 1970. Determining surface roughness and displacement height. *Boundary-layer Meteorology*, **1**, 102–111.

Sternberg, R. W. 1966. Boundary layer observations in a tidal current. *J. Geophys. Res.*, **71**, 2175–2178.

Sternberg, R. W. 1967. Measurements of sediment movement and ripple migration in a shallow marine environment. *Marine Geol.*, **5**, 195–205.

Sternberg, R. W. 1968. Friction factors in tidal channels with differing bed roughness. *Marine Geol.*, **6**, 243–260.

Sternberg, R. W. 1971. Measurements of the incipient motion of sediment particles in the marine environment. *Marine Geol.*, **10**, 113–119.

Sternberg, R. W. 1972. Predicting initial motion and bedload transport of sediment particles in the shallow marine environment. In Swift, D. J. P., Duane, D. B. and Pilkey, O. H. (Eds), *Shelf Sediment Transport: Process and Pattern*, Dowden, Hutchinson and Ross, Stroudsberg, 61–82.

Stienstra, P. 1983. Quaternary sea-level fluctuations on the Netherlands Antilles—possible correlations between a newly composed sea level curve and local features. *Marine Geol.*, **52**, 27–37.

Stride, A. H. 1963. Current-swept sea floors near the southern half of Great Britain. *Quart. J. Geol. Soc. London*, **119**, 175–199.

Stride, A. H. 1974. Indications of long term, tidal control of net loss or gain by European coasts. *Estuar. Coastal Mar. Sci.*, **2**, 27–36.

Stride, A. H. (Ed.) 1982. *Offshore Tidal Sands. Processes and Deposits*, Chapman and Hall, London, 222 pp.

Stride, A. H., Belderson, R. H. and Kenyon, N. H. 1972. Longitudinal furrows and depositional sand bodies of the English Channel. *Mem. B.R.G.M.*, **79**, 233–240.

Stringham, G. E., Simons, D. B. and Guy, H. P. 1969. The behaviour of large particles falling in quiescent liquids. *U.S. Geol. Surv. Prof. Paper*, **562–C**.

Stubblefield, W. L., Lavelle, J. W. and Swift, D. J. P. 1975. Sediment response to the present hydraulic regime on the central New Jersey Shelf. *J. Sediment. Petrol.*, **45**, 337–358.

Stumpf, R. P. 1983. The process of sedimentation on the surface of a salt marsh. *Estuar. Coast. Shelf Sci.*, **17**, 495–508.

Sumer, B. M. and Deigaard, R. 1981. Particle motions near the bottom in turbulent flow in an open channel. *J. Fluid Mech.*, **109**, 311–337.

Sumer, B. M. and Oguz, B. 1978. Particle motions near the bottom in turbulent flow in an open channel. *J. Fluid Mech.*, **86**, 109–127.

Sutherland, A. J. 1967. Proposed mechanism for sediment entrainment by turbulent flows. *J. Geophys. Res.*, **72**, 6183–6194.

Swift, D. J. P. 1976. Coastal sedimentation. In Stanley, D. J. and Swift, D. J. P.(Eds), *Sediment Transport and Environmental Management*, Wiley–Interscience, New York, 255–310.

Symonds, G., Huntley, D. A. and Bowen, A. J. 1982. Two-dimensional surf beat: long wave generation by a time-varying breakpoint. *J. Geophys. Res.*, **87**, 492–498.

Taylor, P. A. and Dyer, K. R. 1977. Theoretical models of flow near the bed and their implications for sediment transport. In Goldberg, E. D., McCave, I. N., O'Brien, J. J. and Steele, J. H. (Eds), *The Sea, Vol. 6*, Wiley–Interscience, New York, 579–601.

Terwindt, J. H. J. 1971. Sand waves in the southern bight of the North Sea. *Marine Geol.*, **10**, 51–67.

Terwindt, J. H. J. and Breusers, H. N. C. 1972. Experiments on the origin of flaser lenticular and sand–clay alternating bedding. *Sedimentology*, **19**, 85–98.

Thimakorn, P. 1980. An experiment on clay suspension under water waves. *Proc. 17th Coastal Eng. Conf.*, 2894–2906.

Thorn, M. F. C. and Parsons, J. G. 1980. Erosion of cohesive sediments in estuaries: an engineering guide. *Proc. 3rd Int. Symp. Dredging Tech.*, 349–358.

Thorne, P. D. 1985. An intercomparison between visual and acoustic detection of seabed gravel movement. *Marine Geol.* In Press.

Thorne, P. D., Heathershaw, A. D. and Troiano, L. 1983/1984. Acoustic detection of seabed gravel movement in turbulent tidal currents. *Marine Geol.*, **54**, M43–M48.

Thorne, P. D., Salkield, A. P. and Marks, A. J. 1983. Application of acoustic techniques in sediment transport research. *Proc. Conf. Acoustics and the Sea Bed. Instit. of Acoustics*, 395–402.

Trask, P. D. 1930. Mechanical analysis of sediments by centrifuge. *Econ. Geol.*, **25**, 581–599.

Tucker, M. J. 1963. Analysis of records of sea waves. *Proc. Inst. Civil Eng.*, **26**, 305–316.

Tucker, M. J., Carr, A. P. and Pitt, E. G. 1983. The effect of an offshore bank in attenuating waves. *Coastal Eng.*, **7**, 133–144.

Tunstall, E. B. and Inman, D. L. 1975. Vortex generation by oscillatory flow over rippled surfaces. *J. Geophys. Res.*, **80**, 3475–3484.

van Andel, T. H. 1981. Consider the incompleteness of geological records. *Nature*, **294**, 397–398.

Vanoni, V. A. and Nomicos, G. N. 1959. Resistance properties of sediment-laden streams. *J. Hydraul. Div. ASCE*, **85**, HY5, 77–107.

van Straaten, L. M. J. U. and Keunen Ph. H. 1958. Tidal action as a cause of clay accumulation. *J. Sediment. Petrol.,* **28**, 406–413.

van Veen, J. 1935. Sand waves in the southern North Sea. *Int. Hydrog. Rev.,* **12**, 21–29.

van Veen, J. 1938. Water movements in the Straits of Dover. *J. du Conseil.,* **13**, 7–36.

Venn, J. F. and D'Olier, B. 1983. Preliminary observations for a model of sand bank dynamics. In Sundermann, J. and Lenz, W. (Eds), *North Sea Dynamics,* Springer-Verlag, Berlin, 472–485.

Vincent, C. E., Young, R. A. and Swift, D. J. P. 1981. Bedload transport under waves and currents. *Marine Geol.,* **39**, 71–80.

Vincent, C. E., Young, R. A. and Swift, D. J. P. 1982. On the relationship between bedload and suspended sand transport on the inner shelf, Long Island, New York. *J. Geophys. Res.,* **87**, 4163–4170.

Vincent, C. E., Young, R. A. and Swift, D. J. P. 1983. Sediment transport on the Long Island shoreface, North American Atlantic Shelf: role of waves and currents in shoreface maintenance. *Continental Shelf Res.,* **2**, 163–181.

Visher, G. S. 1969. Grain size distributions and depositional processes. *J. Sediment. Petrol.,* **39**, 1074–1106.

Walling, D. E. and Webb, B. W. 1981. The reliablity of suspended sediment load data. *Proc. Symp. Erosion and Sediment Transport Measurement. IAHS Publ.,* **133**, 177–194.

Wan, Z. 1982. Bed material movement in hyperconcentrated flow. *Inst. Hydrod. and Hydraul. Eng. Tech. Univ. Denmark. Series Pap.,* **31**.

Wang, H. and Liang, S. S. 1975. Mechanics of suspended sediment in random waves. *J. Geophys. Res.,* **80**, 3488–3494.

Wells, D. R. 1967. Beach equilibrium and second-order wave theory. *J. Geophys. Res.,* **72**, 497–504.

Wells, J. T. and Coleman, J. M. 1981. Physical processes and fine-grained sediment dynamics, coast of Surinam, South America. *J. Sediment. Petrol.,* **51**, 1053–1068.

Wentworth, C. K. 1922. A scale of grade and class terms for clastic sediments. *J. Geol.,* **30**, 377–392.

Werner, F. and Newton, R. S. 1975. The pattern of large-scale bed forms in the Langeland Belt, (Baltic Sea). *Marine Geol.,* **19**, 29–59.

White, B. R. and Schultz, J. C. 1977. Magnus effect in saltation. *J. Fluid Mech.,* **81**, 497–512.

White, S. J. 1970. Plane bed thresholds of fine-grained sediments. *Nature,* **228**, 152–153.

Whitehouse, U. G., Jeffrey, L. M. and Debrecht, J. D. 1960. Differential settling tendencies of clay minerals in saline waters. *Proc. 7th Conf. Clays Clay Mins.,* 1–79.

Wiberg, P. and Smith, J. D. 1983. A comparison of field data and theoretical models for wave current interactions at the bed on the continental shelf. *Continental Shelf Res.,* **2**, 147–162.

Wilkinson, R. H. 1984. A method for evaluating statistical errors associated with logarithmic velocity profiles. *Geo-Marine Letters,* **3**, 49–52.

Wilkinson, R. H., Salkield, A. P. and Moore, E. J. 1984. Photogrammetry in sediment transport studies. In George, D. (Ed.), Underwater Photogrammetry and Television for Scientists, Clarendon Press, Oxford.

Williams, P. B. and Kemp, P. H. 1971. Initiation of ripples on flat sediment beds. *J. Hydraul. Div. ASCE,* **97**, HY4, 505–522.

Willmarth, W. W. and Lu, S. S. 1972. Structure of the Reynolds stress near the wall. *J. Fluid Mech.,* **55**, 65–92.

Wimbush, M. and Munk, W. 1970. The benthic boundary layer. In Maxwell, A. E. (Ed.), *The Sea, Vol. 4,* Wiley–Interscience, New York.

Winkelmolen, A. M. 1969. The rollability apparatus. *Sedimentology,* **13**, 291–305.

Winkelmolen, A. M., van der Knaap, W. and Eijpe, R. 1968. An optical method of measuring grain orientation in sediments. *Sedimentology,* **11**, 183–196.

Wood, P. A. 1977. Control of variation in suspended sediment concentration in the River Rother, West Sussex, England. *Sedimentology*, 24, 437–445.

Wooding, R. A., Bradley, E. F. and Marshall, J. K. 1973. Drag due to regular arrays of roughness elements of varying geometry. *Boundary Layer Meteorology*, 5, 285–308.

Wright, L. D. 1978. River deltas. In Davies, R. A. Jr. (Ed.), *Coastal Sedimentary Environments*, Springer-Verlag, New York, 5–68.

Wright, L. D., Coleman, J. M. and Thom, B. G. 1973. Processes of channel development in a high-tide range environment: Cambridge Gulf–Ord River Delta, Western Australia. *J. Geol.*, 81, 15–41.

Wright, L. D. and Coleman, J. M. 1974. Mississippi River Mouth processes: effluent dynamics and morphologic development. *J. Geol.*, 82, 751–778.

Wright, L. D. and Short, A. D. 1984. Morphodynamic variability of surf zones and beaches: a synthesis. *Marine Geol.*, 56, 93–118.

Wright, L. D. and Sonu, C. J. 1975. Processes of sediment transport and tidal delta development in a stratified tidal inlet. In Cronin, L. E. (Ed.), *Estuarine Research, Vol. II*, Academic Press, New York, 63–76.

Yalin, M. S. 1963. An expression for bedload transportation. *J. Hydraul. Div. ASCE*, 89, HY3, 221–250.

Yalin, M. S. 1964. Geometrical properties of sand waves. *J. Hydraul. Div. ASCE*, 90, HY5, 105–119.

Yalin, M. S. 1977. *Mechanics of Sediment Transport*, Pergamon Press, Oxford, 290 pp.

Yerazunis, S., Bartlett, J. W. and Nissan, A. H. 1962. Packing of binary mixtures of spheres and irregular particles. *Nature*, 195, 33–35.

Yun, C.-X. and Wan, J.-R. 1982. A study of diffusion of upper layer suspended sediments in discharges from the Chang Jiang estuary into the sea, based on satellite imagery. In Kennedy, V. S. (Ed.), *Estuarine Comparisons*, Academic Press, New York, 693–704.

Zabawa, C. F. and Schubel, J. R. 1974. Geologic effects of tropical storm Agnes on Upper Chesapeake Bay. *Maritime Seds.*, 10, 79–84.

Zimmerman, J. T. F. 1981. Dynamics, diffusion and geomorphological significance of tidal residual eddies. *Nature*, 290, 549–555.

Index